IFIP Advances in Information and Communication Technology 445

IFIP – The International Federation for Information Processing

IFIP was founded in 1960 under the auspices of UNESCO, following the First World Computer Congress held in Paris the previous year. An umbrella organization for societies working in information processing, IFIP's aim is two-fold: to support information processing within its member countries and to encourage technology transfer to developing nations. As its mission statement clearly states,

> *IFIP's mission is to be the leading, truly international, apolitical organization which encourages and assists in the development, exploitation and application of information technology for the benefit of all people.*

IFIP is a non-profitmaking organization, run almost solely by 2500 volunteers. It operates through a number of technical committees, which organize events and publications. IFIP's events range from an international congress to local seminars, but the most important are:

- The IFIP World Computer Congress, held every second year;
- Open conferences;
- Working conferences.

The flagship event is the IFIP World Computer Congress, at which both invited and contributed papers are presented. Contributed papers are rigorously refereed and the rejection rate is high.

As with the Congress, participation in the open conferences is open to all and papers may be invited or submitted. Again, submitted papers are stringently refereed.

The working conferences are structured differently. They are usually run by a working group and attendance is small and by invitation only. Their purpose is to create an atmosphere conducive to innovation and development. Refereeing is also rigorous and papers are subjected to extensive group discussion.

Publications arising from IFIP events vary. The papers presented at the IFIP World Computer Congress and at open conferences are published as conference proceedings, while the results of the working conferences are often published as collections of selected and edited papers.

Any national society whose primary activity is about information processing may apply to become a full member of IFIP, although full membership is restricted to one society per country. Full members are entitled to vote at the annual General Assembly, National societies preferring a less committed involvement may apply for associate or corresponding membership. Associate members enjoy the same benefits as full members, but without voting rights. Corresponding members are not represented in IFIP bodies. Affiliated membership is open to non-national societies, and individual and honorary membership schemes are also offered.

Hongxiu Li   Matti Mäntymäki
Xianfeng Zhang (Eds.)

Digital Services and Information Intelligence

13th IFIP WG 6.11 Conference
on e-Business, e-Services, and e-Society, I3E 2014
Sanya, China, November 28-30, 2014
Proceedings

Springer

Volume Editors

Hongxiu Li
Matti Mäntymäki
University of Turku
Turku School of Economics
Rehtorinpellonkatu 3, 20520 Turku, Finland
E-mail: {hongxiu.li, matti.mantymaki}@utu.fi

Xianfeng Zhang
Hainan Normal University
School of Information Science and Technology
No. 99 Longkun South Road, Haikou, 571158, China
E-mail: xfzhang@gmail.com

ISSN 1868-4238   ISSN 1868-422X
ISBN 978-3-662-52621-7   ISBN 978-3-662-45526-5
DOI 10.1007/978-3-662-45526-5
Springer Heidelberg New York Dordrecht London

Typesetting: Camera-ready by author, data conversion by Scientific Publishing Services, Chennai, India

Printed on acid-free paper

Springer is part of Springer Science+Business Media (www.springer.com)

Preface

Since its inception in 1998, the I3E Conference has brought together researchers and practitioners of all aspects of the e-world. The I3E conference focuses on e-business, e-services, and e-society. The I3E conference is truly multi disciplinary covering areas from computer science to information systems and service science. The 2014 I3E conference was the 13th consecutive I3E conference and the second time the conference was organized in China.

The theme of the 2014 I3E conference was "Digital Services and Information Intelligence." Services contribute more than 60% of GDP globally (OECD) and have become a major source of economic growth. The economic role of services will continue to grow in developing economies such as China, India, and Brazil. As pointed out by Vargo and Lusch (2004; 2008), "All economies are service economies." The shift of consumers and enterprise personnel from users to co-creators of value calls for a re-evaluation of the existing design and development approaches, methodologies, and tools to create and distribute services. The second component of the conference theme, "Information Intelligence," points out the importance of developing processes and approaches to effectively manage big data and the associated analytics to turn data into intelligence, and to identify ways to measure the benefits derived from using and analyzing big data at individual, organizational, and societal levels.

As a result, digital services and informational intelligence constitute an umbrella that covers the increasing prevalence of integrating Internet, enterprise applications, user experience, networking, and mobile service demands to create new and enhanced services. The global trend of focusing on service-related activities in all industries pushes the frontier of service innovation. Creating digital services that can make a global impact calls for an understanding of the economic, technical, and social aspects of service development and innovation.

Making a successful conference requires resources and commitment. We would like to thank the authors for their submissions. We also wish to thank the reviewers for ensuring the academic standard of the conference. We wish to extend our thanks to Hainan Normal University for hosting the conference and the keynote speakers, Professor Craig Standing, Professor Jari Salo, Professor Helge Hoivik, and Mr. Yao Chen, for their contributions to the conference. Finally, we would like to thank everyone involved in organizing the conference.

The conference received financial support from the National Science Foundation of China, the International Federation for Information Processing (IFIP),

and the Finnish Foundation of Economic Education (Liikesivistysrahasto). This
generous support played an important role in making the 2014 I3E a successful,
internationally renowned scientific venue.

We hope you enjoy reading the conference proceedings.

September 2014   Xianfeng Zhang
                                                    Hongxiu Li
                                                 Matti Mäntymäki
                                                      Jari Salo

Organization

General Co-chairs

Reima Suomi	University of Turku, Finland
Lihua Wu	Hainan Normal University, China

Steering Committee

Winfried Lamersdorf	University of Hamburg, Germany
Wojciech Cellary	Poznan University of Economics, Poland
Matti Mäntymäki	University of Turku, Finland

Program Committee Co-chairs

Hongxiu Li	University of Turku, Finland
Matti Mäntymäki	University of Turku, Finland
Jari Salo	University of Oulu, Finland
Xianfeng Zhang	Hainan Normal University, China

General Secretaries

Xianfeng Zhang	Hainan Normal University, China
Hongxiu Li	University of Turku, Finland

Honorary Local Organizing Chair

Changri Han	Hainan Normal University, China

Local Organizing Chairs

Shengquan Ma	Hainan Normal University, China
Chun Shi	Hainan Normal University, China
Bin Wen	Hainan Normal University, China

Financial Chair

Haiyan Fu	Hainan Normal University, China

Publications Chair

Yuping Zhou   Hainan Normal University, China

Technical Support Chair

Jianping Feng   Hainan Normal University

Contact Coordinator

Haixia Long   Hainan Normal University, China

Organizing Committee

Caixia Chen   Hainan Normal University, China
Xiaowen Liu   Hainan Normal University, China
Yan Lv   Hainan Normal University, China
Xiao Shen   Hainan Normal University, China
Xiangjun Wang   Hainan Normal University, China
Yu Zhang   Hainan Normal University, China
Junyu Zhang   Hainan Normal University, China

I3E 2014 Keynote Speakers

Jari Salo   University of Oulu, Finland
Craig Standing   Edith Cowan University, Australia
Helge Hoivik   Oslo and Akershus University College, Norway
Yao Chen   Hainan Tourism Development Commission,
                                     China

I3E 2014 Program Committee

Esma Aimeur   University of Montreal, Canada
Joao Paulo Almeida   Federal University of Espírito Santo, Brazil
Hernán Astudillo   Universidad Técnica Federico Santa María,
                                     Chile
Khalid Benali   Loria - Nancy Université, France
Salima Benbernou   Université Paris Descartes, France
Djamal Benslimane   Lyon 1 University, France
Markus Bick   ESCP-EAP Europäische
                                     Wirtschaftshochschule Berlin, Germany
Melanie Bicking   University of Koblenz-Landau, Germany

Herve Verjus	Université de Savoie - LISTIC - Polytech'Savoie, France
Hans Weigand	Tilburg University, The Netherlands
Robert Woitsch	BOC, Austria
Jun Xue	Xi'an University of Posts and Telecommunications, China
Hiroshi Yoshiura	University of Electro-Communications, Japan
Steffen Zimmermann	University of Innsbruck, Austria
Xianfeng Zhang	Hainan Normal University, China
Yu Zhang	Hainan Normal University, China

I3E 2014 Conference Organizers

        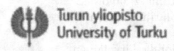

Hainan Normal University, China   University of Turku, Finland

I3E 2014 Conference Sponsors

International Federation for Information Processing   National Natural Science Foundation of China

Hainan Province Computer   e-Business Education and   e-Commerce Committee of
Federation   Research Network for Euro-   China Information Economics
                             Asian Collaboration   Society

Table of Contents

Digital Services

Digital Society

Digital Business

Customer is King? A Framework to Shift from Cost- to Value-Based Pricing in Software as a Service: The Case of Business Intelligence Software

Aaron W. Baur[*], Antony C. Genova, Julian Bühler, and Markus Bick

Department of Business Information Systems,
ESCP Europe Business School Berlin, Germany
{abaur,agenova,jbuehler,mbick}@escpeurope.eu

Abstract. With a shift from the purchase of a product to the delivery of a service, cloud computing has revolutionized the software industry. Its cost structure has changed with the introduction of Software as a Service (SaaS), resulting in decreasing variable costs and necessary amendments to the software vendors' pricing models. In order to justify the gap between the software's price and the incremental cost of adding a new customer, it is essential for the vendor to focus on the added value for the client. This shift from cost- to value-based pricing models has so far not been thoroughly studied. Through literature review and expert interviews, a conceptual model for customer-centric SaaS pricing, especially Business Intelligence & Business Analytics tools, has been developed. The model has then been initially validated by discussions with the top five software players in this realm and builds a strong basis for further theoretical inquiry and practical application.

Keywords: Software, Pricing, Business Intelligence, Cloud Computing, SaaS, E-Business Models.

1 Introduction

One distinct iconoclasm of Jason Maynard, analyst at Credit Suisse, precisely describes the revolution that has happened in the software industry in the last several years: "Traditional software is dead" [1]. Cloud computing, virtualization and Software as a Service (SaaS) switch the delivery of software from physical distribution and installation on local hardware to a provision over the internet. While a complete virtualization is unlikely, the future of computing will lay in a level state: A harmony between the different extremes of pure *on-premise* and pure *over-the-cloud* delivery [2].

SaaS delivery ignited the gradual estrangement from perpetual licenses, with a traditional focus on a large sales force, up-front payments, physical product delivery and a frequent and tedious updating process. Now, vendors can update their products

---

[*] Corresponding author.

H. Li et al. (Eds.): I3E 2014, IFIP AICT 445, pp. 1–13, 2014.

'on-the fly', receive a steady stream of revenue, and focus on a closer client relationship and greater penetration within the client's organizations [3].

These technological and business model changes also largely influence how vendors can set and communicate their pricing policy. Due to the decreased variable costs of vendors, there is a gap between the software's price and the incremental costs of adding a new customer. This misalignment is perceived as unfair and therefore criticized by customers, as studies show [4]. Customers are highly sensitive to the pricing techniques used by vendors, and ironically, it is through the pricing strategies themselves that the company can avoid the customer to focus only on the price as a choice parameter [5]. That's why understanding the client is a key characteristic, as the pricing should be designed upon the variables that the buyer will use in measuring value realization [6]. This is what the customer is willing to pay based on the actual benefit. The result should be a win-win scenario, where customers see the value of the software reflected in their business processes and the vendors benefit from recurring payments [3].

Consequently, choosing the right pricing model is of great importance for software vendors as to attract and retain customers and keep competitors at bay. In order to justify such a 'cost-price gap' and to focus on the added value for the customer, pricing models are now increasingly taking into consideration a customer-centric mind set, by associating price perceptions to product configurations [7].

Analyzing the real value the software represents for the customer needs to be in the center of thought. Hence, the price of the software must be aligned to the customer's value realization, i.e. the shift from cost-based software pricing to a more dynamic value-based software pricing [8]. In the latter case, the price is continuously adapted to the market and it is demand driven, based on a deep knowledge of the customers [9].

Numerous studies attempt to analyze pricing techniques in the SaaS age (e.g. [10,11,12]. However, there has not yet been an analysis of pricing techniques and their correlation with customer value realization specifically applied to Business Intelligence & Business Analytics (BI&BA) solutions for companies. BI&BA tools may be the keys to dealing with today's data glut and customers pose huge expectations in the performance and quality of these software suites [13]. They are of pivotal importance in the management of a company, and it is the software for which the virtualization process has been among the most challenging due to the complexity of the tools [14]. Additionally, it is an area of the software industry with fierce competition and therefore necessitates a customer-centric approach; their pricing models thus present an ideal object of study.

Hence, the focus of this paper is on the effects of this radical change on the software supplier's business model and the resulting customer relationship. This serves to reduce the identified research gap of a missing application of value-based pricing on BI&BA tools. This paper studies the available literature in this field and, together with semi-structured interviews, develops a conceptual model of customer-centric SaaS BI&BA pricing as a research result.

The remainder of this paper is structured as follows: Chapter two presents the research methodology applied, while chapter three introduces common pricing concepts of software products in light of the shift from cost- to value- based pricing.

Chapter four shortly presents the BI&BA industry and the five biggest vendors. Results & implications of the research, i.e. findings in regard to software pricing concepts and a customer-centric framework are then presented in chapter five. Finally, we give a conclusion about a new value proposition applicable to the B2B software industry and sum up the paper in chapter six.

2   Theoretical Background

2.1   Pricing of Software Products

As Patrick Heffron ([10], p. 3) puts it, "understanding software pricing is challenging even for the most savvy business people and seasoned technology veterans". Therefore, we will give a short overview of software pricing determinants and dynamics in this chapter.

In a holistic and general view, pricing depends on three important variables, namely costs, customer and competition, which in literature have been referred to as the Three Cs of Pricing [15]. First, the cost structure of software vendors has changed with the advent of SaaS [16]. In contrary to traditional economic theory, where prices were set on the measurement of "replications", i.e. based on the incremental cost of each additional product [17], such a theory has been put into question within the software industry. Here, replication costs of each additional software license sold are practically inexistent [18]. For example, the hosting, management and recovery of systems (including a 99.9% availability) now take up a much higher percentage of costs than physical reproduction and distribution or customization to different hardware specifications of clients. Expenses are amortized once the number of users grows, which explains the initial difficulty of SaaS providers to achieve profitability [19]. Second, the customers' perception of a price constitutes the top ceiling, above which the vendor should not price the product [20]. According [15], correctly judging this perception is especially challenging for companies in the high-tech and software business. Finally, the competition serves as a benchmark to compare the prices set according to the other two variables [21].

Within these Three Cs of Pricing, concepts can be structured in cost- and value-based models. In cost-based pricing models, price is determined by the production and delivery costs of the service. Relevant examples include flat and user-based pricing [22], [11], usage-based pricing [23,24,25,26], and performance-based pricing [27], [22], [11].

On the other hand, value-based pricing models help software vendors to set the price according to the value received by the customers, and not primarily to their willingness to pay [28], [22]. Important and widely-used forms are penetration pricing [4], [22], skimming pricing [4], [23], [29], and hybrid pricing [22], [11].

What could be noticed, due to high complexity, software vendors in the past have often adopted the "intuitive" approach: They took their development costs as a basis and then subjectively assessed the product in the market and set the price accordingly, with no objective scientific rational [17]. This is especially true for "disruptive" offerings like new BI tools, since comparables are often missing [30]. Even if such a method has been popular in the past, obviously it lacks effectiveness, and its outcomes are basically random [31].

Therefore, new pricing models had to be introduced to be less random and to better justify the gap between the cost of an additional product and the price paid by the customer. For this purpose, vendors try to examine the pricing strategy from the customer's perspective, and "assign a price that is monetary equivalent of the value the customer perceives in the product while meeting profit and return on investment goals" ([11], p.1). They are hence moving towards a customer-value-based approach which has the highest potential of appealing to the customers and effectively positioning the product in the market [19], [32].

Among interesting contributions to software pricing research that incorporate those ideas and approaches are the general software framework of [33], the modified cloud service version of [34] and the customer-value based pricing model of [11]. Another widely-known representative that sees this customer-value creation as the bottom of each software price setting is the strategic pricing pyramid of [30], which is based on a large scale antecedent study in different software settings (see Figure. 1).

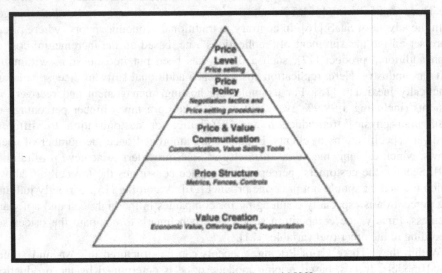

Fig. 1. Strategic Pricing Pyramid [30]

Although the other drivers in this model, i.e. price structure, price & value communication, price policy and last not least the absolute and relative price level are important as well, the basis is represented by the value creation. Value creation, i.e. the satisfaction of the users' needs, and in such a way building customer loyalty in the long-term and attracting new buyers, needs to serve as the core of price determination.

2.2   Business Intelligence and Business Analytics

As described previously, through gaining business advantage from data, these tools help manage information, leading executives to more informed decisions [35]. A wide array of solutions is available, and new players are constantly entering the market [36]. However, BI solutions are expensive to implement and maintain, and require powerful infrastructure. Hence, cloud computing provides the necessary IT capabilities and business agility to cut costs and achieve much needed economies of scale. SaaS solutions provide a more flexible model that aligns better with client's business objectives [37]. The possibility to have an on-demand BI solution allows companies to benefit from the services on a subscription basis, with no need for long capital requests. Even when resources would be available, BI via SaaS may still be preferable when time to market is an issue [38].

According to recent reports, IBM, SAP, Microstrategy, Oracle and Microsoft constitute the global top 5 BI software solution vendors [39,40,41].[1] Therefore, these companies serve as the sample for our study.

In terms of transparency of pricing, the vendors follow very diverse strategies. SAP as the global market share leader in BI&BA does not disclose price lists whatsoever. SAP's rationale is that "... there are so many different solutions available, and they oftentimes are tailored scenarios for specific customer needs. It hasn't really made sense to publish a software price list. We want to ensure each customer or prospective customer gets product pricing that is tailored to their specific use case" (Interview with SAP's spokesman Evan Welsh, as reported by [42]. Oracle, in contrast, is well-known for its transparency and happily publishes price lists, even though virtually no customer will ever pay these (high) prices. The other vendors fall somewhere between these extremes. Overall, many customers are resentful of their vendors' licensing practices, and need more transparency about the pricing models in general, and not only concerning the exact price. Especially with these tools often being key to the enterprises' success rather than just assuming the role of a nice-to-have functionality. That is why enterprises should treat BI software and service issues not only from a technical point of view, but also from a business perspective.

The final part of this paper is therefore going to address this problem, by suggesting a value-based pricing model that is dynamic and customizable, but at the same time transparent.

3   Research Methodology

In order to thoroughly analyze the pricing models applied in the BI&BA SaaS industry, a two-phase approach has been applied: An exploratory phase that includes a literature review, qualitative expert interviews and the design of the model; and a confirmatory phase to validate the model in a dialogue with representatives of the five most important BI software vendors.

---

[1] http://www.ibm.com
http://www.sap.com
http://www.microstrategy.com
http://www.oracle.com
http://www.microsoft.com

In the *first phase*, the available literature in the field of software pricing in BI and BA was identified and screened. For this purpose, both scientific and more practice-oriented sources have been used. In the former case, selected large-scale and reputable digital libraries in IT, engineering, business administration and related fields have been searched. In concrete terms, these sources are EBSCO Business Source Complete, the ACM Digital Library, Web of Science, IEEE Xplore, and ScienceDirect [43]. In the latter case, trade magazines, industry reports, IT/technology magazines, reviews, blogs, market research publications, and company websites have been harnessed. This was deemed necessary to also include the practitioner's view of the topic and to stay on top of developments in the fast paced software market. Search terms were comprised of a variety of queries including *software pricing*, *pricing models* and *value perception* in conjunction with *SaaS*, *Cloud*, and others. Based on these literature findings, interviews were carried out to get further insights into the topic. In total, nine qualitative semi-structured expert interviews [44] have been conducted. Interviewees were senior software industry experts and analysts, both from consultancies and market research firms. The interviews lasted between 57 and 92 minutes, with a median of 77 minutes. After an introduction to the topic, several blocks of questions regarding the changing technological environment, the diffusion of SaaS, customer perception and satisfaction, pricing variables in relation to customers' values and importance of this value realization, among others, were asked. An interview guideline was followed that nevertheless left enough room for the interviewees to set their own emphases. The interviews were then transcribed and coded using MAXQDA 11 software.[2] The literature and interview findings were then used to develop a novel conceptual software pricing model.

In the *second phase,* this resulting model was then discussed, validated and refined with representatives of the leading five international BI&BA software solution providers. These representatives were contacted via email and phone and the discussions held with one executive at a time at the CeBIT 2014 in Hannover, Germany. The discussions gave a first confirmation of the validity of the model.

4   Results and Implications

After having analyzed the literature regarding pricing techniques applied to software products, having conducted interviews, drawn up a conceptual model and having discussed this issue with the five biggest BI software vendors, we now want to present our most important findings.

Since software pricing has become increasingly complex ("constantly changing labyrinth of pricing", [34], p. 127), one of the most important factors for a pricing model is simplicity ("the vision is that less is more", [45], p. 7). The customers need to understand immediately how their value creation is represented in the software pricing ("Firstly, be boring. Secondly, license your software as your customers expect it be licensed – fit in with their business model", [46], p. 58). It is therefore vital to

---

[2] http://www.maxqda.com

attract the customer also on a mental level, acting as a sort of partner who embraces the client's business model.

Additionally, in order to achieve an effective value co-creation it is important to have a flexible model, which reflects the client's need to iteratively address its potentially changing costs [47], [9]. In such a way, it is possible to lock in the customers offering mutually beneficial impacts, which must be clearly understood and perceived in the marketplace. The software vendor needs to leverage the SaaS cost efficiency to deliver the clients a stronger market positioning [47]. Since "users need clarity without surprise" [7], it will pay off to have an open, fair and transparent model.

The ideas from literature have then been cross-checked in the interviews. The opinion of experts working in and with the sector, who are accustomed to the requirements of companies using BI software products, reaffirmed the findings in respect to the importance of customer value realization and perception in this software sector.

While of course there has not been one unison answer to the question *how should the ideal customer-centric pricing model look like?*, the results underpin one common theme: the model has to be customizable and flexible, in order to better represent the needs of each client. This underlines the impression that clients are not that focused on the exact price, but on the adaptability of the pricing scheme to their business model.

However, a dynamic concept that allows the client to reassess the price is not suitable for every vendor's business model. The biggest vendors in the industry deal with large multinational clients who most of the times have had the same contract for many years and do not change it regularly. Hence, they might not wish to offer the client the possibility to reassess the price after one year, since that would potentially result in a loss in revenue. However, software vendors' customer relationships are often perceived as weak, and the clients' satisfaction rate is frequently influenced negatively by causes that do not involve the qualities of the product but the complementary service. Therefore, giving the clients the possibility to reassess the price and change the variables could be a way of facing customers' dissatisfaction.

The interviews also revealed that in the future, software vendors will tend to offer a more complete product, with different functions interfaced as a whole bundle. This would mean that the biggest actors in the market continue to buy smaller companies to increase their offering and beat the competition. Thus, it will become more difficult for small software vendors to survive and remain in the market. Here again, offering the client the possibility to reassess the price of the software after one year - and communicating on this flexibility - could be an effective way for new actors in the industry to gain loyal customers and survive in the market.

What is interesting and somehow unexpected, is that there has neither been a clear preference for one particular pricing model, nor one specific technique of a value-based pricing. Instead, software vendors should try to adapt their strategy to the *client's* business model. A concern that was agreed upon, however, was that such flexibility, and the investigation of the client's needs, business model and willingness to pay, require patience, time and personnel as well as financial resources.

In light of the results of the literature and empirical research, a customer-centric framework to represent the effort of software vendors to support the value creation of their clients has been designed. This serves as a conceptual model. Furthermore, it represents the role of pricing models in ensuring the clients' loyalty, which is of pivotal importance in the SaaS business model.

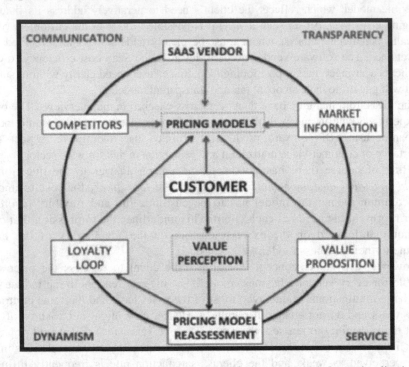

Fig. 2. Customer-centered value proposition of software pricing models [own visualization]

The model develops around the customer, since it is the client who plays the most important role in a value-based pricing model. As a matter of fact, as it emerged in all the interviews, the vendor needs to truly understand the client's needs and perceptions.

The model is then divided into two phases, characterized by the two different colors in the figure. At first, the vendors need to attract the customers to their offerings, through a transparent display of the pricing model. This does not mean that the vendor has to publish the exact price of the offering online, but on the contrary share the process that leads to the price with the customer. This gives the client the opportunity to choose between pricing models based on different variables, so that the client is aware of the final result of his or her choices. However, as it emerged in the literature review (e.g. [34]), it is not realistic to suppose that the price derives only from the value realization of the client. Hence, the model depicts the different variables that lead to the price designation. These variables include primarily the client but also the competitors' offering, the specific market information (including the geographical market, the different industries and sectors that the offering is targeting), and finally the vendor itself, through the cost structure of the product.

In this way, the conceptual model displays a more harmonized pricing framework, which does not focus on the client's value only during the price designation phase, but also and most importantly in the subsequent stage. Finally, the first stage also includes communication, a keyword that previous pricing models have ignored completely. This empirical research has confirmed that communicating the pricing model is essential in value-based pricing models since the value realization often does not take place solely due to a lack of clients' understanding. It is a concept that needs to be directly linked to pricing techniques.

The second phase of the conceptual model illustrates the post-purchase stage, which is focused on two other key words, service and dynamism.

The first concept is self-explanatory, since SaaS involve the offering of a service instead of a product. However, through the interviews and discussions with the top five BI vendors, it became apparent how important it is for the vendor to offer a full service. This full service is not only limited to the functioning of the software itself, but also encompasses guidance of the customer and a focus on relationship management.

The client has chosen the software because of the value proposition of the vendor. However, the conceptual model also illustrates the follow-up of this proposition, with the value perception that is essential to reassess the price in collaboration with the vendor. As it repeatedly emerged in the interviews, the customers usually value the trial period as an essential tool to understand the product, and to decide whether it meets the needs of their businesses.

Within the model, the same concept of a trial is included, but applied to the after-purchase phase. Hence, the client has a period of time to see whether the variables he has chosen as a fundament for the price of the service are the ones that best depict his or her business and best represent the value realization. In this way, the clients can then reassess the price according to their business model needs, through a pricing concept that is dynamic and allows for alterations. In contrast with previous pricing models, the dynamism here concentrates on the client's needs and not on the vendor's strategy (as in the traditional value-based pricing models, namely *penetration pricing* and *skimming pricing*).

The reassessment of the price aims to shift the client's investigation from an ex-ante to an ex-post stage. As a matter of fact, the vendor needs to keep a good relationship with the client, with a frequent exchange of information, and through this dialogue the vendor can save resources, time and money from investigating the client's needs before reaching him.

Finally, in the model, this leads to a loyalty loop, as in McKinsey's consumer decision journey model [48]. Since the clients see their needs fulfilled, and the value exploited, they do not seek other software vendors and remain loyal to the service offered. This is important in light of recent research publications, which have been analyzed in the previous chapters. They revealed that customers in the majority of cases show a low level of client satisfaction, and are not pleased with the vendor relationship vendors [39,40,41].

Hence, it can be seen as crucial to switch to and constantly strive for a customer-centered value proposition.

5   Conclusion

After the software revolution and the introduction of Software as a Service (SaaS), the end customer has become essential in the definition of the overall strategy of a software vendor. Therefore, the client's needs and expectations have to be reflected in the Business Intelligence & Business Analytics (BI&BA) software supplier's pricing strategy even more. Our research aim therefore was to develop a customer-centric conceptual model for the pricing of SaaS BI&BA software. To reach this goal, a thorough literature study both in scientific and practice-oriented publications has been conducted, followed by a round of nine semi-structured expert interviews. With this data pool, a conceptual model that consolidates and compresses prior models has been developed. To validate this conceptual model, we discussed it with representatives of the five leading BI&BA software vendors.

In the course of the study it became clear that each software pricing model has its own unique strengths and weaknesses; each customer achieves value in a different way. Therefore, taken together, the conceptual model developed here caters to these two aspects and is intended to show the need for flexibility and scalability on behalf of the software vendor, in order to meet the clients' expectations.

We think the contribution of this research is twofold:

First, in the scientific community, there is very little pricing research in the particular but important case of Business Intelligence software. Here, the conceptual model can help stimulating the discussion immediately in order to ignite further necessary research. It may contribute to focus more on the value-perception and the interaction between customer and vendor, and less on technical issues, which have already been thoroughly researched.

Second, the findings and the model can help practitioners in rethinking their pricing method. Higher market pressure and increased competition in this field of the software industry make customer satisfaction and loyalty ever more important. Thus, having a strategic and customer-centric look on the current pricing practice can lead to lower customer churn rates, higher customer satisfaction and more pricing flexibility.

However, our study also contains a number of limitations: On the one hand, the research has been conducted without investigating the technical characteristics of the software products under examination. This would have allowed a deeper understanding of the cost structure, and hence the possible pricing techniques to apply. On the other hand, the findings have only been discussed with the five biggest BI&BA software vendors and small-and mid-sized companies have been ignored. As this particular share of the software market is very dynamic, partly due to constant start-up activity, a generalization of the findings may be problematic at this point of time of our research.

Future research may address these shortcomings and should potentially also take into account the well-described vendor/client lock-in effects of SaaS applications which frequently occur. In addition, the conceptual model is just a starting-point to initiate more research in terms of putting the client center-stage when it comes to value delivery, value communication and value pricing in the software and BI&BA industry.

References

1. The Economist: Universal Service? Proponents of "software as a service" say it will wipe out traditional software (April 20, 2006)
2. Carraro, G., Chong, F.: Software as a service (SaaS): An enterprise perspective. Microsoft Corporation (2006),
   http://msdn.microsoft.com/en-us/library/aa905332.aspx
3. Gruman, G., Morrisson, A., Retter, T.: Software Pricing Trends: How vendors can capitalize on the shift to new revenue models. PriceWaterhouseCoopers (2007),
   http://www.pwc.com/en_us/us/technology-innovation-center/assets/softwarepricing_x.pdf
4. Rohitratana, J., Altmann, J.: Impact of pricing schemes on a market for software as a service and perpetual services. Future Generation Computer Systems 28(8), 1328–1359 (2012)
5. Bertini, M., Wathieu, L.: How to stop customers from fixating on price. Harvard Business Review, 85–91 (2010)
6. Bontis, N., Chung, H.: The evolution of software pricing: from box licenses to application service provider models. Internet Research: Electronic Networking Applications and Policy 10(3), 246–255 (2000)
7. Schneider, S.: The dirty little secret of software pricing. RTI Whitepaper (2012),
   http://www.rti.com/whitepapers/Dirty_Little_Secret.pdf
8. Baker, W.: Software license and maintenance pricing principles – Best practices and case studies. In: Soft Summit Conference, Santa Clara, CA (2004),
   http://www.softsummit.com/library/presentations/2006/HomayounHatami_McKinseyandCompany.pdf
9. Lehmann, S., Buxmann, P.: Pricing strategies of software vendors. Business & Information Systems Engineering 6, 452–462 (2009)
10. Heffron, P.: Understanding Software Pricing Models. Esna Technologies (2013),
    http://ease.esna.com/Sys/Document/Open/70t000000000031000a
11. Harmon, R., Demirkan, H., Hefley, B., Auseklis, N.: Pricing strategies for information technology services: a value-based approach. In: 42nd Hawaii International Conference on System Sciences. IEEE Press, New York (2009)
12. Choudhary, V.: Software as a service: Implications for investment in software development. In: 40th Hawaii International Conference on System Sciences. IEEE Press, New York (2007)
13. Swoyer, S.: Analytics in the cloud. The Challenges and Benefits. TDWI (2013),
    http://tdwi.org/research/2013/11/tdwi-ebook-analytics-in-the-cloud-the-challenges-and-benefits.aspx
14. Van Der Lans, R.: Data Virtualization for Business Intelligence Systems: Revolutionizing Data Integration for Data Warehouses. The Morgan Kaufmann Series on Business Intelligence. Morgan Kauffman, Waltham (2012)
15. Mohr, J., Sengupta, S., Slater, S.: Marketing of High-Technology Products and Innovations. Pearson Education, New Jersey (2005)
16. Churakova, I., Mikhramova, R.: Software as a Service: Study and Analysis of SaaS Business Model and Innovation Ecosystems (2010),
    http://lib.ugent.be/fulltxt/RUG01/001/459/665/RUG01-001459665_2011_0001_AC.pdf

17. O'Connor, S.: How indigenous software companies price their product and service offerings: An exploratory investigation (2009), http://repository.wit.ie/1453/1/How_indigenous_software_comp anies_price_their_product_and_service_offerings_an_explorato ry_investigation.pdf

18. Ojala, A., Tyrvainen, P.: Business models and market entry mode choice of small software firms. Journal of International Entrepreneurship 4, 60–81 (2006)

19. Desisto, R., Paquet, R.: Learn the Economic Advantages of a Pure SaaS Vendor. Gartner (2007), https://www.gartner.com/doc/537208/learn-economic-advantages-pure-saas

20. Marn, M., Roegner, E., Zawada, C.: Pricing new products. McKinsey Quarterly 3(XXX), 40–49 (2003)

21. Shipley, D., Jobber, D.: Integrative pricing via the pricing wheel. Industrial Marketing Management 30, 301–314 (2001)

22. Harmon, R., Raffo, D., Faulk, S.: Value-based pricing for new software products: Strategy insights for developers. In: Portland International Conference on the Management of Engineering and Technology. IEEE Press, New York (2004)

23. Singh, J.: Software Pricing Strategies (2010), http://www.slideshare.net/Alistercrowe/software-pricing-strategies

24. Ojala, A.: Comparison of different revenue models in SaaS. In: Prakash, E. (ed.) Proceedings of 5th Computer Games, Multimedia & Allied Technology Conference, pp. 120–123 (2012)

25. Waters, B.: Software as a service: A look at the customer benefits. Journal of Digital Asset Management 1, 32–39 (2005)

26. Choudhary, V., Tomak, K., Chaturvedi, A.: Economic Benefits of Software Renting. Journal of Organizational Computing and Electronic Commerce 8(4), 277–305 (1998)

27. Keranen, T.: Value based pricing of ICT services at Citiusnet Ltd. (2010), http://publications.theseus.fi/handle/10024/17291

28. Monroe, K.: Pricing: Making Profitable Decisions. McGraw-Hill/Irwin, New York (2002)

29. Monroe, K.: Buyers' Subjective Perceptions of Price. Journal of Marketing Research 10, 70–80 (1973)

30. Kittlaus, H.B., Clough, P.N.: Software Product Management and Pricing: Key Success Factors for Software Organizations. Springer, Heidelberg (2009)

31. Hinterhuber, A.: Customer value-based pricing strategies: Why companies resist. Journal of Business Strategy 29(4), 41–50 (2008)

32. Cavusgil, S.T.: Pricing for global markets. Columbia Journal for World Business 31(4), 66–78 (1986)

33. Iveroth, E., Westelius, A., Petri, C.-J., Olve, N.-G., Coster, M., Nilsson, F.: How to differentiate by price: Proposal for a five-dimensional model. European Management Journal, 109–123 (2012)

34. Laatikainen, G., Ojala, A., Mazhelis, O.: Cloud Services Pricing Models. In: Herzwurm, G., Margaria, T. (eds.) ICSOB 2013. LNBIP, vol. 150, pp. 117–129. Springer, Heidelberg (2013)

35. Muntean, M., Muntean, C.: Evaluating a Business Intelligence Solution. Feasibility Analysis Based On Monte Carlo Method, ECECSR Journal 2, 85–102 (2013)

36. Baur, A.W., Breitsprecher, M., Bick, M.: Catching Fire: Start-Ups in the Text Analytics Software Industry. In: Americas Conference on Information Systems (AMCIS), Savannah (August 2014)

37. Thompson, W., Van der Walt, J.: Business Intelligence in the cloud. SA Journal of Information Management 12, 1 (2010)
38. Mitchell, R.: Is SaaS a good fit for BI? Computerworld, pp. 22–25 (February 22, 2010)
39. Schlegel, K., Sallam, R., Yuen, D., Tapadinhas, Y.: Magic Quadrant for Business Intelligence and analytics platforms. Gartner (2013), http://www.gartner.com/technology/reprints.do?id=1-1DZLPEP&ct=130207&st=sb
40. Business-Software Report: Top 10 Business Intelligence Software Report, Comparison of the leading Business Intelligence Software Vendors. Business-Software (2013), http://c3330831.r31.cf0.rackcdn.com/top_10_bi.pdf
41. Dresner, H.: Wisdom of crowds embedded intelligence market study. Dresner Advisory Services (2013), http://www.dundas.com/wisdom-of-crowds-embedded-bi-study
42. Wailgum, T.: Why doesn't SAP publish its price list? ASUG News (2011), http://www.asugnews.com/article/why-doesnt-sap-publish-its-price-list
43. Chen, H., Chiang, R.H.L., Storey, V.C.: Business Intelligence and Analytics: From Big Data to Big Impact. MIS Quarterly 36(4), 1165–1188 (2012)
44. Hair, J.F., Celsi, M.W., Money, A.H., Samouel, P., Page, M.J.: Essentials of Business Research Methods, 2nd edn. M. E. Sharpe, Armonk (2011)
45. Luoma, E.: Examining Business Models of Software-as-a-Service Firms. In: Altmann, J., Vanmechelen, K., Rana, O.F. (eds.) GECON 2013. LNCS, vol. 8193, pp. 1–15. Springer, Heidelberg (2013)
46. Davidson, N.: Don't just roll the dice – A useful short guide to software pricing. Simple Talk Publishing (2009)
47. Kauffman, R.J., Ma, D.: Cost Efficiency Strategy in the Software-as-a-Service Market: Modeling Results and Related Implementation Issues. In: Altmann, J., Vanmechelen, K., Rana, O. (eds.) GECON 2013. LNCS, vol. 8193, pp. 16–28. Springer, Heidelberg (2013)
48. Court, D., Elzinga, D., Mulder, S., Vetvik, O.J.: The consumer decision journey. McKinsey Quarterly (2009), http://www.mckinsey.com/insights/marketing_sales/the_consumer_decision_journey

Weibo or Weixin? Gratifications for Using Different Social Media

Chunmei Gan[1,*] and Weijun Wang[2]

[1] School of Information Management, Sun Yat-sen University, Guangzhou, China
chunmei_gan@163.com
[2] School of Information Management, Central China Normal University, Wuhan, China
wangwj@mail.ccnu.edu.cn

Abstract. Social media has experienced great changes in recent years. Various social media platforms emerge and develop significantly. Why users choose to use some particular social media becomes a major concern. Through adopting uses and gratifications theory, this study aims to explore gratifications sought from using two popular social media: Weibo and Weixin. Data was collected by eighteen in-depth interviews and content analysis was conducted for data analysis. Results show that, seven gratifications could be sought from using Weibo: information seeking, social interaction, entertainment, pass time, self-expression, information sharing and social networking, while five gratifications from Weixin usage: private social networking, social interaction, convenient communication, high-quality information provided and information sharing. In addition, three general gratifications for usage of both Weibo and Weixin were identified: information sharing, social networking and social interaction. Implications of this study are also discussed.

Keywords: instant messaging, social networking service, social media, gratification, uses and gratifications theory, motivation.

1 Introduction

The use of social media has become more widespread and continues to grow significantly in recent years. Social media is "a group of Internet-based applications that build on the ideological and technical foundations of Web2.0, and that allow the creation and exchange of user generated content" [1]. Social media enables users to create and share information, and to communicate with each other. Typical social media platforms include: instant messaging services like Weixin and QQ, social networking services like Renren and Facebook, microblog services like Weibo and Twitter.

CNNIC statistical report mentioned that, as the No.1 Internet application, instant messaging had a continuously rising utilization ratio (from 82.9% in 2012 to 86.2% in 2013), while microblog and other communication-type applications had a continuously declining utilization ratio (for microblog, from 54.7% in 2012 to 45.5% in 2013) [2].

---

* Corresponding author.

H. Li et al. (Eds.): I3E 2014, IFIP AICT 445, pp. 14–22, 2014.

Also, CNNIC research report finds that, there are quick changes for social applications. Social networking sites located in the first line previously; with the emergence of microblog (also "Weibo") sites, number of users of social networking has dramatically dropped; however, the emergence and development of Weixin has brought about a complicated competition environment for microblog, as some functions of microblog have been substituted or transferred [3]. Then one question arises: why do users choose to use some particular social media? Specially, for Weibo and Weixin users, why do some users use Weibo, while some use Weixin, and some use both?

The extant studies provide us a new perspective to consider the research questions. That is, when users choose to use Weibo, Weixin or both, they could seek for special gratifications to meet their requirements. So, what type of gratifications do users seek for when using Weibo and Weixin? To answer the question, based on the work of [4] and [5], we try to adopt uses and gratifications theory (U&G theory) as our theoretical background, and use in-depth interview as the method of data collection, so as to explore the general and specific gratifications sought from using Weibo and Weixin.

The paper is structured as follows. First, uses and gratifications theory will be introduced as the theoretical background and related studies will be elaborated. Second, research methodology will be described, including research design and data collection. Then findings about gratifications from Weibo and Weixin will be showed in Section 4. And finally, results will be discussed; implications, limitations and future research will also be presented.

2   U&G Theory and Related Studies

2.1   U&G Theory

U&G theory was first put forward by Katz " [6]. This theory explores users' usage motivation from the perspective of individuals, and puts forward that users are active and can meet their requirements to use some particular media with special motivations [6]. Furthermore, Katz considers users' media usage behavior as a chain of causation, i.e., "social factors + psychological factors - expectations for media - usage of media - gratifications sought" [6].

U&G theory has the following ideas: individuals are active media users, individuals use some media for particular aims, individuals meet their social and psychological gratifications through using some media, media should connect users' gratifications with their usage of the media when spreading information [6]. The theory aims to understand users' motivations to use media, to explain how users seek their gratifications through media usage, and to show relationships among users' cognitive needs, motivations and their behavior.

For usage of traditional media, contents that media spread and experience brought to users by media usage lead to users' gratifications sought from media usage [7]. With the emergence and development of various new media, uses and gratifications theory is further improved, referring to, such as, interaction and verified gratifications. In addition, researchers adopt uses and gratifications theory to investigate different types of gratifications users seek from using different new media.

2.2   Related Studies

Researchers have already made many efforts to explore reasons on why users use social media or related communication technologies. For example, Whiting and Willians [4] identified ten gratifications for using social media: social interaction, information seeking, pass time, entertainment, relaxation, communicatory utility, convenience utility, expression of opinion, information sharing, and surveillance/ knowledge about others. Ku et al. [5] compared gratifications from usage of three CMC technologies (SNS, IM, and e-mail); results revealed four general gratifications: relationship maintenance, information seeking, amusement, and style; and showed specific gratifications: sociability gratification sought from using instant messaging and social networking sites, gratification of kill time sought from using instant messaging. Through 77 surveys and 21 interviews, Quan-Haase and Young [8] found six gratifications users obtained from Facebook: pastime, affection, fashion, share problems, sociability, and social information; furthermore, differences were revealed that Facebook is more related to having fun and knowing about the social activities occurring in one's social network, while instant messaging focused more on relationship maintenance and development. Raacke and Bonds-Raacke [9] showed that many college students used friend-networking sites to meet gratifications such as keeping in touch with friends, posting/looking at pictures.

In addition, researchers have applied U&G theory to explain motivations for using various social media. Lo and Leung [10] revealed gratifications sought from IM usage: peer pressure/entertainment, relationship maintenance, free expression and sociability. Raacke and Bonds-Raacke [9] found gratifications for using social networking sites: keep in touch with old/current friends, post/look at pictures, make new friends, locate old friends, learn about events, post-social functions, feel connected, share information, academic purposes and dating. Pai and Arnot [11] also identified four gratifications for adopting social networking sites: belonging, hedonism, self-esteem, and reciprocity. The above-mentioned findings by [4] and [5] also adopted uses and gratifications for their investigations.

3   Research Method and Data Collection

3.1   Research Method

Applying uses and gratifications theory, we try to explain the reasons why users choose different social media (Weibo and Weixin) although they provide some overlap functions. Three stages were adopted for the research design to achieve the research aim.

First, literature review was conducted so as to acquire a preliminary list of gratifications for using Weibo and Weixin. On the basis of gratifications identified by [4] and [5], we further reviewed articles related to usage of different social media, and established a preliminary list of gratifications: relationship maintenance, information seeking, social interaction, entertainment, and killing time.

Second, we conducted in-depth interview to collect data to validate and modify the preliminary list of gratifications. In-depth interview is suitable because the aim of the current research is to explore why users choose to use a particular social media tool.

After data collection, we adopted content analysis for analyzing the data. The researchers independently read and reviewed responses from interviewees, and extracted possible gratifications based on the preliminary list of gratifications. For those gratifications that did not appear on the list, we listed them as a new group. Also, we discussed the responses that we did not agree with each other. After data analysis, we got the final list of gratifications to answer the research questions.

Table 1. Descriptive statistics for the sample (N = 2 + 18)

No	Gender	Grade	Major	Experience of using Weibo	Experience of using Weixin
*	Female	Sophomore	Lirary science	2 years	1 year
*	Male	Sophomore	Lirary science	3 years	2 years
A	Female	Sophomore	Information management & information system	Less than 1 year	Less than 1 year
B	Female	Sophomore	Library science	3 years	2 years
C	Male	Sophomore	Archives science	5 years	4 years
D	Male	Sophomore	Information management & information system	2 years	2 years
E	Male	Sophomore	Library science	3 years	1 year
F	Female	Sophomore	Library science	4 years	3 years
G	Male	Sophomore	Information management & information system	3 years	3 years
H	Female	Sophomore	Archives science	2 years	Less than 1 year
I	Female	Sophomore	Archives science	5 years	4 years
J	Female	Junior	Information management & information system	5 years	4 years
K	Male	Junior	Library science	1 year	2 years
L	Male	Junior	Archives science	1 year	1 year
M	Female	Junior	Information management & information system	2 years	Less than 1 year
N	Male	Junior	Library science	3 years	2 years
O	Male	Junior	Archives science	4 years	3 years
P	Female	Junior	Library science	3 years	3 years
Q	Male	Junior	Information management & information system	1 year	1 year
R	Female	Junior	Archives science	4 years	3 years

Note: The first two samples with * were invited for pre-interivews.

3.2   Data Collection

The interviewees were university students in their second or third year from one university in south China. According to Lincoln and Guba [12], the number of interviewees for interviews should be at least 12. So we choose 18 students as samples. Two pre-interviews were conducted before the final interviews so as to check out the questions. All interviews were conducted Face - to - Face in one of the researcher's office in one month, and each interview lasted for 45 – 60 minutes.

Table 1 shows the descriptive statistics for interview samples. Among 18 interviewees, 9 were female and 9 were male. For their study grades, 9 were sophomore and 9 were junior. All of them were from three majors: library science, archives science and information management & information system. For experience for using Weibo, 5 students had 3 years, 3 students had 1, 2, 4, and 5 years respectively, and only 1 student had less than 1 year. For experience for using Weixin, 5 students had 3 years, 4 students had 2 years, and 3 students had less than 1 year, 1 and 4 years respectively.

4   Results

4.1   Gratifications from Weibo

Seven gratifications were identified with respect to the use of Weibo. They will be discussed as follows.

(1) Information seeking. 83.33% (15) interviewees reported using Weibo to seek various information, such as information about social events and news, history events, work and studies. Interviewee C said that "Weibo can provide different types of information, I like topics such as 'hot topics', 'hot post in 24-hour' and 'specially focused topics', so as to quickly seek information on what happened." Interviewee A further referred to use Weibo for "seeking information about studying aboard, such as introduction of universities and majors, daily study and life on some particular universities."

(2) Social interaction.72.22% (13) interviewees mentioned using Weibo for social interaction, i.e., interacting with various users ranging from acquaintances, new friends and even famous people. Interviewee J said that "Weibo can help me quickly know what happens on my friends and communicate with them; also, I can pay attention to those who share similar interests with me." Also interviewee D mentioned that "I can keep in touch with others, talk anything with them, even we do not meet each other Face-to-Face. Weibo expands my social life online."

(3) Entertainment.72.22% (13) interviewees reported using Weibo for entertainment. Interviewee B mentioned that "Weibo brings me much fun. I can always find interesting videos or posts. I enjoy it." And interviewee Q said that "I like playing games in Weibo platform, especially with friends."

(4) Pass time. 66.67% (12) interviewees used Weibo for passing time. Some responses from interviewees were "I read posts from Weibo when waiting for bus", "I log in Weibo when class is boring", "I use Weibo when I can't sleep during night".

(5) Self-expression. 55.56% (10) interviewees mentioned using Weibo for self-expression that they can freely publish whatever they want to, even with one or two words, or just some expression. Interviewee R said that "I like to write down what I am thinking on things that I see or experience. Weibo provides various forms for me to do so, such as words, pictures, music or videos." While interviewee G mentioned that "I like making comments to others' posts I am interested in. It is a good place to express whatever you want to."

(6) Information sharing. 55.56% (10) used Weibo for sharing information. Some responses were "I would like to share my ideas through posts or comments", "When I read some interesting posts, I will share them with others", "Through sharing, I could make advertisement for myself."

(7) Social networking. 44.44% (8) reported using Weibo for social networking. Interviewee P mentioned that "I think Weibo is a good social networking tool. It not only makes us keep in touch with old friends, also give us opportunities to know others in the world." Interviewee E further said that "I use Weibo to maintain my social networks through sharing, commenting, or playing together with my online friends."

4.2 Gratifications from Weixin

With respect to Weixin, five gratifications were identified as follows.

(1) Private social networking. This type of gratification was mentioned by 88.89% (16) interviewees. Some responses were "Compared to Weibo, Weixin is more private. Only my friends can read my posts", "(Weixin) is a private social networking platform, not an open one like Weibo", "I communicate with my real friends by Weixin, share ideas with them. So I can protect my privacy. I think this advantage is the most one that we choose to use Weixin".

(2) Social interaction. 77.78% (14) interviewees reported use Weixin for social interaction. Some responses were "I use Weixin to interact with my friends that I have already know in the real life", "Weixin is convenient for me to communicate with my friends. Most of my friends and classmates use Weixin, and we interact with each other through it".

(3) Convenient communication. 66.67% (12) interviewees used Weixin due to convenient communication. Interviewee F said that "Weixin is convenient for us to communicate with each other. There are messages and voices. I like the real-time speaking function. It is just like you are talking with somebody using the mobile phone, but without any costs." Interviewee M further mentioned that "It is indeed a convenient communication tool. It is mobile, so no matter where you are, you can use it to communicate with your friends. Expect for words, voices are more intuitive. Even your friends are not online when you talk with them, they can hear from you after logging in and replying to you quickly."

(4) High-quality information provided. 55.56% (10) interviewees mentioned the high-quality information provided by Weixin. Some responses were "There is not so much rubbish information. Only those I am interested in and focus on can be provided for me", "It is not the same as using Weibo. Weibo always pushes information that I

do not like, such as advertisements. When I use Weixin, I only pay attention to that useful information. And other rubbish information will not come to me if I do not follow them", "The information has high-quality, sometimes I think that I am reading a good prose. And there are no advertisements, less rubbish information".

(5) Information sharing. 50% (9) reported to use Weixin for information sharing. Some responses were "I use Weixin because I can share information with my friends: my current state, what I am doing at the moment, good articles that I ever read", "When I read some interesting posts, I always share them with my friends. I just click the "share" button and it is done".

5   Discussion, Implications and Limitations

5.1   Discussion

Considering Weibo and Weixin as popular social media platforms, users choose to use them for meeting different requirements. For example, for Weibo usage, users seek for diversified information, perceive enjoyment. While for using Weixin, users keep close connections with friends, acquire more valued information. Gratifications sought from Weibo or Weixin further foster users' regular and continuance usage. Thus, we attempt to adopt U&G theory to explore why users choose to use different social media (Weibo or Weixin), to explain what types of gratifications users can seek from using these social media.

Our research indicates some major findings with respect to the choice of instant messaging services. Results show that users acquire various gratifications to varying degrees when using Weibo or Weixin [5]. For the usage of Weibo, users can receive gratifications of information seeking (83.33%), social interaction (72.22%), entertainment (72.22%); pass time (66.67%), self-expression (55.56%), information sharing (55.56%) and social networking (44.44%). With Weixin usage, gratifications of private social networking (88.89%), social interaction (77.78%), convenient communication (66.67%), high-quality information provided (55.56%) and information sharing (50%) are sought. In addition, results show the general gratifications for usage of both Weibo and Weixin: information sharing, social networking and social interaction. More details about the findings will be discussed below.

For gratifications identified for Weibo and Weixin, results suggest that multiple gratifications can be sought from using Weibo and Weixin. With respect to Weibo, the results are partially consistent with the findings of [4-5]. However, our results also show that users use Weibo for social networking. For Weixin use, gratifications of social interaction, convenient communication and information sharing sought were partially accord with what [4] identified. In addition, private social networking and high-quality information provided are special gratifications that sough from Weixin.

For general gratifications from both Weibo and Weixin, social interaction, information sharing and social networking were identified. As popular social media tools, Weibo and Weixin play important roles in fostering users to interact with others, to share information and to maintain their relationships [10] [14].

5.2 Implications

This research has some implications for academia and practitioners in several ways. For academic, this research utilizes uses and gratifications theory to explain the usage of Weibo and Weixin. The findings could further strengthen the relevance and suitability of uses and gratifications theory in social media research. Also, results suggest specific gratifications for Weibo or Weixin, and general gratifications for both of Weibo and Weixin. For practitioners, this research provides a rich understanding of why users choose different social media. These findings could help them better understand users' different preferences and choose suitable social medial platforms so as to meet users' gratifications and effectively achieve their own marketing purposes.

5.3 Limitations and Future Research

The current study has several limitations that should be considered in the future research. First, the study only focuses on the specific sample of Chinese university students. Therefore, the results may not be generalizable to all social media users. Future research needs to explore other types of social media users, such as working staff. Second, this study conducts an exploratory research, and adopts interviews as the main method for data collection. Future research should consider quantitative research and collect data from more samples.

Acknowledgements. This work was supported by National Natural Science Foundation of China (71073066 & 71271099) and Central China Normal University (Program of excellent doctoral dissertation cultivation grant).

References

1. Kaplan, A.M., Haenlein, M.: Users of the world, unite! The challenges and opportunities of social media. Business Horizons 53(1), 59–68 (2010)
2. China Internet Network Information Center. Statistical report on Internet development in China, http://www1.cnnic.cn/IDR/ReportDownloads/
3. China Internet Network Information Center. Research report on user behavior of social applications in China, http://www.cnnic.net.cn/hlwfzyj/hlwxzbg/sqbg/201312/t20131225_43545.htm
4. Whiting, A., Williams, D.: Why people use social media: a uses and gratifications approach. Qualitative Market Research: An International Journal 16(4), 362–369 (2013)
5. Ku, Y.-C., Chu, T.-H., Tseng, C.-H.: Gratifications for using CMC technologies: A comparison among SNS, IM, and e-mail. Computers in Human Behavior 29, 226–234 (2013)

6. Katz, E., Blumler, J.G., Gurevitch, M.: Utilization of Mass Communication by the Individual. In: Blumler, J.G., Katz, E. (eds.) The Uses of Mass Communications: Current Perspectives on Gratifications Research, pp. 19–32. Sage Publications, Beverly Hills (1974)
7. Stafford, T.F., Stafford, M.R.: Identifying motivations for the use of commercial web sites. Information Resources Management Journal 14(1), 22–30 (2001)
8. Quan-Haase, A., Young, A.L.: Uses and gratifications of social media: A comparison of Facebook and instant messaging. Bulletin of Science, Technology and Society 30(5), 350–361 (2010)
9. Raacke, J., Bonds-Raacke, J.: MySpace and Facebook: Applying the uses and gratifications theory to exploring friend-networking sites. CyberPsychology and Behavior 11(2), 169–174 (2008)
10. Lo, O.W.Y., Leung, L.: Effects of gratification-opportunities and gratifications-obtained on preferences of instant messaging and e-mail among college students. Telematics and Informatics 26(2), 156–166 (2009)
11. Pai, P., Arnott, D.C.: User adoption of social networking sites: Eliciting uses and gratifications through a means-end approach. Computers in Human Behavior 29(3), 1039–1053 (2013)
12. Lincoln, Y., Guba, E.: Naturalistic inquiry. Sage, New York (1985)
13. Park, N., Kee, K.F., Valenzuela, S.: Being immersed in social networking environment: Facebook groups, uses and gratifications, and social outcomes. CyberPsychology and Behavior 12(6), 729–733 (2009)

A Bibliometric Analysis of Social Media Research from the Perspective of Library and Information Science

Chunmei Gan[1,*] and Weijun Wang[2]

[1] School of Information Management, Sun Yat-sen University, Guangzhou, China
chunmei_gan@163.com
[2] School of Information Management, Central China Normal University, Wuhan, China
wangwj@mail.ccnu.edu.cn

Abstract. A bibliometric analysis was conducted on social media research in journals under the subject category "Information Science & Library Science" of the Social Science Citation Index. 646 articles were retrieved using the term "social media" as a keyword to search parts of titles, abstracts or keywords of publications. The research performance and trends were analyzed with descriptors of types and languages, characteristics, countries, journals, authorships and author keywords. Results showed that, social media research steadily increased from the period of 2002 to 2013 and the annual publication output in 2012 and 2013 were almost half of the total. A total of 9,851 pages, 29,433 cited references, 1,540 authors and 3,740 citations were identified in all 646 articles, with the average per article of 15.25 pages, 45.46 cited references, 2.38 authors and 5.79 citations. Analysis of countries and journals suggested an uneven distribution of publications on national and journal levels. The USA attained a leading position by contributing the largest share of articles. UK, Spain and China were the other three top productive countries in total publications. 73.53% of the total articles were published in 25 journals with impact factors ranging from 0 to 5. More than half (51.24%) journals had an impact factor between 1 and 3. Journal of Health Communication with 2.079 IF had published the most articles. The most commonly used author keywords appeared in the articles were "social media", "social network", "Internet", "communication", "Web 2.0", "blog", "Twitter", "Facebook" and "virtual community".

Keywords: Social media, bibliometric, SSCI, research performance, trends.

1   Introduction

Social media develops rapidly over the past years, and plays an increasingly important role in a variety of fields, such as business, education and daily life for organizations or individuals. There is a growing consensus that social media is fundamentally changing the way individuals communicate, consume and collaborate [1]. Recent reports show that, 65% of Australians use social media [2], while 43.8% of Internet users use social media in China [3].Also, 73% of online American adults use a social

---

* Corresponding author.

H. Li et al. (Eds.): I3E 2014, IFIP AICT 445, pp. 23–32, 2014.

networking site of some kind [4]. In addition, reports reveals that, marketers place very high value on social media, a significant 86% of marketers indicate that social media is important for their business [5].

Social media research has experienced a considerable increase due to the mentioned popular use of social media technologies and their impacts. Great efforts have been made around the world to better understand and utilize social media. A number of reviews were published within the social media research fields (e.g. [1, 6-8]). However, there were no attempts to provide a more quantitative assessment of the current status and trends of this research thus far.

Scientometrics that measuring the contribution of scientific publications within a given topic could represent current research trends and be used to identify focuses of future. It uses statistical and mathematical tools to map out data and patterns of bibliographical records pertaining to a network of scientific documents. Through bibliometric research of literature, the next research trend could be predicted [9]. Bibliometric studies that evaluate status and progresses of a specific topic based on publication records have a long research strand and have been applied to evaluate performance of publications in many areas.

The purpose of the present research, through a comprehensive bibliometric analysis, is to analyze the status and trends of social media research from the perspective of Library and Information Science (LIS) until the year of 2013, so as to help researchers better understand the panorama of global social media research and predict dynamic direction of research.

2   Research Methods

The data used in this study was obtained from the database of SSCI published by Thomson Reuters Web of Science, Philadelphia, PA, USA. As a strictly selected abstract database, Web of Science has been long recognized as the most authoritative scientific and technical literature indexing tool that can provide the most important areas of science and technology research [10]. To retrieve papers dealing with social media research in LIS discipline, the term "social media" was used as a search query phrase to search the titles, abstracts and keywords of referenced publications under the category of "Information Science & Library Science". A total of 768 publications were identified in the SSCI databases for the period 2002-2013.

In addition, the impact factor (IF) of a journal was determined for each document as reported in the year 2012 by ISI's Journal Citation Reports (JCR). Citation counts of all the papers obtained on March 10th, 2014 when the SSCI search process for this study was conducted. And publications originating from England, Scotland, North Ireland and Wales were grouped under the UK (United Kingdom) heading.

We then performed a bibliometric analysis based on the following descriptors: types and languages of publications, characteristics of scientific output, publication distribution by countries, distribution of outputs in journals, authorships as well as frequencies of author keywords, so as to study the worldwide research activity on social media from the perspective of LIS in general and determine the research trends and performance. The statistical analysis was conducted by the combined use of BibExcel [11] and MS Excel. Also, for analysis of author keywords, following the formation of the basic word list, all words devoid of meaning (i.e., "if", "and", "with", etc.) were removed from the list, and the alternate forms of words were grouped together.

3   Results and Discussion

3.1   Types and Languages of Publications

There were 768 papers on social media research under category of LIS in the ISI Web of Science database. 6 document types were identified. The most frequent document type was journal articles (646), comprising 84.11% of the total publication, followed by book review (50, 6.51%) and editorial material (43, 5.6%). Other less significant document types were review (27, 3.52%), proceedings paper (25, 3.26%) and news item (2, 0.26%), which accounted for 7.03% in total.

Since original articles were the most-frequently used type, also following the conventions used in other bibliometric studies, journal articles were used for further analysis, while other types of publications were then removed.

As for publishing language, English remains the dominant language, making up 92.41% (597) of all the published articles. This result was consistent with the fact that English is the prevalent academic language as most journals indexed by SSCI are published in English. Other publication languages included Spanish (34), Hungarian (7), Portuguese (4), German (3) and French (1).

3.2   Characteristics of Scientific Output

The articles devoted to social media research from the perspective of LIS were summarized in Table 1, showing some basic characteristics of all 646 articles. The number of total pages (NP) for all 646 articles was 9,851, and the average page per article (NP/NA) was 15.25. Further, among the 646 articles, the maximum pages were 52, and the minimum was 1. The article with 52 pages was "researchers and practitioners talk about users and each other: making user and audience studies matter", written by Dervin B and Reinhard C D, and published in "Information Research - An international electronic journal". Both of the two articles with 1 page were published in "Econtent" journal, written by Scott D M in 2007 and 2009, and titled with "social media debate" and "after thought armed with social media", respectively.

Table 1. Characteristics of articles outputs

	NA	NP	NP/NA	NR	NR/NA	NAU	NAU/NA	TC	TC/NA
Total	646	9,851	15.25	29,433	45.56	1,540	2.38	3,740	5.79
Max	/	52	/	160	/	17	/	139	/
Min	/	1	/	0	/	1	/	0	/

Note: NA number of total articles, NP number of total pages, NP/NA average page per article, NR number of cited references, NR/NA average cited references per article, NAU number of total authors, NAU/NA average author per article, TC times cited, TC/NA times cited per article, Max The maximum, Min The minimum, / no data

The total number of cited references (NR) for all 646 articles was 29,433, and the average cited references per article (NR/NA) were 45.46. In addition, the maximum cited reference was 160, while the minimum was 0. The article with 160 cited references was titled "online social movements", a review paper published in "Annual Review of Information Science and Technology" journal in 2011, written by Hara N and Huang B Y from Indiana University. Also, there were 15 articles with 0 cited references. The further analysis of these 15 articles revealed that their topics were intensively related to social media applications in different areas, such as business, marketing and library. Meanwhile, 8, 5, 1 and 1 article(s) were published in journals "Econtent", "Online", "Learned Publishing" and "Profesional de la Información", respectively.

For authors, there were 1,540 authors in total (NAU) for all 646 articles, and the number of average authors per article (NAU/NA) was 2.38, indicating the collaboration trend for current research. Moreover, the article with the most authors was written by 17 authors, of which 6 from Korea and 11 from Pakistan, and published in "Journal of Health Communication".

Additionally, there were 3,740 citations (TC) in the total of 646 publications for an average of 5.79 citations per article (TC/NA), indicating the high impact or visibility of social media research from LIS. The most frequently cited article was "acceptance of Internet-based learning medium: the role of extrinsic and intrinsic motivation", written by Lee M K O, Cheung C M K and Chen Z from Hong Kong, China. This article was published in "Information & Management" journal, and empirically investigated extrinsic and intrinsic motivators affecting students' intention to use an Internet -based learning medium. It has been cited 139 times since published in 2005. Meanwhile, among the most times cited articles, 129 (19.97%) articles were cited by 2,970 (79.41%) times, following the Pareto principle.

In addition, Figure 1 shows the publication years of all 646 articles. We could find that research on social media has risen yearly from 2003 to 2013 and aroused more and more attention in recent years. The annual number of publication output has grown rapidly from 16 in 2002 to 174 in 2013. Articles published in the year of 2012 and 2013 accounted for almost half of the total publications (45.2%).

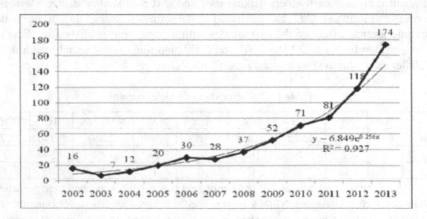

Fig. 1. Publications of each year

Also, as shown in Fig. 1, the progression in the number of articles each year was further studied. There were significant correlations between the publication years and the number of articles published in those years. The growth pattern of the progression was simulated by one exponential function $y = 6.849e^{0.256x}$, with coefficients of determination $R^2=0.927$.

3.3   Publication Distribution by Countries

LIS scholars from 54 countries have contributed in social media research. Table 2 presented the top 15 countries/territories that were most productive, suggesting a geographic inequality. Among 17 countries/territories ranked top 15, there were 9 European countries, 5 Asian countries, 2 and 1 from North America and Africa, respectively. Moreover, USA ranked first with 297 (39.23%) publications, followed by UK (63, 8.32%) and Spain (49, 6.47%). China (38, 5.02%) ranked the fourth in total publications.

Table 2. Top 15 countries/territories with most articles

Country	Articles	Percentage%	Country by region	Country	Articles	Percentage%	Country by region
USA	297	39.23%	North America	Finland	15	1.98%	Europe
UK	63	8.32%	Europe	Sweden	15	1.98%	Europe
Spain	49	6.47%	Europe	Singapore	14	1.85%	Asia
China	38	5.02%	Asia	Israel	13	1.72%	Asia
Canada	33	4.36%	North America	South Africa	9	1.19%	Africa
South Korea	29	3.83%	Asia	Ireland	8	1.06%	Europe
Netherlands	24	3.17%	Europe	Italy	8	1.06%	Europe
Australia	23	3.04%	Europe	Taiwan	8	1.06%	Asia
Germany	21	2.77%	Europe	Total	667	88.11%	/

3.4   Distribution of Output in Journals

All 646 articles were published in a wide range of 78 journals. Table 3 lists the top 16 journals (25 journals in total) with the greatest number of published articles, comprising 73.53% of the total 646 articles. There was a high concentration of social media publications in these top journals, which follows the Zipf's law and is consistent with observation in other fields. Journal of Health Communication ranked first with 55 published articles, followed by Journal of Computer-Mediated Communication with 50 published articles.

Impact factor and journal rank of each journal are also shown in Table 3. Impact factors of all the 25 journals were ranged from 0 to 5 according to 2012 JCR Social Science Edition. The journal with the highest impact factor was MIS Quarterly (4.659, ranked 1st of 85 journals in LIS) with 8 published articles, followed by Scientometrics (2.133, ranked 7th) and Journal of Health Communication (2.079, ranked 8th).

Table 3. The top 16 journals with the greatest number of articles

Journal	P	P%	IF	Journal rank
Journal of Health Communication	55	8.51%	2.079	8/85
Journal of Computer-Mediated Communication	50	7.74%	1.778	13/85
Government Information Quarterly	39	6.04%	1.910	11/85
Social Science Computer Review	37	5.73%	1.303	24/85
Journal of the American Society for Information Science and Technology	32	4.95%	2.005	10/85
Profesional de la Información (Spanish)	29	4.49%	0.439	53/85
Information Society	28	4.33%	1.114	30/85
Information Systems Research	20	3.10%	2.010	9/85
Aslib Proceedings	19	2.94%	0.432	54/85
Telematics and Informatics	16	2.48%	/	/
Journal of Documentation	15	2.32%	1.138	29/85
Online Information Review	13	2.01%	0.939	37/85
Information Research - An International Electronic Journal	11	1.70%	0.520	49/85
Library Hi Tech	11	1.70%	0.621	46/85
Libri	10	1.55%	0.368	64/85
Journal of Information Science	10	1.55%	1.238	26/85
Scientometrics	10	1.55%	2.133	7/85
Information & Management	10	1.55%	1.663	15/85
Electronic Library	10	1.55%	0.667	44/85
Ethics and Information Technology	9	1.39%	0.846	40/85
International Journal of Information Management	9	1.39%	1.843	12/85
Journal of Knowledge Management	8	1.24%	1.474	20/85
MIS Quarterly	8	1.24%	4.659	1/85
Econtent	8	1.24%	0.127	79/85
Journal of Management Information Systems	8	1.24%	1.262	25/85

Note: P total publications, P% share in publications, IF impact factor

Shares of articles in different ranges of impact factor were further analyzed (Table 4). There were 9 journals with impact factor between 0 and 1, 10 journals between 1 and 2, 4 journals between 2 and 3, and 1 journal between 4 and 5. In addition, journals with impact factor between 1 and 3 accounted for more than half (51.24%) of the total publications, which indicated that qualities of publications in social media research from LIS scholars were high.

Table 4. Shares of publications in different ranges of impact factor

IF	J	P	P%	IF	J	P	P%
0<IF≦1	9	120	18.58	3<IF≦4	0	0	0
1<IF≦2	10	214	33.13	4<IF≦5	1	8	1.24
2<IF≦3	4	117	18.11	N/A	1	16	2.48
Total	25	475	73.53				

Note: IF impact factor, J number of journals, P total publications, P% share in publications, N/A no data

3.5 Patterns of Authorships

There were 201 articles with single author, while 445 articles with multiple authors. Moreover, 1,448 authors appearing 1,540 times were identified for all the 646 articles. 67 authors published at least 2 articles, and 1,381 author published one article. Table 5 lists frequency of times authors appeared of the total publications.

Table 5. Frequency of times authors appeared

T	F	F%	T	F	F%	T	F	F%
9	1	0.07	4	3	0.21	2	54	3.73
6	1	0.07	3	8	0.55	1	1381	95.37

Total: 1448

Note: T Times appeared, F frequency, F% share of frequency

We also list the most productive authors published at least 4 articles in Table 6. The most productive author in social media research from LIS was Park H W (12 articles) from Yeungnam University in South Korea, followed by Bonson E (6 articles) from University of Huelva in Spain, Rice R E (5 articles) from University of California, Santa Barbara in USA and Thelwall M (5 articles) from University of Wolverhampton in UK.

Table 6. The most productive authors

Author name	Articles	Author name	Articles	Author name	Articles
Park HW	12	Flores F	4	Marcella R	4
Bonson E	6	Yuan YC	4	Bertot JC	4
Rice RE	5	Jaeger PT	4	Total	52
Thelwall M	5	Baxter G	4		

3.6 Frequencies of Author Keywords

The technique of statistical analysis of keywords may reflect directions of research. Especially, authors' keywords analysis could offer the information of research trends as viewed by researchers [9]. Examination of author keywords revealed that, 1,649 keywords were identified among the total 646 publications. Table 7 shows the frequency of different number of keywords appeared in different times. Among them, 1,386 (84.05%) keywords appeared only once, and 147 (8.91%) appeared twice. These words maybe reflected a lack of continuity in research and a wide difference in research focuses [12].

Table 7. Frequency of keywords appeared in different times

Times appeared	Number of key-words	Frequency %	Times appeared	Number of key-words	Frequency %	Times appeared	Number of key-words	Frequency %
1	1,386	84.05	9	2	0.12	21	2	0.12
2	147	8.91	10	3	0.18	22	1	0.06
3	55	3.34	12	1	0.06	23	1	0.06
4	18	1.09	13	2	0.12	35	1	0.06
5	9	0.55	14	1	0.06	38	1	0.06
6	6	0.36	15	1	0.06	45	1	0.06
7	2	0.12	18	1	0.06	69	1	0.06
8	5	0.30	19	1	0.06	142	1	0.06

We also presented the top 15 most active author keywords in Table 8. Other than the term "social media" used for searching, the most frequently used keywords was "social network", which is highly accorded with that fact that social network has attracted much attention from scholars. Also, "Internet" and "communication" were the other two most active words. This is consistent with the fact that social media is a type of new Internet application and is convenient for people to communicate with each other. In addition, "Web 2.0", "blog", "Twitter", "Facebook" and "virtual community" were paid more attention to because they are typical applications of social media and used mostly according to the current statistic reports. Furthermore, social media is applied into different areas, as "library" (21), "political participation" (13) and "e-government" (12) were also appeared in the list.

Table 8. Top 15 most active author keywords

Author keyword	Times	Author keyword	Times	Author keyword	Times
social media	142	Facebook	21	political participation	13
social network	69	library	21	e-government	12
Internet	45	virtual community	19	Information retrieval	10
communication	38	social interaction	18	knowledge management	10
Web 2.0	35	social capital	15	innovation	10
Blog	23	social network analysis	14	Total	550
Twitter	22	digital divide	13		

4   Conclusion

In this study, we provided a supplement evaluation on the global research status and trends in social media studies from the perspective of LIS, utilizing scientometrics approach by summarizing types and languages, characteristics, geographic distributions, journals, patterns of authorships and frequencies of author keywords. The findings of this study could help LIS researchers better understand the performance of social media research in the world, as well as direct for future research.

The study showed that social media research in LIS started in 2002 and has a steady growth in the scientific outputs, following one exponential function in terms of increasing number of annual publications. Also, publications of the year of 2012 and 2013 accounted for almost half of the total publications (45.2%), suggesting social media research has attracted increasing interests from LIS scholars in recent years.

There were 768 papers related to social media research under category of LIS, of which 6 document types were identified and journal articles made up of the majority (84.11%). All 646 articles had a total of 9,851 pages, 29,433 cited references and 1,540 authors. The average for one article was 15.25 pages, 45.46 cited references and 2.38 authors. Meanwhile, all articles got 3,740 citations with an average of 5.79 citations per article. And 129 (19.97%) articles were cited by 2970 (79.41%) times, which follows the Pareto principle.

The research output distributed unevenly over 54 countries contributed in social media research. The USA attained a leading position by contributing the largest share of articles. UK, Spain and China were the other three top productive countries in total publications.

All 646 articles were published in 78 journals, of which 73.53% were published in top 16 journals with impact factors ranging from 0 to 5, and followed the Zipf's law. The journal with the most published articles was Journal of Health Communication with 2.079 IF, ranking 8th of 85 journals in LIS. Also, MIS Quarterly and Scientometrics were the top 2 journals with the highest IF among the top 16 journals. Also, more than half (51.24%) journals had an impact factor between 1 and 3.

201 articles were written by single author and 445 articles by multiple authors. 1,448 authors were identified with 1,540 times appeared, of which 67 authors published at least 2 articles and 1,381 author published one. The most productive author was Park H W (12 articles), Bonson E (6 articles), Rice R E (5 articles) and Thelwall M (5 articles).

1,649 keywords were presented among the total 646 publications, of which 1,386 (84.05%) and 147 (8.91%) keywords appeared only once and twice, respectively. The most commonly used author keywords appeared in the articles were "social media" and "social network", indicating that topics related to social network have received clearly increasing interests. In addition, "Internet", "communication", "Web 2.0", "blog", "Twitter", "Facebook" and "virtual community" were the most active author keywords, which reflected the current research focuses and might become the future research angles.

Acknowledgements. This work was supported by National Natural Science Foundation of China (71073066 & 71271099) and Central China Normal University (Program of excellent doctoral dissertation cultivation grant).

References

1. Wu, J.J., Sun, H.Y., Tan, Y.: Social media research: A review. Journal of Systems Science and Systems Engineering 22(3), 257–282 (2013)
2. Sensis. Yellow Pages Social Media Report (2013),
   http://about.sensis.com.au/IgnitionSuite/uploads/docs/
   Yellow%20Pages%20Social%20Media%20Report_F.PDF
3. The Kantar China Social Media Impact Report (2014),
   http://cn.kantar.com/media/633232/2014.pdf
4. Social Media Update (2013),
   http://pewinternet.org/Reports/2013/Social-Media-Update.aspx
5. 2013 Social Media Marketing Industry Report,
   http://www.socialmediaexaminer.com/SocialMediaMarketingIndus
   tryReport2013.pdf
6. Bullard, S.B.: Writing on the Wall: Social Media-The First 2,000 Years. Journalism & Mass Communication Quarterly 91(1), 182–183 (2014)
7. Gao, J.: Producing the Internet: Critical Perspective of Social Media. International Journal of Communication 8, 165–167 (2014)
8. Mesquita, A.: Social Media for Academics: A Practical Guide. Online Information Review 37(6), 987–988 (2013)
9. Garfield, E.: Citation Indexing for Studying Science. Nature 227(5259), 669–671 (1970)
10. Boyack, K.W., Klavans, R., et al.: Mapping the backbone of science. Scientometrics 64(3), 351–374 (2005)
11. Persson, O., Danell, R., Schneider, J.W.: How to use Bibexcel for various types of bibliometric analysis. In: Åström, F., Danell, R., Larsen, B., Schneider, J. (eds.) Celebrating scholarly communication studies: A Festschrift for Olle Persson at his 60th Birthday, pp. 9–24. International Society for Scientometrics and Informetrics, Leuven (2009)
12. Chuang, K.Y., Huang, Y.L., Ho, Y.S.: A bibliometric and citation analysis of stroke-related research in Taiwan. Scientometrics 72(2), 201–212 (2007)

Understanding the Influence of Electronic Word-of-Mouth on Outbound Tourists' Visit Intention

Ping Wang

Turku School of Economics, University of Turku, Turku, Finland
ping.wang@utu.fi

Abstract. The valence of online user-generated reviews is an increasing important antecedents affecting tourist's decision with the pervasion of Web 2.0 and information and communication technology. The purpose of this study is to explore the influence of electronic word-of-mouth (eWOM) on outbound tourists' intention to visit a destination through a dual-process perspective. A research model was proposed based on the dual-route theory of elaboration likelihood model (ELM) and theory of planned behavior (TPB). The research model was empirically tested with the data collected among the Chinese outbound tourists. The research results indicate that, tourist's attitude towards a destination was positively influenced by argument quality of eWOM, and intention to recommend the destination before travel was positively influenced by attitude towards destination and source credibility of destination related eWOM. Outbound tourists' intention to visit a destination was positively influenced by argument quality, attitude towards destination, and WOM intention. Several practical and theoretical implications are also discussed in the paper.

Keywords: Electronic word-of-mouth (eWOM), travel decision, online travel information, visit decision, elaboration likelihood model, theory of planned behavior

1 Introduction

With the advancement of information and communication technology (ICT) and advent of Web 2.0 in recent years, tourists are enabled to communicate virtually and share travel experience, reviews online [1]. Increasing amounts of user-generated-content (UGC) were spawned in the form of online travel experience or online reviews that pertain to personal experience with particular service or products which is also described as electronic word-of-mouth (eWOM) [2]. EWOM are available to a multitude of people via Internet and Web 2.0 platforms, which is likely to generate persuasive effects on target audiences, such as potential customers [3]. As an information-intensive industry, tourism was deeply influenced by eWOM [4-7]. Electronic word-of-mouth have facilitated travel information searching behavior, and influenced tourists' travel planning behavior [8, 9]. Tourists nowadays are increasingly using eWOM to inform themselves about travel-related services and products, for travel planning and travel related purchasing [10-12].

H. Li et al. (Eds.): I3E 2014, IFIP AICT 445, pp. 33–45, 2014.

As eWOM has strongly affected tourism industry especially in tourist's decision making process, a considerable amount of literature has been focused on eWOM in tourism [13-16]. These studies mainly focused on two streams: one stream focused on exploring the factors influencing eWOM generating motivation [2, 9, 17], while the other stream concentrated on the impacts of eWOM on target behavior [18, 19]. Notwithstanding, our understanding about how eWOM influence tourists' visit intention is still scant. What is the most effective information influencing tourists' attitude towards destination, and furthermore their visit intention, and how? Thus, researches aiming at solving this puzzle are needed.

A structural model based on well-tested theories was proposed in this research to explore the influence process of eWOM on outbound tourists' intention to visit a destination. Empirical data was collected among Chinese outbound tourists via structural questionnaire survey. This study contributes to the understanding of how destination related electronic word-of-mouth impact outbound tourists' leisure travel destination decision-making with empirical evidence. Furthermore, the influence route before initial attitude formation toward a destination is also explored.

The remainder of this article is structured as follows: In section 2, theory of elaboration likelihood is introduced, and incorporated to the theory of planned behavior holding that behaviours are influenced by intentions, intentions by attitudes [20, 21]. Nine hypotheses are proposed. Research method is delineated in section 3, including data collection, descriptive analysis of the sample characters, and measurement model. In section four, structural equation model is used to analyse data with AMOS22.0. In section five, we give implications for both academic and practice. At the end of the paper in section six, limitations of current research and future research suggestions were discussed.

2   Research Model and Hypotheses

2.1   Elaboration Likelihood Model

Elaboration likelihood model (ELM) originates from social and consumer psychology, which suggests that attitude changes in two distinct routes: central route and peripheral cue [22]. The basic idea of ELM suggests that the way individuals are persuaded varies according to the extent to which they are willing to engage in elaboration of the persuasive issue. In the central route, individuals think more critically about issue-related arguments in an informational message and scrutinize the both pros and cons as related to those arguments prior to forming their target behavior [23]. In contrast, in peripheral route to attitude change, a relatively less cognitive effort is needed. A person may simply rely on cues related to target behavior, like prior users, prior related personal experience, and credibility. The central route is message-related argument oriented, while the peripheral route is process cues-oriented.

'Elaboration' is defined as 'the extent to which the individuals engage in information contained in the communication, and mentally modify or process the issue' [24]. The term 'likelihood' illustrates whether elaboration is likely or unlikely [24], and is used to describe individuals would add something of their own to a given information in a concrete communication [22, 23]. In the context of current study, people in high likelihood of having an outbound travel are more likely to be persuaded by potential destination related electronic word-of-mouth through a central route, what's more, they think more critically about the destination related-message of eWOM. On the contrary, people in a low likelihood would refer to peripheral cues, such as the credibility of the information source. It is suggested that individuals' degree of elaboration forms a spectrum of message-related thinking, which could vary from little to high extent. What's more, the attitude change may happen at any extent of elaboration [23], that is, both central route and peripheral route can lead to attitude change, as shown in Figure 1.

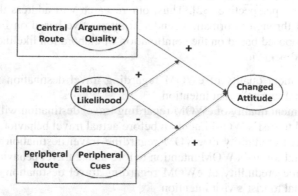

Fig. 1. Elaboration likelihood model

In the context of tourism of this study, source credibility refers to the credibility of website or online community in which eWOM of travel destination is disseminated. Argument quality refers to the quality of content/message contained in the travel destination-related eWOM. Source credibility and argument quality of social media material are empirically tested to be related with attitude formation in the context such as outbound studying destination choice [25] and adoption of accommodation related online review [4]. Thus, we propose that argument quality and source credibility of destination-related eWOM have positive influence on tourists' attitude towards the destination. Hypotheses 1 and 2 are:

H1: Argument quality of destination-related eWOM will positively influence tourists' attitude towards leisure travel destination.

H2: Source credibility of destination-related eWOM will positively influence tourists' attitude towards leisure travel destination.

2.2   Incorporating Behavior Intention to ELM

Intentions to perform different kinds of behaviors can be predicted from attitude towards the behavior in one of the most researched and reputational psychological theory—theory of planned behavior [20]. Several prior studies applied theory of planned behavior to explore the relationship between tourists' attitude and behavior intention [26-28]. Ajzen (1991) also assumes that a person's likelihood behavior can be dictated to his/her available resources. Therefore, we incorporate intention into the elaboration likelihood model. Concretely, the tourists' behavior intention in our study is discussed with the terms of intention to visit a destination and WOM intention. Word-of-Mouth intention refers to tourist intention to recommend a destination to others according to their knowledge gained from destination-related eWOM before their own travel.

Argument quality describes the persuasive strength embedded in the eWOM related to a leisure travel destination. The dependent variables of intention imitated from the theory of planned behavior (TPB) are intention to visit an outbound leisure travel destination, and the intention to recommend the destination (WOM intention). Prior research has focused on predicting WOM adoption from information receiver's perspective [29-31], few researches focused on exploring WOM intention from information sender's perspective [32]. Thus, our research will address the intention to distribute WOM through information sender's perspective based on ELM. Hypotheses 3 to 6 are proposed based on the combination of elaboration likelihood model and theory of planned behavior.

H3: The argument quality of eWOM regarding travel destination will positively influence potential tourist's visit intention.

H4: The argument quality of eWOM regarding travel destination will positively influence potential tourist's WOM intention before actual travel behavior.

H5: The source credibility of eWOM regarding travel destination will positively influence potential tourist's WOM intention before actual travel behavior.

H6: The source credibility of eWOM regarding travel destination will positively influence potential tourist's visit intention.

After information searching behavior, an initial destination image will come into formation in the mind of potential tourists [33]. We also propose that positive attitude will lead to word-of-mouth before practical travel experience, and strengthen visit intention.

H7: Tourist's attitude towards destination will positively influence visit intention.

H8: Tourist's attitude towards destination will positively influence WOM intention.

H9: Tourist's WOM intention will positively influence visit intention.

Figure 2 presents the research model and hypotheses in our study:

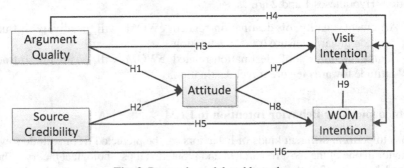

Fig. 2. Research model and hypotheses

3   Research Method

3.1   Data Collection

Prior to our formal survey in China, pilot studies for both English and Chinese versions of questionnaire have been done during May 5th to 15th 2014. Some English-speaking experts and professionals in tourism were interviewed, including academic researchers in University of Turku and the Finnish Tourist Board, to test the logistics, validity, and construct of the questionnaire. The questionnaire was then translated into Chinese after that, and proofread by some bilinguals capable of Chinese and English to ensure the consistency of our questionnaire. The pilot survey of Chinese questionnaire was done among 15 Chinese people to make sure the language and logic of questionnaire in Chinese. The final version was achieved after all the interviews, pilot survey and discussion with experts.

Our data were collected through online survey due to its advantage of faster, cheaper, and easier use for both participants and researchers compared to off-line survey. As our target group is those who are interested in Finland and most potential tourists to Finland in the near future, questionnaire was distributed through the official account of VisitFinland on Wechat and Weibo—two of the most popular social media in China. In this way, potential respondents will mostly be the followers of Visit-Finland. As being the followers of official account of Finnish Tourist Board Weibo and Wechat, it's true that they already paid attention to Finland as a travel destination. 959 responses were collected in two weeks after our questionnaire distribution. For our study in this paper, only respondents who have never travelled to Finland, and were influenced by online travel experience or reviews during their destination decision process, were included. Our final samples for analysis consist of 195 respondents.

3.2   Sample Characteristics

The demographic profile of our sample is presented in Table 1. More than 80% are between the age of 20 and 40 years old, and nearly a half of them are singles. Most (89.7%) of them received relatively high education from university with a bachelor degree or above. 79% of them have a monthly income of more than 5,000 Yuan, and more than 40 % of them earn more than 10,000 Yuan per mouth, which confirmed their likelihood of taking an outbound travel in recent three years. In addition, nearly 87% of the respondents have outbound travel experience in recent three years, and 60% of them have travelled abroad for at least three times in recent three years.

3.3   Measurement

Five factors were included in our model: argument quality, source credibility, attitude towards destination, WOM intention, and visit intention. Each construct was measured with multiple items adapted from previous literatures to validate content. Necessary modifications have been done according to our current research context.

Table 1. Sample characteristics (N=195)

Sample characteristic	Category	Frequency	Percent %
Gender	Male	4	27.7
	Female	141	72.3
Age	Less than 20	6	3.1
	20-30	94	48.2
	31-40	76	39.0
	41-50	15	7.7
	51-60	4	2.1
Education	Under bachelor degree	22	11.3
	Bachelor degree	130	66.7
	Master degree	40	20.5
	Doctoral degree	3	1.5
Family size	Single	95	48.7
	Couple without children	44	22.6
	Couple without children at home	5	2.6
	Couple with children at home	48	24.6
	Other	3	1.5
Income per mouth	Less than 5,000	41	21.0
	5,001-10,000	69	35.4
	10,001-20,000	51	26.2
	20,001-30,000	20	10.3
	More than 30,000	14	7.2
Outbound Travel Frequency (in recent three years)	1	95	13.8
	2	44	17.9
	3	5	26.2
	4	48	8.2
	5 or more	3	33.8

4   Data Analysis and Results

The data were analysed using structural equation model with SPSS Amos 22.0. First, the reliability and validity of the scales used in our measurement model were tested. Model fit was also examined.

4.1   Reliability and Validity

The validation of our measurement model was investigated in both convergent validity and discriminant validity. We investigated convergent validity by computing all factor loadings, composite reliabilities, Cronbach's Alpha and the average variance extracted (AVE). As shown in Table 3, all the factor loadings exceeded the recommended value of 0.70, composite reliabilities exceeded 0.8, and all AVEs exceeded 0.5, which shows a good convergent validity was achieved [34]. All Cronbach's Alpha values are higher than 0.7, which implies good reliability [35].

Table 2. Convergent validity: Factor loading, composite reliability, Cronbach's Alpha and AVE

Item	Mean (N=195)	Factor Loading	Composite Reliability	Cronbach's Alpha	AVE
AQ1	4.09	0.8867			
AQ2	3.73	0.8151	0.8790	0.7952	0.7081
AQ3	3.90	0.8208			
SC1	3.89	0.7844			
SC2	3.84	0.8909	0.8855	0.8591	0.7211
SC3	3.99	0.8686			
ATT1	4.51	0.8828			
ATT2	4.52	0.9361	0.9143	0.8591	0.7809
ATT3	4.58	0.8289			
WOM1	4.29	0.9356			
WOM2	4.46	0.8730	0.9381	0.9012	0.8349
WOM3	4.27	0.9313			
INT1	4.23	0.9142			
INT2	4.35	0.9539	0.9538	0.9273	0.8731
INT3	4.45	0.9346			

In addition, we compared the square root of the AVE and factor correlation coefficients to examine the discriminant validity of our constructs. As shown in Table 4 below, all factors' latent variable correlations with other factors are smaller than the square root of its AVEs, which shows a good discriminant validity of our scales [34, 36].

Table 3. Discriminant validity: latent variable correlations and the square roots of AVEs (Square root of AVEs in the main diagonal)

	AQ	SC	ATT	WOM	INT
AQ	**0.8415**				
SC	0.5987	**0.8492**			
ATT	0.1750	0.1480	**0.8837**		
WOM	0.1281	0.1768	0.6563	**0.9137**	
INT	0.2400	0.1525	0.5903	0.6747	**0.9344**

4.2   Hypotheses Testing

Before exploring the path coefficient, the goodness of model fit was tested in AMOS. The results of model fit were shown in Table 5. According to the rules of thumb, the values of GFI, NFI,CFI all proved a satisfactory structural model fit to the data [37].

Table 4. Goodness-of-fit measures for original model

Chi-square	Degrees of freedom	GFI	NFI	CFI
199.890 (.000)	81	0.891	0.853	0.937

Notes: p-value in parentheses; GFI=goodness of fit; NFI=normal fit index; CFI=comparative fit index.

Figure 3 graphically presents our path coefficient results and variance explained. Six out of nine hypotheses were supported in our results (see Table 6). The results indicate that: tourist's attitude towards destination was positively influenced by argument quality (β=0.173, p<0.01); word-of-mouth intention was positively influenced by attitude (β=0.720, p<0.001) and source credibility (β=0.104, p<0.01); visit intention was positively influenced by argument quality (β=0.192, p<0.01), attitude towards destination (β=0.149, p<0.05), and WOM intention(β=0.620, p<0.001). It is estimated that predictors in our model of visit intention can explain 58.5 percent of its variance, 52.7 percent of WOM intention, and 3.5 percent of attitude.

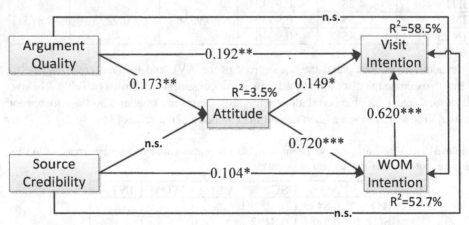

Note: n.s.: not significant; ***p<0.001; **p<0.01; *p<0.05

Fig. 3. Results of analysis

Table 5. Summary of hypotheses testing

Hypothesis	Result
H1: Argument quality→ Attitude towards destination (+)	**Supported**
H2: Source credibility → Attitude towards destination (+)	Not Supported
H3: Argument quality→ Visit intention (+)	**Supported**
H4: Argument quality→ WOM intention (+)	Not Supported
H5: Source credibility → WOM intention (+)	**Supported**
H6: Source credibility → Visit intention (+)	Not Supported
H7: Attitude→Visit intention (+)	**Supported**
H8: Attitude→ WOM intention(+)	**Supported**
H9 WOM intention→ Visit intention (+)	**Supported**

5   Conclusions and Discussion

5.1   Main Results

The results of this study demonstrate a combination of dual process attitude change theory—elaboration likelihood model—with the theory of planned behavior. The empirical results indicate that tourist's attitude towards travel destination is positively influenced by argument quality rather than source credibility of eWOM. However, argument quality of eWOM will only influence tourists' own attitude and visit intention, but not their recommendation behavior. Additionally, it's not argument quality but source credibility of eWOM that would lead to tourist's recommendation intention even before their own travel experience. Contrarily, tourist's visit intention is influenced by eWOM through a central route of argument quality, but the influence of source credibility is not significant as implicated by our empirical results. Consistent with prior research results, visit intention and recommendation intention were positively influenced by attitude [38].

5.2   Practical Implications

This research contributed to tourism industry for giving implications on better understanding tourists' decision making behavior. As an information intensive industry, nowadays increasing amount of new information and communication technology (ICT) has been penetrating into nearly every tourism sector [39]. As to tourism stakeholders, especially for destination marketers, this study makes great value in their understanding of what types of information positively motivate potential tourist attitude towards destination, and finally visit intention.

Our empirical findings also indicated that during tourist's destination decision making process, argument quality of online information sources is highly valued for the formation of positive initial image and attitude towards travel destination. It means that, in tourist's destination decision making process, tourist will think critically about destination-related arguments and scrutinize the relative merits and relevance of those arguments prior to forming a positive attitude to travel destination, and further visit intention. In this case, eWOM will influence tourist's destination decision through a central route.

However, tourist's recommendation behavior was influenced by source credibility and attitude towards destination rather than argument quality. It implies that individuals' recommendation behavior is determined more by the credibility of the source they obtained, but requires less scrutinize of the information and content contained in the eWOM itself. Thus, eWOM will influence tourist's recommendation behavior through a peripheral route, which is cues-oriented.

5.3   Theoretical Implications

This research also has theoretical implications in IS user behavior research. As illustrated, our research model synthesized the dual-process theory and the theory of planned behavior in social psychology literature by incorporating behavior intention into ELM. The model was empirically tested in the context of electronic word-of-mouth in tourism. These two models complete each other. Theory of planned behavior illustrates the relationship between attitude and intention, but does not explain the how the influence process before attitude formation. Elaboration likelihood model supplement TPB in the influence process before the very initial attitude change. The combination of these two models shows good model fitness with our data. Thus, our research contributes a predictive model of influence process of IS on user's intention in the context of tourism.

6   Limitation and Future Research

Some limitations exist in this research. Exploring outbound tourist behavior is challenging, especially for an empirical research not only due to the difficulties in data collection, but also in the uncertainty influence of multi-culture values. Thus, data collection could conduct through wider channels in future study. In addition, the impact of culture values should be taken into consideration in future research.

We explored potential tourist's visit intention through a general dual influence process perspective, which limits the explanation power of attitude in our model. Therefore, research in the future could decompose attributes of argument quality of eWOM to identify the decomposed attribute of argument quality.

Acknowledgement. I would like to state my gratitude to Katarina Wakonen and Kristiina Hietasaari in Finnish Tourist Board and VisitFinland representatives in China for the data collection, the participants for pilot study, as well as Hongxiu Li, Reima Suomi for all the assistance in conducting this study.

This research is supported by China Scholarship Council, and part of the achievements of the China MOE Project of Humanities and Social Sciences (No.13YJC630228), and the National Natural Science Foundation of China (No. 71362027).

References

1. Kim, M.J., Lee, C.K., Chung, N., Kim, W.G.: Factors Affecting Online Tourism Group Buying and the Moderating Role of Loyalty. Journal of Travel Research 53(3), 380–394 (2014)
2. Bronner, F., de Hoog, R.: Vacationers and eWOM: Who Posts, and Why, Where, and What? Journal of Travel Research 50(1), 15–26 (2011)
3. Shih, H.P., Lai, K.H., Cheng, T.C.E.: Informational and Relational Influences on Electronic Word of Mouth: An Empirical Study of an Online Consumer Discussion Forum. International Journal of Electronic Commerce 17(4), 137–165 (2013)
4. Filieri, R., McLeay, F.: E-WOM and Accommodation: An Analysis of the Factors That Influence Travelers' Adoption of Information from Online Reviews. Journal of Travel Research 53(1), 44–57 (2014)
5. Kozak, M.: Holiday taking decisions - The role of spouses. Tourism Management 31(4), 489–494 (2010)
6. Jalilvand, M.R., Samiei, N.: The impact of electronic word of mouth on a tourism destination choice Testing the theory of planned behavior (TPB). Internet Research 22(5), 591–612 (2012)
7. Li, H., Yong, L.: Post adoption behaviour of e-service users: an empirical study on Chinese online travel service users. In: ECIS 2011 Proceedings, Paper 56 (2011)
8. Cox, C., Burgess, S., Sellitto, C., Buultjens, J.: The role of user-generated content in tourists' travel planning behavior. Journal of Hospitality Marketing & Management 18(8), 743–764 (2009)
9. Li, H., Liu, Y.: Understanding post-adoption behaviors of e-service users in the context of online travel services. Information & Management (2014)
10. Tham, A., Croy, G., Mair, J.: Social Media in Destination Choice: Distinctive Electronic Word-of-Mouth Dimensions. Journal of Travel & Tourism Marketing 30(1-2), 144–155 (2013)
11. Standing, C., Tang-Taye, J.P., Boyer, M.: The Impact of the Internet in Travel and Tourism: A Research Review 2001-2010. Journal of Travel & Tourism Marketing 31(1), 82–113 (2014)
12. Amaro, S., Duarte, P.: Online Travel Purchasing: A Literature Review. Journal of Travel & Tourism Marketing 30(8), 755–785 (2013)
13. Cantallops, A.S., Salvi, F.: New consumer behavior: A review of research on eWOM and hotels. International Journal of Hospitality Management 36, 41–51 (2014)
14. Litvin, S.W., Ronald, E.: Goldsmith, and Bing Pan, Electronic word-of-mouth in hospitality and tourism management. Tourism Management 29(3), 458–468 (2008)
15. Iang, J.X., Gretzel, U., Law, R.: Do Negative Experiences Always Lead to Dissatisfaction? - Testing Attribution Theory in the Context of Online Travel Reviews. Information and Communication Technologies in Tourism 297–308 (2010)
16. Jeong, E., Jang, S.: Restaurant experiences triggering positive electronic word-of-mouth (eWOM) motivations. International Journal of Hospitality Management 30(2), 356–366 (2011)
17. Munar, A.M., Jacobsen, J.K.S.: Motivations for sharing tourism experiences through social media. Tourism Management 43, 46–54 (2014)
18. Cheung, C.M.K., Lee, M.K.O., Rabjohn, N.: The impact of electronic word-of-mouth - The adoption of online opinions in online customer communities. Internet Research 18(3), 229–247 (2008)

19. Verhagen, T., Nauta, A., Feldberg, F.: Negative online word-of-mouth: Behavioral indicator or emotional release? Computers in Human Behavior 29(4), 1430–1440 (2013)
20. Ajzen, I.: The theory of planned behavior. Organizational Behavior and Human Decision Processes (50), 179–211 (1991)
21. Ajzen, I.: Nature and operation of attitudes. Annual Review of Psychology 52, 27–58 (2001)
22. Petty, R.E., Cacioppo, J.T., Schumann, D.: Central and peripheral routes to advertising effectiveness: The moderating role of involvement. Journal of Consumer Research 10(2), 135 (1983)
23. Bhattacherjee, A., Sanford, C.: Influence processes for information technology acceptance: An elaboration likelihood model. MIS Quarterly 30(4), 805–825 (2006)
24. Priester, J.R., Petty, R.E.: The influence of spokesperson trustworthiness on message elaboration, attitude strength, and advertising effectiveness. Journal of Consumer Psychology 13(4), 408–421 (2003)
25. Shu, M., Scott, N.: Influence of Social Media on Chinese Students' Choice of An Overseas Study Destination: An Information Adoption Model Perspective. Journal of Travel & Tourism Marketing 31(2), 286–302 (2014)
26. Lam, T., Hsu, C.H.C.: Predicting behavioral intention of choosing a travel destination. Tourism Management 27(4), 589–599 (2006)
27. Hsu, C.H.C., Huang, S.: An Extension of the Theory of Planned Behavior Model for Tourists. Journal of Hospitality & Tourism Research 36(3), 390–417 (2012)
28. Song, H., You, G.-J., Reisinger, Y., Lee, C.-K., Lee, S.-K.: Behavioral intention of visitors to an Oriental medicine festival: An extended model of goal directed behavior. Tourism Management 42, 101–113 (2014)
29. Cheung, M.Y., Luo, C., Sia, C.L., Chen, H.P.: Credibility of Electronic Word-of-Mouth: Informational and Normative Determinants of On-line Consumer Recommendations. International Journal of Electronic Commerce 13(4), 9–38 (2009)
30. Jin, S.A.A., Phua, J.: Following Celebrities' Tweets About Brands: The Impact of Twitter-Based Electronic Word-of-Mouth on Consumers' Source Credibility Perception, Buying Intention, and Social Identification With Celebrities. Journal of Advertising 43(2), 181–195 (2014)
31. Fang, Y.H.: Beyond the Credibility of Electronic Word of Mouth: Exploring eWOM Adoption on Social Networking Sites from Affective and Curiosity Perspectives. International Journal of Electronic Commerce 18(3), 67–101 (2014)
32. Shih, H.-P., Lai, K.-H., Cheng, T.C.E.: Informational and Relational Influences on Electronic Word of Mouth: An Empirical Study of an Online Consumer Discussion Forum. International Journal of Electronic Commerce 17(4), 137–165 (2013)
33. Fodness, D., Murray, B.: Tourist information search. Annals of Tourism Research 24(3), 503–523 (1997)
34. Fornell, C., Larcker, D.F.: Structural equation models with unobservable variables and measurement error: Algebra and statistics. Journal of Marketing Research, 382–388 (1981)
35. Nunnally, J.C., Ira, H.: Bernstein, and Jos MF ten Berge, Psychometric theory, vol. 226. McGraw-Hill, New York (1967)
36. Haas, J.S., Kaplan, C.P., Gerstenberger, E.P., Karlikowske, K.: Changes in the Use of Postmenopausal Hormone Therapy after the Publication of Clinical Trial Results. Ann. Intern. Med. 140, 184–188 (2004)

37. Gefen, D., Straub, D., Boudreau, M.-C.: Structural equation modeling and regression: Guidelines for research practice. Communications of the Association for Information Systems 4(1), 7 (2000)
38. Wang, J., Ritchie, B.W.: Understanding accommodation managers' crisis planning intention: An application of the theory of planned behaviour. Tourism Management 33(5), 1057–1067 (2012)
39. Zelenka, J.: Information and Communication Technologies in Tourism - Influence, Dynamics, Trends. E & M Ekonomie a Management 12(1), 123–132 (2009)

The Role of Trust towards the Adoption
of Mobile Services in China: An Empirical Study

Shang Gao[1,2,*] and Yuhao Yang[2]

[1] Department of Computer and Information Science,
Norwegian University of Science and Technology, Norway
shanggao@idi.ntnu.no
[2] School of Business Administration,
Zhongnan University of Economics and Law, Wuhan, China
vincentyoo@foxmail.com

Abstract. This research aims to study the role of trust towards the adoption of mobile services in China. This study examined users' adoption of general mobile services by extending TAM with additional trust-related constructs. Based on the literature review from previous research, a research model with 10 research hypotheses is proposed in the study. This research model is empirically evaluated using survey data collected from a sample of 373 subjects. Seven research hypotheses are positively significant supported, while three research hypotheses are rejected in this study. The results indicate that both perceived reputation and perceived structural assurance directly affects the consumers' trust in mobile services. However, perceived environment risk does not have positive effect on consumers' trust in mobile services. Another interesting finding is that consumers' trust in mobile services does not have direct positive effect on users' intention to use mobile services.

Keywords: Mobile Services, Trust, Technology Acceptance Model, Technology Adoption.

1   Introduction

Along with the popularity of mobile devices and advances in wireless technology, mobile services have become more and more prevalent. Despite the rapid global diffusion of mobile devices, some mobile services have experienced much slower uptake from consumers [17]. A mobile service is a term used to describe software that runs on a mobile device. Mobile services are designed to educate, entertain, and assist users in their daily lives.

While there has been an increasing availability of mobile services, limited attention has been given to user adoption of mobile services, particularly with newly developed advanced mobile services (e.g., advanced mobile services). Prevalence of mobile services depends not only on technology advancement, but also on user adoption. Most research about the adoption of mobile services was mainly based on the

---

* Corresponding author.

H. Li et al. (Eds.): I3E 2014, IFIP AICT 445, pp. 46–57, 2014.

technology acceptance model (TAM) (e.g.,[9, 19, 28, 38] [21]). Some previous research (e.g., [10, 18]) indicated that traditional adoption theories (e.g., TAM) failed to explain the adoption of mobile services. It is believed that current research has some limitations in explaining consumers' trust in mobile services and their feelings attached to using them. Therefore, we believe that there is a need to include the trust element into the mobile services adoption research. The problem that we want to address in this research is to investigate the role of trust towards the adoption of mobile services in terms of a research model.

Based on analysis of prior literature on technology diffusion and trust, a research model is developed to investigate the adoption of mobile services. To operationalize the research model, a measurement instrument is used to measure each of the constructs. The objective of this paper is to empirically examine how well the proposed research model is able to explain mobile services adoption. The research model is analyzed using Partial Least Squares (PLS) analysis.

The remainder of this paper is organized as follows: the literature review is provided in Section 2. The research model and hypotheses are presented in Section 3. The method and process of our empirical study are described in Section 4. This is followed by a discussion of the findings and limitations of this study in Section 5. Section 6 concludes this research and suggests directions for future research.

2   Literature Review

The literature about technology adoption and trust is discussed in this section.

2.1   Technology Adoption Research

An important and long-standing research question in information systems research is how we can accurately explain user adoption of information systems [13]. Several models have been developed to test the users' attitude and intention to adopt new technologies or information systems. These models include the Technology Acceptance Model (TAM) [11], Theory of Planned Behavior (TPB) [1], Innovation Diffusion Theory (IDT) [33], Unified Theory of Acceptance and Use of Technology (UTAUT) [35]. Among the different models that have been proposed, TAM, which is the extension of the Theory of Reasoned Action (TRA) [15], appears to be the most widely accepted model. TAM focus on the perceived usefulness (PU) and perceived ease of use (PEOU) of a system and has been tested in some domains of E-business and proved to be quite reliable to predict user acceptance of some new information technologies, such as intranet [26], electronic commerce [32], and online shopping [22].

However, TAM's limitations relative to extensibility and explanation power have been noted [6]. Many researchers have suggested that TAM needs to be extended with additional variables to provide a stronger model [27]. Some researchers have also indicated that the major constructs of TAM cannot fully reflect the specific influences of technological and usage-context factors that may alter users' acceptance [31]. Therefore, PU and PEOU may not fully explain people's intention to adopt mobile services. We believe that TAM has limitations when investigating users' adoption of mobile services, which is also confirmed by prior research work in [37]. Moreover, although UTAUT unifies more factors and consolidates the functions of the technology acceptance model with the constructs of eight prominent models in IS adoption research, this increases the complexity, so that it is more complicated to test its applicability. While the acceptance and adoption of IT services has been one of the most prevailing IS research topics (e.g.,[12]), the pervasiveness of mobile services raises new questions in exploring the adoption of mobiles services, such as what are the key factors driving the adoption of mobile services, and how do usage context affect users' adoption of mobile services.

All the findings above motivate the development of a research model, which is described in next section.

2.2   The Role of Trust in Mobile Services

Trust is a crucial enabling factor in relations where there is uncertainty, interdependence, risk, and fear of opportunism [29]. For example, because of the absence of proven guarantees that the mobile services providers will not engage in harmful opportunistic behaviors, such as privacy violations and unauthorized use of credit card information, some users generally stay away from mobile services providers they do not trust.

In the domain of mobile services, trust can be defined as a user's belief or faith in the degree to which a specific service can be regarded to have no security and privacy threats [16]. In other words, trust means believing in mobile services now with an expectation that it will be risk free and you will be getting intangible benefits in certain ways at an unspecific time in the future. For mobile services providers, cultivating users' trust is a time-consuming process. Trust is hard to gain, but easy to lose.

Several factors can influence users' trust in mobile services. First, many users are not familiar with mobile services and mobile technology. Second, unfamiliarity with mobile services providers may make users' perception on the activities involved in mobile services as risky. For example, the lack of physical presentation of the service providers and the inability to feel and inspect the desired real products in a mobile transaction may make users feel vulnerable. Third, mobile services are also confronted with other challenges (e.g., privacy and security issues). Most users are concerned about the security of mobile services and mobile services providers' ability to protect unauthorized access to personal information. As indicated in [25], security and privacy issues are critical to the success of consumers' trust building in Internet shopping. We believe that this principle can apply to mobile services as well. Compared to traditional electronic services, users tend to perceive mobile services as riskier in nature and are more reluctant to adopt them. Fourth, corporate branding and reputation could be used to engender trust in mobile services. Last but not least, legislations and governmental policies on mobile services may also affect users' trust in mobile services.

2.3   Trust in Technology Adoption Research

Trust is defined as the willingness of a party to be vulnerable to the actions of another party based on the expectation that the latter one will perform particular actions [29]. After having identified the concept of trust as a key success factor in e-commerce [30], it makes trust become a much concerned forefront direction in the area of acceptance research. Bélanger explored trust in e-government services [5] and Gefen et al. studied the role of trust in online shopping [23]. Further, Watzdorf et al. found that the impact of trust on intention to use is insignificant in mobile emergency applications [36]. Last but not least, Chandra illustrated that building sufficient trust in mobile payment system is imperative for its adoption [8]. The difference in results probably attributes to different samples and objects. In this research, we would like to further explore the importance of trust towards mobile services adoption in China.

3   Research Model and Hypotheses

This study examines the acceptance of general mobile services in China. The proposed research model (see Figure 1) is an extension of TAM. In addition to perceived ease of use and perceived usefulness, the model includes trust related elements as additional factors to study the relating to user adoption of mobile services.

This study is based on a model [8] which focused on consumers' adoption of Mobile payment. Chandra's model is an extension of TAM model, and the model included the perceived usefulness (PU), perceived ease of use (PEOU), adoption intention of M-Payment system (AI). In addition to that the model indicated a positive influence of PU on AI and PEOU on PU, the influence of PEOU on AI is insignificant. In our research, we still want to confirm the influence of PEOU on AI as we change the research target into Chinese and research subject into mobile services, thus we formulate the following rescarch hypotheses.

H1a: The perceived usefulness has a positive impact on the intention to use mobile services.

H1b: The perceived ease of use has a positive impact on the intention to use mobile services.

H1c: The perceived ease of use has a positive impact on the perceived usefulness.

As proposed in [8], the two dimensions of consumer trust are trust in mobile service provider and trust in technology facilitated by mobile service provider characteristics and mobile technology characteristics respectively. Perceived reputation (PR) of the mobile services provider is identified a major category of trust in mobile services provider. Perceived environment risk (PER) and perceived structural assurance (PSA) are identified another two categories of trust in technology facilitated. Then, we modulate these three variables to our research model.

Perceived reputation (PR): The extent to which consumers believe in the mobile services provider's competency, honesty, and benevolence [14].

Perceived environment risk (PER): Security-related risks faced by consumers while using mobile services, including privacy risk, financial risk [2].

Perceived structural assurance (PSA): The institutional environment (here mobile technology including the mobile Internet and mobile platform) that all structures like guarantees, regulations, and promises are operational for safe, secure, and reliable services [39].

Hence, we formulate the following hypotheses.

H2: Perceived reputation of the mobile services provider has a positive impact on the level of consumer trust in mobile services.

H3a: Perceived environment risk has a negative impact on the level of consumer trust in mobile services.

H3b: Perceived structural assurance has a positive impact on the level of consumer trust in mobile services.

H3c: Perceived structural assurance has a negative impact on the Perceived environment risk in mobile services.

H4a: Consume trust in mobile services has a positive impact on the perceived usefulness of mobile services.

H4b: Consume trust in mobile services has a positive impact on the perceived ease of use of mobile services.

H4c: Consume trust in mobile services has a positive impact on the intention to use mobile services.

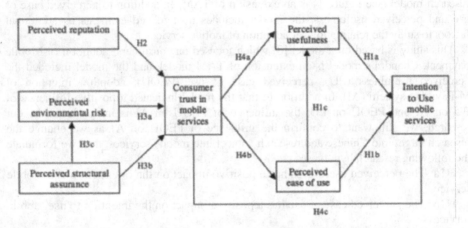

Fig. 1. Research Model

4   An Empirical Study with the Research Model

In this research, the research model was examined through the use of general mobile services in China.

4.1   Mobile Commerce in China

According to the report from the Chinese Ministry of Information Industry, the number of mobile phone subscribers in China exceeded 1 billion in March 2012 as more Chinese people began to consider mobile phones as an everyday necessity. China has the largest mobile phone market in the world. However, the mainstream usage of mobile phones in China focuses on phone calls, SMS, instant messaging services, contact services, and purchasing ring tongs [20]. Social networking services on mobile devices are becoming more and more popular in young generation in China. However, not many researches are focused on the adoption of mobile services in China. Therefore, we believe that it is still worth to develop a research model to measure the adoption of mobile services in China.

4.2   Instrument Development

The validated instrument items [2, 39] [11, 18] [12] from previous research were used as the foundation to create the instrument for this study. Developed from the literature, the measurement questionnaire consisted of 24 items. In order to ensure that the instrument better fit this empirical study, some minor words changes were made to ensure easy interpretation and comprehension of the questions. A seven point Likert scale, with 1 being the negative end of the scale (strongly disagree) and 7 being the positive end of the scale (strongly agree), was used to examine participants' responses to all items in this part.

4.3   Samples

The data for this study were collected through self-administered questionnaires by inhabitants of the biggest city in the central part of China. The survey was distributed in terms of paper-based questionnaires individually from January 1st 2014 to March 15, 2014. 435 completed questionnaires were collected, among which 373 of them were valid questionnaires (i.e., valid respondent rate 85.7%). Among the participants, 66.2%of the participants used Android mobile devices, and 22.3% of the participants used iPhones. 41% of the participants were male, and 59% were female. Moreover, 52.1% of the participants were below 22 years old.

4.4   Measurement Model

To test the reliability and validity of each construct in the mobile service acceptance model, the Internal Consistency of Reliability (ICR) of each construct was tested with Cronbach's Alpha coefficient. For the purposes of testing the research hypotheses, data were analyzed using the structural equation modeling (SEM).

   In this study, we examined goodness-of-fit of the measurement model by using six widely-used fit indices: the chi-square/degrees of freedom (x2/df), the goodness-of-fit index (GFI), the adjusted goodness-of-fit index (AGFI), the comparative fit index (CFI), the normed fit index (NFI), and the root mean square error of approximation (RMSEA). The fitness measures are shown in Table 1.

   Table 1 shows that all the fitness measures are within acceptable range. Therefore, we consider the measurement model is acceptable, and the measures indicate that the model fit the data.

   Convergent validity was assessed through composite reliability (CR) and the average variance extracted (AVE). Bagozzi and Yi [4] proposed the following three measurement criteria: factor loadings for all items should exceed 0.5, the CR should exceed 0.7, and the AVE of each construct should exceed 0.5. As shown in Table 2, all constructs are in acceptable ranges except the construct perceived environmental risk (PER). Further, the Cronbach's Alpha values range from 0.732 to 0.887. All the constructs are above 0.700. Consequently, the scales are deemed acceptable to continue. As for discriminant validity, the variances extracted by the constructs are more than the squared correlations among variables.

Table 1. Fit indices for the measurement model

Measures	Recommended Criteria	Measurement Model	Suggested by Authors
Chi-square/d.f.	<3.0	3.266	Hayduk (1988) [24]
GFI	>0.9	0.844	Scott (1995) [34]
AGFI	>0.8	0.807	Scott (1995) [34]
CFI	>0.9	0.893	Bagozzi and Yi (1988)[3]
NFI	>0.9	0.853	Bentler (1990) [7]
RMESA	<0.08	0.078	Bagozzi and Yi (1988)[3]

Table 2. Factor loadings, composite reliability, and AVE for each construct

Construct	Item	Factor Loading	Composite Reliability	AVE	Cronbach's Alpha
PR	PR1	0.78	0.905	0.761	0.843
	PR2	0.85			
	PR3	0.77			
PER	PER1	0.64	0.036	0.218	0.732
	PER2	0.94			
	PER3	0.53			
PSA	PSA1	0.82	0.901	0.752	0.832
	PSA2	0.75			
	PSA3	0.80			
Trust (TU)	TU1	0.84	0.912	0.725	0.868
	TU2	0.87			
	TU3	0.88			
	TU4	0.53			
PU	PU1	0.60	0.881	0.713	0.793
	PU2	0.83			
	PU3	0.84			
PEOU	PEOU1	0.75	0.917	0.689	0.887
	PEOU2	0.75			
	PEOU3	0.76			
	PEOU4	0.80			
	PEOU5	0.83			
Intention to use (IU)	IU1	0.80	0.871	0.694	0.770
	IU2	0.81			
	IU3	0.60			

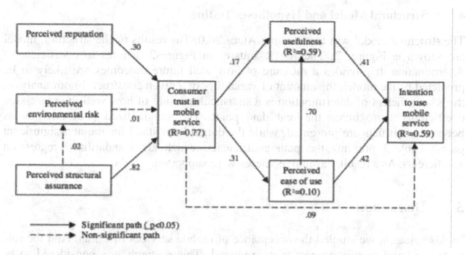

Fig. 2. Mobile services Acceptance Research Model—Results

Table 3. Test of hypotheses based on path coefficient

Hypothesis	Path Coefficient	Hypothesis Result
H1a: The perceived usefulness has a positive impact on the intention to use mobile services.	0.47***	Supported
H1b: The perceived ease of use has a positive impact on the intention to use mobile services.	0.44***	Supported
H1c: The perceived ease of use has a positive impact on the perceived usefulness.	0.63***	Supported
H2: Perceived reputation of the mobile services provider has a positive impact on the level of consumer trust in mobile services.	0.26***	Supported
H3a: Perceived environment risk has a negative impact on the level of consumer trust in mobile services.	0.01	Rejected
H3b: Perceived structural assurance has a positive impact on the level of consumer trust in mobile services.	0.73***	Supported
H3c: Perceived structural assurance has a negative impact on the Perceived environment risk in mobile services.	0.02	Rejected
H4a: Consume trust in mobile services has a positive impact on the perceived usefulness of mobile services.	0.14***	Supported
H4b: Consume trust in mobile services has a positive impact on the perceived ease of use of mobile services.	0.29***	Supported
H4c: Consume trust in mobile services has a positive impact on the intention to use mobile services.	0.09	Rejected

*$p<0.05$; **$p<0.01$; *** $p<0.01$

4.5   Structural Model and Hypotheses Testing

The structural model was tested using Amos 20.0. The results of the structural model are shown in Figure. 2. The R^2 (R square) in Figure 2 denotes to coefficient of determination. It provides a measure of how well future outcomes are likely to be predicted by the model, the amount of variability of a given construct. In our analysis, the R^2 coefficient of determination is a statistical measure of how well the regression coefficients approximate the real data point. The standardized path coefficients between constructs are presented, while the dotted lines stand for the non-significant paths. Table 3 presents the path coefficients, which are standardized regression coefficients. As a result, seven hypotheses were supported.

5   Discussion

In this research, we studied the acceptance of mobile services in China. And the role of trust played in this process was examined. This research was considered to be beneficial for academic research since it extends and enhances the understanding of adoption and mobile services in China. The findings demonstrated the appropriateness of the research model and hypotheses for measuring mobile services adoption.

The positive impact of perceived reputation on trust was significant. Among the numerous services, providers with good reputation can decrease the perceived vulnerability of users and a good impression based on previous experience can encourage the users to make an attempt to new services from the providers. This result is consistent with previous technology adoption studies in online shopping [22, 23]. For those providers with successful services before, they have more advantages in new services promotion than other market entrants.

The other factor, which has a positive influence on trust, was perceived structure assurance, which refers to mobile Internet and operating platform technologies here. Due to a safety structure assurance, users believed that they would not suffer attack and potential loss from hackers, and they would be compensated if that happened [2].

However, perceived environment risk does not have positive effect on consumers' trust in mobile services. According to the descriptive results, perceived environment risk has an average value of 5.13 on a seven point Likert scale. But the influence of perceived environment risk on users' trust was insignificant. The possible explanation might be that users are confidence with their qualified skills to avoid fraud services, and the mobile service has some schemes to protect users' security and privacy issues. Trust has both positive influence on perceived ease of use and perceived usefulness significantly.

Further, another interesting finding of this study was that trust did not have direct positive effect on the intention to use mobile services. This is contradictory with some previous research (e.g.,[1, 24]). As our research aimed to investigate users' adoption of general mobile services in China, the results may vary from a specific mobile service to another mobile service. The research results also confirmed the positive influence of perceived usefulness and perceived ease of use on the intention to use mobile services. It revealed the importance of ease of use and usefulness of mobile services toward users' adoption of mobile services.

However, we were also aware of some limitations. Firstly, we only tested the research model and research hypotheses with general mobile services in China. Therefore, the generalizability of the results to a specific mobile service remains to be determined. Secondly, the subjects in this study were mainly university students. This sample may not be fully representative of the entire population in China.

6   Conclusion and Future Research

This research was designed to study the role of trust towards the adoption of mobile services in China. To our best knowledge, only a few studies were concerned with identifying trust-related factors that affect the adoption of mobile services in China. This study examined users' adoption with general mobile services by extending TAM with additional trust-related constructs. A research model with 10 research hypotheses was proposed in the study. Seven research hypotheses were positively significant supported, while three research hypotheses were rejected in this study. The results indicated that both perceived reputation and perceived structural assurance directly affects the consumers' trust in mobile services. But, consumers' trust in mobile services does not have direct positive effect on user's intention to use mobile services.

Continuing with this stream of research, we plan to examine the applicability of the research model with a specific mobile service, such as mobile information service or mobile payment. Future research is also needed to empirically verify the research model with larger samples across the world. Furthermore, some mediating factors (e.g., gender) may provide fresh insights and offer new directions for future research.

Acknowledgement. This research is supported by the Fundamental Research Funds for the Central Universities, China (Project No. ZNUFE. 2012065).

References

1. Ajzen, I.: The theory of planned behavior. Organizational Behavior and Human Decision Processes 50(2), 179–211 (1991)
2. Aldridge, A., White, M., Forcht, K.: Security considerations of doing business via the Internet: cautions to be considered. Internet Research 7(1), 9–15 (1997)
3. Bagozzi, R.P., Yi, Y.: On the evaluation of structural equation models. Journal of the Academy of Marketing Science 16(1), 74–94 (1988)
4. Bagozzi, R.P., Yi, Y.: Specification, evaluation, and interpretation of structural equation models. Journal of the Academy of Marketing Science 40(1), 8–34 (2012)
5. Bélanger, F., Carter, L.: Trust and risk in e-government adoption. The Journal of Strategic Information Systems 17(2), 165–176 (2008)
6. Benbasat, I., Barki, H.: Quo Vadis TAM. Journal of the Association for Information Systems 8(4), 211–218 (2007)
7. Bentler, P.M.: Comparative fit indexes in structural models. Psychological Bulletin 107(2), 238 (1990)

8. Chandra, S., Srivastava, S.C., Theng, Y.-L.: Evaluating the role of trust in consumer adoption of mobile payment systems: An empirical analysis. Communications of the Association for Information Systems 27, 561–588 (2010)
9. Chen, L.-D.: A model of consumer acceptance of mobile payment. Int. J. Mob. Commun. 6(1), 32–52 (2008)
10. Dahlberg, T., Mallat, N., Ondrus, J., et al.: Past, present and future of mobile payments research: A literature review. Electronic Commerce Research and Applications 7(2), 165–181 (2008)
11. Davis, F.D.: Perceived usefulness, perceived ease of use and user acceptance of information technology. MIS Quarterly 13, 319–340 (1989)
12. Davis, F.D., Bagozzi, R.P., Warshaw, P.R.: User acceptance of computer technology: a comparison of two theoretical models. Manage. Sci. 35(8), 982–1003 (1989)
13. DeLone, W., McLean, E.: Information Systems Success: The Quest for the Dependent Variable. Information Systems Research 3 (1) (1992)
14. Doney, P.M., Cannon, J.P.: An Examination of the Nature of Trust in Buyer-Seller Relationships. Journal of Marketing 61(2), 35–51 (1997)
15. Fishbein, M., Ajzen, I.: Belief, Attitude, Intention and Behavior: An Introduction to Theory and Research. Addison-Wesley (1975)
16. Gao, S., Krogstie, J., Gransæther, P.A.: Mobile Services Acceptance Model. In: The Proceedings of International Conference on Convergence and Hybrid Information Technology. IEEE Computer Society (2008)
17. Gao, S., Krogstie, J., Siau, K.: Adoption of mobile information services: An empirical study. Mobile Information Systems 10(2), 147–171 (2014)
18. Gao, S., Krogstie, J., Siau, K.: Developing an instrument to measure the adoption of mobile services. Mobile Information Systems 7(1), 45–67 (2011)
19. Gao, S., Moe, S.P., Krogstie, J.: An empirical test of the mobile services acceptance model. In: 2010 Ninth International Conference on Mobile Business and 2010 Ninth Global Mobility Roundtable (ICMB-GMR), pp. 168–175. IEEE (2010)
20. Gao, S., Zang, Z., Gopalakrishnan, S.: A study on distribution methods of mobile applications in China. In: The Proceedings of Seventh International Conference on Digital Information Management (ICDIM), pp. 375–380. IEEE Computer Society (2012)
21. Gao, S., Zang, Z., Krogstie, J.: The Adoption of Mobile Games in China: An Empirical Study. In: Liu, K., Gulliver, S.R., Li, W., Yu, C., et al. (eds.) ICISO 2014. IFIP AICT, vol. 426, pp. 368–377. Springer, Heidelberg (2014)
22. Gefen, D.: TAM or Just Plain Habit: A Look at Experienced. Online Shoppers. Journal of End User Computing 15(3), 1–13 (2003)
23. Gefen, D., Karahanna, E., Straub, D.W.: Trust and TAM in Online Shopping: An Integrated Model. MIS Quarterly 27(1), 51–90 (2003)
24. Hayduk, L.A.: Structural equation modeling with LISREL: Essentials and advances. JHU Press (1988)
25. Hoffman, D.L., Novak, T.P., Peralta, M.: Building consumer trust online. Commun. ACM 42(4), 80–85 (1999)
26. Horton, R.P., Buck, T., Waterson, P.E., et al.: Explaining intranet use with the technology acceptance model. Journal of Information Technology 16, 237–249 (2001)
27. Legris, P., Ingham, J., Collerette, P.: Why do people use information technology?: a critical review of the technology acceptance model. Inf. Manage. 40(3), 191–204 (2003)
28. Luarn, P., Lin, H.-H.: Toward an understanding of the behavioral intention to use mobile banking. Computers in Human Behavior 21(6), 873–891 (2005)

29. Mayer, R., Davis, J., Schoorman, D.: An Integrative Model of Organizational Trust. The Academy of Management Review 20(3), 709–734 (1995)
30. McKnight, D.H., Chervany, N.L.: What trust means in e-commerce customer relationships: an interdisciplinary conceptual typology. International Journal of Electronic Commerce 6, 35–60 (2002)
31. Moon, J.-W., Kim, Y.-G.: Extending the TAM for a World-Wide-Web context. Inf. Manage. 38(4), 217–230 (2001)
32. Pavlou, P.A.: Consumer Acceptance of Electronic Commerce: Integrating Trust and Risk with the Technology Acceptance Model. Int. J. Electron. Commerce 7(3), 101–134 (2003)
33. Rogers, E.M.: The diffusion of innovations. Free Press, New York (1995)
34. Scott, J.E.: The measurement of information systems effectiveness: evaluating a measuring instrument. ACM SIGMIS Database 26(1), 43–61 (1995)
35. Venkatesh, V., Morris, M.G., Davis, G.B., et al.: User Acceptance of Information Technology: Toward a Unified View. MIS Quarterly 27(3), 425–478 (2003)
36. von Watzdorf, S., Ippisch, T., Skorna, A., et al.: The Influence of Provider Trust on the Acceptance of Mobile Applications: An Empirical Analysis of Two Mobile Emergency Applications. In: 2010 Ninth International Conference on Mobile Business (ICMB-GMR), pp. 329–336 (2010)
37. Wu, J.-H., Wang, S.-C.: What drives mobile commerce? An empirical evaluation of the revised technology acceptance model. Inf. Manage. 42(5), 719–729 (2005)
38. Yang, K.C.C.: Exploring factors affecting the adoption of mobile commerce in Singapore. Telemat. Inf. 22(3), 257–277 (2005)
39. Zucker, L.G.: Production of trust: Institutional sources of economic structure, 1840–1920. Research in Organizational Behavior (1986)

The Use of Lean Principles in IT Service Innovation: Insights from an Explorative Case Study

Yiwei Gong[1, 2,*] and Marijn Janssen[2]

[1] School of Information Management, Wuhan University,
430072 Wuhan, Hubei, P.R. China
[2] Faculty of Technology, Policy and Management, Delft University of Technology,
Jaffalaan 5, 2628 BX Delft, The Netherlands
{Y.Gong,M.F.W.H.A.Janssen}@tudelft.nl

Abstract. IT service innovation is often dependent upon the relationship with outsourcing vendors. In such situations innovation is a result of knowledge accessing and utilization between outsourcing provider and user. As a management thinking, the application of Lean principles can facilitate knowledge accessing and utilization to enable IT service innovation for the customer. Based on the knowledge-based view of the firm, we developed a conceptual framework to describe how Lean can drive IT service innovation within IT outsourcing relationships. This framework is used to analyze the use of Lean principles in an explorative case study of a service organization and its two IT outsourcing providers. The framework and the case study show that IT service innovation is an ongoing process. A clear strategic direction and learning environment are critical to achieve it. Applying Lean principles facilitates the learning behaviors and allows smooth communication on innovative ideas and in this way drive innovation.

Keywords: Service Innovation, Lean, IT Service, IT Outsourcing, Knowledge-based View.

1 Introduction

Along with the growth of the services sector, there is a need for methods to enhance innovation in services within a short time frame. While traditional product innovation emphasizes the design of tangible and relatively static products, services are often intangible and customers are involved in the service delivery process [1, 2]. Grönroos [3] identified three basic characteristics of services: 1) services are processed using a series of activities (a business process) rather than things, 2) services are to some extent produced and consumed simultaneously, and 3) the customer participates in the service delivery process. Consequently, an approach that can be used to enhance innovativeness in production and manufacturing environments might need to be adapted to the idiosyncratic nature of the services domain [4]. We adopt a multi-dimensional definition of service innovation from [5] (p. 494) "A service innovation is a new

---

* Corresponding author.

H. Li et al. (Eds.): I3E 2014, IFIP AICT 445, pp. 58–69, 2014.

service experience or service solution that consists of one or several of the following dimensions: new service concept, new customer interaction, new value system/business partners, new revenue model, new organizational or technological service delivery system."

In recent years, Lean has become popular and sometimes even a strategic imperative for organizations to improve their performance. Lean can be used by organizations to turn passive and defensive organizational cultures into proactive and open cultures that promotes organizational learning and innovation [6-8]. Although Lean has been around for decades, practitioners and scholars primarily applied and studied Lean primarily in manufacturing environments. Because of its apparent success, Lean has gradually expanded itself to other domains such as healthcare [9], government [10] and the service industries [11-13], including IT services [14, 15]. Despite its broad acceptance it is hardly been applied for innovation.

Innovation in IT services is becoming an increasingly complex issue. On the one hand, IT enables innovation and the adoption of new IT technologies often need to be complemented by organizational change activities [16]. On the other hand, IT outsourcing is a common practice in the execution and operation of IT services. A key issue in service innovation is therefore the management of the outsourcing relationships and knowledge sharing between organizations [17]. Prior innovation approaches (e.g. service blueprinting [1]) was used to help organizations improve their service offerings process in the design or implementation stages of services [18]. A service innovation approach that can address the management of IT outsourcing relationships and knowledge sharing between organizations for its use in IT service context has not yet been reported.

In this paper, we propose a Lean IT service innovation framework based on the knowledge-based view (KBV) of the firm. The framework is then used to analyze an insurance company and the relationships with its two IT outsourcing suppliers. The analyses provide insights of the use of Lean for IT service innovation.

2 Lean Driven Innovation

Innovation concerns the generation of a new idea and the way to implement it into new product, process or service [19]. The earlier literature interprets innovation mainly from a product or process perspective [e.g. 20, 21]. While *product innovation* is about the introduction of new product aiming at satisfying new customer needs, *process innovation* is concerned with introducing new elements into an organization's operations such as input materials, task specifications, work and information flow mechanisms, and equipment used to provide a product or service [22]. More recent literature also discusses innovations from administration or management perspectives. *Administrative innovation* is the innovation that relates to strategies, structure, systems, or people in the organization [23]. *Management innovation* is defined as the invention and implementation of a management practice, process, structure, or technique that is new to the state of the art and is intended to further organizational goals [24].

According to the nature of change, innovation can be classified as incremental or radical [19]. *Incremental innovations* improve existing products or processes within the existing structures and strategies, while *radical innovations* give rise to revolutionary changes in strategies and structures and even culture [23]. A radical innovation is likely to happen in a long time frame and many radical innovations take more than five years before starting to pay back [25].

Although innovation has been a subject of scientific research for over a half century, only the last decade scientific literature started to report the connection between Lean and innovation. However, to date the number of publications is still very small. According to Hoppmann, Rebentisch, Dombrowski and Zahn [26 Exhibit 1], up to 2011 only 27 publications can be found about Lean innovation or product development (PD) and half of these publications fall into the PD domain. Existing literature regards Lean driven innovations as either incremental or radical innovations in different forms. In the manufacturing industry Smeds [27] argued that the reorganization of manufacturing according to Lean principles can trigger a radical techno-organizational change towards a Lean enterprise which includes a new structure, strategy and culture. Byrne, Lubowe and Blitz [28] reported on the effect of using Lean Six Sigma management to derive radical innovations. In the research and development (R&D) domain, Schuh, Lenders and Hieber [29] revealed that the implementation of Lean thinking in innovation management facilitates incremental process and product innovations. Besides the manufacturing industry, applying Lean in supporting innovation was found in healthcare and pharmaceutical industry to achieve incremental process innovation [7, 30].

3   Knowledge-Based View in IT Service Innovation

Lean provides companies a way of working that enables them to introduce and exploit service innovations continuously. Based on KBV, we propose a Lean IT service innovation framework which helps organizations to understand how Lean facilitates service innovation in an IT outsourcing relationship.

The KBV considers knowledge as the most strategically significant resource of a firm and the source of competitive advantage resides in the application of the knowledge rather than in the knowledge itself [31]. It is an outgrowth of the resource-based theory (RBT) of the firm which was initially proposed by Penrose [32]. The KBV postulates that organizations render their services offered by using their knowledge. This knowledge is embedded in and carried through multiple entities including organizational culture and identity, routines, policies, systems, documents, and individual employees [33]. The KBV is also able to analyze the efficiency characteristics of inter-organizational cooperation, including outsourcing [34]. The precondition for using the KBV in analyzing outsourcing relationships is that the major motivation for outsourcing is to gain access to knowledge which can be used to innovate.

While accessing knowledge can be considered as the learning activities of the organizations in an IT outsourcing relationship, knowledge utilization are the organizational activities to transform knowledge assets into goods or services [35]. Knowledge utilization in the service process is the mechanism built within the IT outsourcing relationship that enables IT outsourcing suppliers and clients to share and offer the required expertise which can be used to internalize and facilitate internal innovation.

The KBV suggests that innovating IT outsourcing relationships need specific mechanisms to support efficient knowledge accessing and utilization. The evidence reported by existing literature and the theoretical body of knowledge on Lean indicates that Lean is able to underpin the aforementioned two important components and results in different kinds of innovation outcomes. Figure 1 shows a framework conceptualizing Lean IT service innovation from the perspective of the service organization. The framework describes how IT service innovations can be achieved in cooperation with IT outsourcing suppliers. In this framework, knowledge accessing and utilization are underpinned by four Lean principles (numbered in the figure). Principles are often used to guide decision-making [36]. Lean principles are widely used to guide the implementation of Lean thinking [37, 38]. The four Lean principles presented in Figure 1 are selected as they concern specific areas where IT outsourcing relationships should pay attention to in order to avoid waste and focus on creating value in the knowledge accessing and utilization.

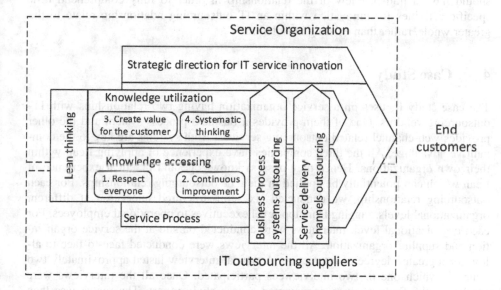

Fig. 1. A conceptual framework of Lean IT service innovation

In the framework, knowledge accessing is facilitated by the Lean principle "respect everyone" and "continuous improvement". *Respect everyone* is about involving all people with different backgrounds, insights, skills and experiences that make a distinct contribution to reach the specified goals of an IT service innovation. Involvement drives people to learn and to improve, unlocks creative potential, ultimately enhancing innovativeness. People are the driving force behind IT outsourcing relationships and have to be respected and nurtured to give them a sense of belonging. People should be provided with plenty of opportunities to hone their skills that are at the heart of both an organization's competitiveness and the success of any IT outsourcing relationship. Similarly, *continuous improvement* concerns the continuous effort of people to shift from reactive firefighting to proactive problem solving, and having the aim of relentlessly improving the execution and efficiency of processes in an ongoing manner. By focusing on continuous improvement, waste is gradually eliminated causing everyone in the IT outsourcing relationship to focus more on better customer services and innovation.

Knowledge utilization is facilitated by the Lean principles "creating value for the customer" and "systematic thinking". *Creating value for the customer* concerns everyone knowing what the customer desires and demands. This means people begin with the final result in mind and focus on doing the right things in order to reach this desired result. By clearly hearing the voice of the customer without noise, unclarities or ambiguities, the supplier knows what the customer considers to add value and what is considered waste, thereby allowing the IT outsourcing relationship to innovate. *Systematic thinking* focuses on viewing the interconnected processes that make up the entire value stream, while being aware of the cause-and-effect interdependencies that either add value or create waste. All parties involved in an IT outsourcing relationship should have a holistic view of the relationship in order to fully comprehend how specific activities for innovation within an IT outsourcing relationship relate to the greater whole, rather than existing in isolation.

4   Case Study

The case study focuses on a service organization having two relationships with IT-outsourcing vendors. One of them provides a business process system and the other provides web-channel related outsourcing services. Both systems are key to the competitive advantage. All the three companies have experience of applying Lean within their own organizations. Hence they possess knowledge and practical experience of Lean which can potentially be applied in their IT outsourcing relationships. For each outsourcing relationship, we conducted six semi-structured interviews at different organizational levels, ranging from top-level executives to operational employees. For each organizational level, interviews were conducted on site at the service organization and supplier organization. All the interviews were conducted face-to-face to allow for a greater degree of interaction. A typical interview lasted approximately two hours in which one or two interviewees participated. Once all the interviews were conducted the findings were documented a case study report. The report was then provided to the interviewees and asked whether it contained any mistakes. The framework we introduced in the last section is used for describing the case. The rest of this section briefly presents the research finding of this case study.

4.1   Strategic Direction for IT Service Innovation

Company A is an insurance group of operating companies with different labels and brands in several European countries. Company A covers a broad range of insurance services and products, distributed to the market through a broad mixture of distribution channels. The company is a front-runner in experimenting and exploiting digital customer services and distribution channels. In the view of Company A, IT service innovation can be accomplished by adopting new technologies, by enabling new customer-oriented services, utilizing a mix of service distribution channels and by improving the performance of business processes. Company A selected two IT outsourcing suppliers which have a long history of cooperation with Company A to acquire required technical expertise. Table 1 provides an overview of these IT outsourcing relationships.

Table 1. An overview of the IT outsourcing relationships studied

IT-Outsourcing Supplier	Outsourcing domain	IT outsourcing content and motivation
B	Business process systems	Co-development of an insurance ERP system which can support multiple digital channels for customer interaction
C	Service delivery channels	Development of reusable website components for every operating company of Company A to allow them building up their own website quickly by configuring components

For a long time, Company A had separate IT departments in each operating company. The IT productivity was considered to be too low and the IT departments spent a significant amount of capital to build customized solutions. The strategy of Company A was to centralize IT organization to integrate operations in order to simplify its IT landscape and use module-based and standard solutions to allow quick creation of new business processes and the deployment of new services. In the centralized IT environment, Company A decided to use the standard solution from Supplier B for core insurance and banking systems, next to the reusable, modular frontend components from Supplier C.

4.2 Business Process System Innovation

In the co-development project between Company A and Supplier B, both companies were aiming to develop a standard solution for the insurance industry. Such a partnership itself is already viewed as an innovative cooperation by some interviewees. Company A is the first insurance company adopting the insurance solution of Supplier B together with other ERP components from Supplier B across the entire company. In this sense, Company A plays an important role for Supplier B to improve its insurance solution. Looking from the other side, Supplier B can offer Company A not just the latest IT technology, but also knowledge and experience acquired from IT projects implemented at other market players in the insurance industry. In their collaboration, Company A is involved as of requirements engineering up until final validation of the solution that was built by Supplier B.

Knowledge Accessing
Both companies attach a great deal of weight to the professional development of their employees. Supplier B has plenty of internal training resources to allow its people to learn, including the development of soft skills referring to collaboration and cultural

aspects. Building together with Supplier B, Company A also has an internal training program on the products of Supplier B. In addition, consultants from Supplier B are co-located at Company A to facilitate the deployment of the solution and bring back the feedback from Company A to the development center of Supplier B. There is a lot of discussion between the two companies on identifying the opportunities of improvement and innovation. The use of Lean techniques such as Kaizen and A3 thinking for problem solving on an outsourcing relationship level was being discussed. Both companies know that people need to adopt innovation and have to incorporate these in their daily behaviors to ensure continuous improvement. Management at a tactical level carefully balances the efforts on routine work/delivery and innovation. They promote proactive innovation in a modest but sustainable way so that it has fewer interruptions on the current processes and consequently less resistance.

Knowledge Utilization
The chief architect and people at the strategic level of Company A decide whether a business process is designed well enough and which processes are important enough to be placed on the roadmap for the system development. By sharing this roadmap with Supplier B, Company A shares a clear vision on the transformation. In the software development, Company A is closely involved with Supplier B in the development of the insurance solution. Occasionally, Company A's experts and architects are invited to the development center of Supplier B for joint solution design sessions.

People at the tactical level of Company A hold a plan-build-run meeting every two weeks with people from Supplier B. The meeting provides updates on all the issues that they previously decided to monitor. Issues they focus on are related to safeguarding the service delivery or creating innovative ideas. On an operational level, there is a collaborative contract to support operational people of Company A with quick problem-solving and addressing any findings for improvement and innovation.

However, despite that people have learned value stream mapping during trainings, we found that neither of the companies uses this technique in their outsourcing relationship. Some interviewees indicated to believe that both company have optimized procedures, but neither organization wants to change its internal processes for an optimized cross-organizational value stream. This implies that a higher level of openness and a clear business model for value sharing are required to achieve common systematic thinking in IT outsourcing relationships to enable cross-organizational optimization.

4.3 Service Delivery Channels Innovation

In the outsourcing relationship for the development of website components, Supplier C provided software development outsourcing services to Company A through onshore and offshore teams. A parallel 'software development factory' structure was introduced to provide Company A with flexible and sufficient software development capacity. Two factories similar in size are working in parallel: one factory belongs to Supplier C, whereas the other factory belongs to Company A. In addition, a small team of software architects from Supplier C are working on-site at Company A. The two parallel factories are treated in an equal manner, meaning that for any random order, a set of requirements of to-be developed software components will be sent to either one of these factories. A deliverable, being one or more software components developed by either the factory of Company A or the factory of Supplier C, will be sent to a unified quality gate of Company A to be tested.

Knowledge Accessing

The strategic direction is to create a competitive but open and healthy learning environment between the two parallel factories. This learning environment stimulates the people involved to improve their software development skills and to learn from their peers working for the other factory.

The business strategy of Company A demands each operating company to have its own website as a service delivery channel to its customers. The best way to achieve this is to develop reusable components and build websites based on those components rather than to develop each website individually. As a consequence, Company A decided that the reusable components have to be built on the Microsoft .NET framework. A major challenge for Company A was that its software development personnel had to switch to this new development environment, and that people had to learn new technical skills. Within this outsourcing relationship, having a peer factory with strong software development skills at Supplier C results in competition between the factory of Company A and the factory of Supplier C. This results in a competitive culture in which both factories strive to outperform each other. As a consequence, people working in both development factories attempt to earn the respect and recognition of their peers for their craftsmanship, accompanied by a strong willingness to learn and to improve.

Both organizations recognize the benefit of such a competitive environment. Every quarter both factories visit each other to facilitate learning. Each visit typically lasts several days, depending on the subjects and purposes, and provides sufficient time for communication and discussion. Subjects discussed include software engineering and project management. Having sufficient time during a visit allows people to work together on a certain issue in an in-depth manner. Examples of discussions are how the process of developing a software component can be broken down into an action plan in a team meeting, how the most difficult part of the component can be identified, and how to decide upon the most optimal priority of different tasks to be carried out during development. Both organizations also provide or support various software-related technical trainings to employees. Occasionally, Supplier C invites people from Company A to attend its internal trainings on new and related technologies. Those trainings allow both factories to keep up with the latest technologies that are related to their current or future work.

In this case study, we found that board level meetings often involve operational people. During such meetings, people discuss issues of continuous improvement and conduct Kaizen sessions. These Kaizen sessions involve managers from both factories and also the software architects from Supplier C working on-site at Company A. Issues about continuous improvement are also discussed on a lower level. Those issues include, but are not limited to, how capacity forecasting can be improved, how people's skills can be improved, or if their roles and positions should be adjusted.

Both factories also have their own daily start-up meeting. During the bilateral visits, personnel join in the local daily start-up meetings and exchange ideas and experiences. The parallel factories setting allows people to compare different ways of working in designing and analyzing a software component and the way of prioritizing the work in progress. This comparison results in fresh ideas for continuous improvement and experiments and subsequently results in process innovations.

Knowledge Utilization

The knowledge utilization in this IT outsourcing relationship begins with functional requirements engineering. Functional requirements of websites originally come from the operating companies of Company A. The IT department at Company A translates those functional requirements into IT requirements and component-based system design solutions. During this process, software architects from Supplier C are often involved, sometimes leading the design. To evaluate those novel solutions, Microsoft is involved during the review process to ensure the technical feasibility of the solution. Eventually, these design solutions become orders for software components and are sent to one of the two factories.

During the software component development stage, we found that both factories make use of internal value stream mapping. Activities were undertaken to create process maps, but did not cross the border of the factory. In other words, these activities were not performed from an end-to-end perspective, but limited to internal software development processes. Nevertheless, analyzing segments of multiple value streams helped both factories to improve their way of working. The type of innovations achieved mainly improve the efficiency of software development. Those improvements originate from the best practices of the two parallel factories. As the two factories remain independent in their daily operations, knowledge utilization rarely happened in this stage.

5   Discussion and Conclusions

The case study presented in this paper shows that the application of Lean principles facilitate knowledge accessing and utilization in both IT outsourcing relationships to a large extent. Not all the four principles were completely followed in the studied two outsourcing relationships. Systematic knowledge sharing is often found to be difficult to be achieved in IT outsourcing relationships, as the exchange is blocked by the boundary of organizations. This reduces the opportunities of innovations. Higher levels of openness are a requirement for both organizations to facilitate innovation. In addition, the case study shows that service innovations can have various forms. The co-development relationship between Company A and Supplier B can be considered as an innovation in itself where both companies established a strategic alignment to create a new ERP solution. The next step in the innovation process will be to develop this as a standard for the insurance industry. In the outsourcing relationship between Company A and Supplier C, we observed several process innovations resulting in the improvement of software development processes.

IT service innovation is an ongoing process that provide the end-customer with a new service experience or solution. To achieve this, the service organization should establish a clear strategic direction for IT service innovation. Based on this strategic direction and Lean thinking, the service organization can better involve its IT outsourcing suppliers in the efforts for knowledge accessing and utilization. Different learning environment (e.g. co-development or competition) can be established to facilitate the knowledge accessing and utilization for different domains. Lean principles are widely applicable in different contexts to amplify the effect of the learning environment and the efficiency of knowledge utilization.

Traditionally Lean driven innovation focuses on manufacturing and production environment rather than service industries. IT service innovation often co-evolves with IT outsourcing activities and as such innovation is a result of knowledge accessing and utilization between the service organization and its outsourcing suppliers. With a Lean management system, IT service innovation should be viewed as an ongoing process in which innovations at different domains including business process system and service delivery channels are involved. The key principles of Lean driven IT service innovation are 1) respect everyone 2) continuous improvement 3) create value for the customer and 4) systematic thinking. The explorative case study shows that the application of these Lean principles facilitate knowledge accessing and utilization. Applying these principles facilitate the learning behaviors of the people working in the IT outsourcing relationships. Based on a clear strategic direction of IT service innovation and applying Lean principles allowed smooth communication on innovative ideas and drive innovations at different domains of the IT service process. Future research should focus on examining the applicability of those Lean principles to facilitate IT service innovation in different domains besides the insurance industry.

Acknowledgement. The preliminary work of this research is supported by the Lean Education and Research Network (LEArN) at Nyenrode Business University.

References

1. Bitner, M.J., Ostrom, A.L., Morgan, F.N.: Service blueprinting: a practical technique for service innovation. California Management Review 50, 66–94 (2008)
2. Oliveira, P., Hippel, E.V.: Users as service innovators: The case of banking services. Research Policy 40, 806–818 (2011)
3. Grönroos, C.: Service Management and Marketing. A customer relationship management approach. Wiley, Chichester (2001)
4. Ettlie, J.E., Rosenthal, S.R.: Service versus Manufacturing Innovation. Journal of Product Innovation Management 28, 285–299 (2011)
5. Den Hertog, P., Van der Aa, W., De Jong, M.W.: Capabilities for managing service innovation: towards a conceptual framework. Journal of Service Management 21, 490–514 (2010)
6. Dahlgaard, J.J., Dahlgaard-Park, S.M.: Lean production, six sigma quality, TQM and company culture. The TQM Magazine 18, 263–281 (2006)
7. Johnstone, C., Pairaudeau, G., Pettersson, J.A.: Creativity, innovation and lean sigma: a controversial combination? Drug Discovery Today 16, 50–57 (2011)
8. Bhasin, S.: Performance of organisations treating lean as an ideology. Business Process Management Journal 17, 986–1011 (2011)

9. Brandao de Souza, L.: Trends and approaches in lean healthcare. Leadership in Health Services 22, 121–139 (2009)
10. Janssen, M., Estevez, E.: Lean government and platform-based governance - Doing more with less. Government Information Quarterly 30, S1–S8 (2013)
11. Piercy, N., Rich, N.: Lean transformation in the pure service environment: the case of the call service centre. International Journal of Operations & Production Management 29, 54–76 (2009)
12. Staats, B.R., Upton, D.M.: Lean Knowledge Work. Harvard Business Review 89, 100–110 (2011)
13. Swank, C.K.: The lean service machine. Harvard Business Review 81, 123–129 (2003)
14. Poppendieck, M., Poppendieck, T.: Lean software development: An agile toolkit. Addison-Wesley Professional (2003)
15. Staats, B.R., Brunner, D.J., Upton, D.M.: Lean principles, learning, and knowledge work: Evidence from a software services provider. Journal of Operations Management 29, 376–390 (2011)
16. Gallouj, F., Savona, M.: Innovation in services: a review of the debate and a research agenda. Journal of Evolutionary Economics 19, 149–172 (2009)
17. Agarwal, R., Selen, W.: Dynamic Capability Building in Service Value Networks for Achieving Service Innovation. Decision Sciences 40, 431–475 (2009)
18. Bettencourt, L.A., Brown, S.W., Sirianni, N.J.: The secret to true service innovation. Business Horizons 56, 13–22 (2013)
19. Urabe, K.: Innovation and the Japanese management system. In: Urabe, K., Child, J., Kagono, T. (eds.) Innovation and Management International Comparisons. Walter de Gruyter (1988)
20. Dewar, R.D., Dutton, J.E.: The adoption of radical and incremental innovations: An empirical analysis. Management Science 32, 1422–1433 (1986)
21. Rowe, L.A., Boise, W.B.: Organizational Innovation: Current Research and Evolving Concepts. Public Administration Review 34, 284–293 (1974)
22. Afuah, A.: Innovation management: Strategies, implementation and profits. Oxford University Press, New York (1998)
23. Popadiuk, S., Choo, C.W.: Innovation and knowledge creation: How are these concepts related? International Journal of Information Management 26, 302–312 (2006)
24. Birkinshaw, J., Hamel, G., Mol, M.J.: Management Innovation. Academy of Management Review 33, 825–845 (2008)
25. Stamm, B.V.: Managing Innovation, Design and Creativity. John Wiley & Sons (2008)
26. Hoppmann, J., Rebentisch, E., Dombrowski, U., Zahn, T.: A Framework for Organizing Lean Product Development. Engineering Management Journal 23, 3–15 (2011)
27. Smeds, R.: Managing Change towards Lean Enterprises. International Journal of Operations & Production Management 14, 66–82 (1994)
28. Byrne, G., Lubowe, D., Blitz, A.: Using a Lean Six Sigma approach to drive innovation. Strategy & Leadership 35, 5–10 (2007)
29. Schuh, G., Lenders, M., Hieber, S.: Lean Innovation: Introducing Value Systems to Product Development. Portland International Conference on Management of Engineering & Technology. IEEE, Cape Town (2008)
30. Garcia-Porres, J., Ortiz-Posadas, M.R., Pimentel-Aguilar, A.B.: Lean Six Sigma applied to a process innovation in a Mexican health institute's imaging department. In: 30th Annual International Conference of the IEEE Engineering in Medicine and Biology Society. IEEE, Vancouver (2008)

31. Alavi, M., Leidner, D.E.: Review: Knowledge Management and Knowledge Management Systems: Conceptual Foundations and Research Issues. MIS Quarterly 25, 107–136 (2001)
32. Penrose, E.T.: The Theory of the Growth of the Firm. Wiley, New York (1959)
33. Grant, R.M.: Toward a knowledge-based theory of the firm. Strategic Management Journal 17, 109–122 (1996)
34. Grant, R.M.: Reflections on knowledge-based approaches to the organization of production. Journal of Management & Governance 17, 541–558 (2013)
35. Grant, R.M., Baden-Fuller, C.: A Knowledge Accessing Theory of Strategic Alliances. Journal of Management Studies 41, 61–84 (2004)
36. Gong, Y., Janssen, M.: An Interoperable Architecture and Principles for Implementing Strategy and Policy in Operational Processes. Computers in Industry 64, 912–924 (2013)
37. Womack, J.P., Jones, D.T.: Lean Thinking: Banish Waste and Create Wealth in Your Corporation, Revised and Updated. Productivity Press (2003)
38. Shingo: The Shingo Prize for Operational Excellence: Model & Application Guidelines. Jon M. Huntsman School of Business (2012)

Design and Implementation of Enterprise Office Management System Based on PHP

Yuan Hao[1], Laihong Lu[2], and Yuping Zhou[3,*]

[1] S.A.S FRTAO , Marseille, France
yuan.hao@frtao.com
[2] Liaoyuan Coal Mine Machinery Manufacturing Co., Ltd. Liaoyuan, China
469843951@qq.com
[3] The Collage of Information Science and Technology, Hainan Normal University,
Haikou, China
zypnew@qq.com

Abstract. Using a network to manage enterprise office affairs is an important means of improving efficiency of the office management and achieving standardization for the enterprise business. The major technology about using PHP and MySQL to develop an enterprise office management system was illustrated in this paper. The contents including the realization of functional modules were discussed, such as: The foreground web page module used by employees, business management module, design of the database tables and web management technology of administrator. The system has features of full-featured, high-efficiency, advanced technology. The systems provide an important reference platform to help enterprise reduce management costs, improve internal management, and enhance competition in the market for comprehensive competitiveness.

Keywords: PHP, enterprise, office management, system design.

1 Introduction

This system was the enterprise office management system, with full function and good maneuverability. The system was divided into the foreground use part and background management part. Its main function was to manage enterprise's daily affairs, office data management and maintenance. The foreground part was used by employees, therefore its interface design should mainly be user-friendly operation, beautiful interface, efficient and practical. Background management part was used by the system administrator. The system administrator will complete data management and updates in the system in the background. The system mainly used PHP and MySQL as a development tool, and Apache as a server. The network technology was applied in the enterprise management. It was easy for the employee and enterprise managers to operate or use. It provided the convenient, fast and practical software for the enterprise's daily management.

---

* Corresponding author.

H. Li et al. (Eds.): I3E 2014, IFIP AICT 445, pp. 70–78, 2014.

PHP (Hypertext Preprocessor) is a server scripting language, which is designed specifically for the Web. PHP codes can be embedded in an HTML page, which were executed when the page was accessed in each time [1]. PHP has advantages in many aspects, such as: PHP is very fast, with high performance; PHP are open source software, with a very good development space and scalability, and it can also be effective to make horizontal extension for a large number of servers; PHP can be applied on all major operating systems, and its code can be run without any modification in different systems; PHP supports multiple databases, and it can be connected to any database providing ODBC driver; PHP is free, in which a lot of open source codes are available. If modifying or adding new features to the language, it can be free of charge, therefore the cost of using PHP is low [2]. The unique syntax of PHP is the mixture of C, Java, Perl and PHP own syntax [3].

MySQL is a relational database management system. The relational database is the most common type of database, because of its faster access, and more flexible for data expression. MySQL supports structured query language SQL, so data query is simple and quick. For medium-sized enterprises, MySQL function can be fully met. The advantages of MySQL are its small in size, fast, cross-platform, and open-source. It is especially the feature of the open-source that a lot of free resources can be obtained, which causes many small and medium sized websites to choose MySQL as a database support of website background, in order to reduce the overall cost of the website. MySQL is a true multi-user, multi-threaded SQL database server. MySQL is small and exquisite database server software, which is ideal for small application system [4].

Apache server is an open source Web server of Apache Software Foundation. Apache has a good cross-platform and can run on almost any computer platform widely used.

The system was developed with today's most popular PHP and MySQL softwares. Two functions of the foreground use and background management were considered in this system. The database of this system was built scientifically, and each module function was carefully designed. Furthermore, it was a full-featured office management system, with the nice interface and simple operation, in order to satisfy the demands of all aspects of daily office management.

In the era of information, office collaboration in different place and office business in the business trip were often required for the modern enterprise. The network framework of this system was based on the application mode of a wide area network. Therefore, this system can achieve cross-regional business processing [5]. This was also another utility of the system.

2   Study Background

According to the survey, the current many enterprises exist many problems in the office management, such as low office efficiency, big proportion of artificial management, manpower and resource waste, some works still finished in manual, but can be done automatically, etc. Although some companies have bought the office software, the office automation becomes a "stopping, frozen", and office level is still in a backward state due to the backward software version, complicated interface not easy to operate, and poor maneuverability, etc. factors. It was the common desire of the people, and was also the main target of this system design to develop an office software with the humanized interface, enterprise office management level improved, the office efficiency of internal enterprise improved. The choice of development tools was also considered carefully. At present, there are a lot of development systems based on Web technology, but the three main kinds of dynamic technology are PHP, JSP and ASP.NET, etc. [6], each technology has its own characteristics.

PHP is a completely free, and the source code can be downloaded freely. It is a development tool favored by a lot of programmers. PHP is a simple, efficient and dynamic scripting language, with the advantages of cross-platform, powerful database support [7]. As the PHP is independent on the platform, it can run in a relatively high safety system platform [8]. Therefore, it has more advantages for system background construction. PHP (Professional Hypertext Preprocessor) technology and MySQL database have many advantages. Therefore, they are the best combination to develop the dynamic website. There are many software developed using them as a main tool.

PHP outstanding characteristics are fast in performance, a powerful CGI script language, more efficient to use memory, which can take up less memory consumption. PHP has good portability, and Web back-end CGI programs written in PHP can be easily transplanted to different operating systems.

3   System Overall Design

The enterprise office management system is an office management system with a collection of network and computer technology and full functions. It is mainly composed of two parts of the foreground use and background management. The overall structure of the system was shown in Figure 1. The use function module block of the foreground included the notice information, file transfer, attendance, employee information, salary information, online communication and user registration and other parts. Background management function module included the section management, notice management, document management, salary management, online communication management, employee information management, system management, attendance management, user management, system management, and other parts.

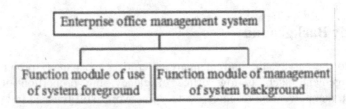

Fig. 1. Overall structure of enterprise office management system

3.1   Foreground Function Design of the System

The notice information modules in the foreground function of the enterprise office management system can complete the notice query, notice release and news alerts, and other functions. The file transfer module can complete the file upload, file access and file deletion, etc. Functions. The employee information module can complete staff recruitment, information inquiry, staff training, and other functions. The business attendance module can complete the functions such as employee attendance, leave management and attendance report. The salary information module can complete the salary query, report and printing, etc. the online communication module can complete the user login and online communication, etc. The user registration module can complete the registration information, account setting and user logged off, etc. The foreground module of the system is the interface used by common users, and the access to the system was restricted. The foreground use function module of the enterprise office management system was shown in Figure 2.

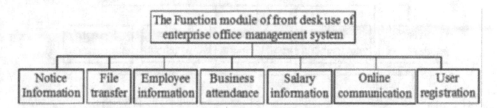

Fig. 2. The Function module of foreground use of enterprise office management system

3.2   Background Function Design of the System

The background function in the enterprise office management system included all data information management involved in foreground functions, such as documents, employees, salary, and etc. information management. In addition, the section information management and system management were included. The management to the section can include the query, add, delete and modify, etc. to section information. The management work of the system background should be done by the system administrator. The system administrator was conferred the management function of the system, therefore, his access right was bigger than that of normal users. The background management function module in the enterprise office management system was shown in Figure 3.

4   Design of the System Data Tables

The development and implementation of the enterprise office management system involves the analysis and design to the database and database tables, including the design of the data entities, attributes, entity relation, etc. Two kinds of operation permissions users were designed in this system: common users and administrators. According to the permission categories, the operating range for different users was limited [9]. System mainly involves the table structure design, the establishment of the link between tables, the establishment of the records in the table, etc. in the relational model. There were 31 table files involved in the database of the system. The structure of each table file table was carefully designed. On the premise of minimum redundancy guaranteed, links between tables were established scientifically to improve the efficiency of system running time, and take up less storage space. The table structure of employee basic information table (employee), department information table(dept), the announcement information table(notice), staff attendance sheet(sign), users of the system table(sysuser) and file information table (file) were given here, as shown in table 1 to table 6. The other table files involved in the system will not be illustrated here.

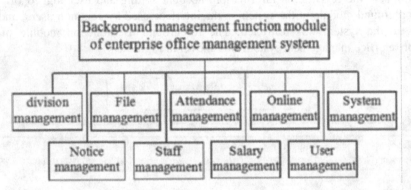

Fig. 3. Background management function module of enterprise office management system

Table 1. Staff basic information table (employee)

Field name	Data type	Length	Whether primary key	Description
emplid	varchar	6	Yes	Working ID
name	varchar	16	No	Name
sex	varchar	6	No	Gender
age	int	4	No	Age
birthday	datetime	8	No	Date of birth
Birthplace	varchar	20	No	Native place
nation	varchar	16	No	nationality
maristatus	varchar	4	No	Marital status
eduback	varchar	20	No	Education background
degree	varchar	12	No	Degree
eduschool	varchar	30	No	School of graduation
title	varchar	10	No	title of a technical post
position	varchar	20	No	Duty

Table 1. *(Continued)*

Field name	Data type	Length	Whether primary key	Description
wage	money	8	No	Basic salary
dept	varchar	30	No	Section
tel	varchar	20	No	Phone
address	varchar	80	No	Address
email	varchar	50	No	E-mail
onlinesta	varchar	4	No	Online status
image	Text	20	No	Photo path

Table 2. Sector information table (dep)

Field name	Data types	Length	primary key or not	Description
deptid	varchar	4	yes	Sector ID
deptname	varchar	30	No	Sector name
memo	Text	50	No	Sector description

Table 3. Notice information table (notice)

Field name	Data types	Length	primary key or not	Description
noticeid	varchar	6	yes	Notice ID
noticetitle	varchar	40	No	Notice title
noticetime	datetime	8	No	Notice time
noticeperson	varchar	20	No	Notice peopole
noticecontent	Text	50	No	Notice contents

Table 4. Attendance table (sign)

Field name	Data types	Length	primary key or not	Description
signid	varchar	6	yes	ID
datetime	datetime	8	No	Date and time
employee Name	varchar	20	No	Employee Name
late	varchar	4	No	Late or not
quit	varchar	4	No	Leave early or not

Table 5. System user table (sysuser)

Field name	Data types	Length	primary key or not	Description
userid	varchar	6	yes	ID
username	varchar	20	No	User name
userpwd	varchar	20	No	User password
logintime	datetime	8	No	Login Time
system	varchar	4	No	System Administrator or not
sign	varchar	4	No	Online Logo

Table 6. File information table (file)

Field name	Data type	Length	Whether primary key	Discription
fileID	int	4	yes	File ID
fileName	varchar	30	no	File name
fileSender	varchar	20	no	Delivery man
filereceiver	varchar	20	no	Recipient
fileTitle	varchar	50	no	File header
fileTime	datetime	8	no	Delivery time
fileContent	Text	30	no	File discription
path	varchar	100	no	File path
recsta	varchar	10	no	Receiving state

The relation among the data tables in the system database was more complex, and more table files were involved, such that the employee information management involves the operation and run of many table. Its link establishment must be accurate to reduce the redundancy of database. Only when created, stored procedure was compiled. Afterwards, recompiling was not needed whenever executing the stored procedure. Therefore, the efficiency of database access can be improved when using the stored procedure [10].

5   Introduction to the System Realization Technology

In the PHP programm, there was a common.inc.php file, in which some constants, configuration files, path of the session, host information, database information, global function, etc. were stored to ensure important information needed for the program at runtime [11]. In the development system, constants were defined and "$magic_quotes = get_magic_quotes_gpc ()" was used in order to increase the readability of the code and ease use. The automatic escape function was shut down, and self-defining functions escape was used. The database encoding and references to the parameters configuration page were completed by the program, therefore, the modularization of the program was realized, which is easy for maintenance and modification in the future.

The development of this system also involves image processing, encoding and decoding, compressed file processing, XML parsing, identity authentication of HTTP, POP3, SNMP, database processing, network interface application, security coding mechanism, and other functions. Using PHP + MySQL technology can solve these above problems well. API function supplied in MySQL in PHP can be used to operate the database. The database management, maintenance, and data retrieval and other various operations can be completed through the MySQL function library.

Much more module designs were involved in the system development, of which each module had personalized design in the design process. Such as: the main function of the background page design in the employee information management was to achieve the management and maintenance to the enterprise staff information, therefore, ordinary users can query, but can't modify. Only admin can modify operations. Next example: file transfer management page design was to save files uploaded and downloaded to the server. The path of transferring files in the server was saved to the database for the receiver to download again. All employees had permission to upload and download files. Online communication design was a complex link, because it was a web page on which the enterprise employee can transfer information online, make real-time communication, with more services demanded in the Web service, stronger required to the interactivity and instantaneity [12]. The attendance management page design was an important means for the enterprise to manage staff. It included signing in, signing out, taking a vacation, attendance statistics, etc. functions, which were used to compare and appraise staff. They were main indicators for the enterprise to evaluate staff performance. The design of the user password Settings page involves two levels of ordinary users and administrators super users. The user password settings page was for enterprise employees to reset the password, in order to ensure the safety of employees account. All employees in the enterprise can reset personal password, but the admin user had permissions to modify any employee's passwords. Employee salary management is an important part of the function, therefore, not only the real-time update should be considered, but also the security and confidentiality of data should be considered.

In short, a variety of different design technology, from hardware to software, was involved in the development and design process of the whole system, therefore, not only the development of the system should be completed well, but also the later system maintenance work should be considered.

6   Discussion and Implication

Today's advanced PHP + MySQL development tools were adopted in a PHP-based enterprise office management system. The system provided more comprehensive management functions, which enables the administrator and ordinary users to operate and manage conveniently. It is a relatively ideal office management platform of the enterprise. The system improves the work efficiency of the office management, reduces the waste of resources, and gets rid of the nuisance manual operation. The deficiency is that the system running speed will be slow down while the video function of online communication was used, and user's personalized printing function remains to be perfect. These will be further improved in the follow-up development process of the system.

References

1. Ma, N.: General Invoicing Management Platform Based on PHP. University of Electronic Science and Technology (2013)
2. Welling, L., Thomson, L.: PHP and MYSQL WEB Development (Wu Xin, etc translation), vol. 4, pp. 44–55. Machinery Industry Press, Beijing (2009)
3. Wu, Y.: Design and Implementation of the E-commerce Website Security System with Delphi and PHP. Hunan University (2013)
4. Gao, Y., Zhu, W.: Study on Credit Enterprise Information Systems. Science and Technology Information 36, 12–12 (2012)
5. Li, H.: The Research and Implementation of.NET-based Collaborative Office System Chongqing University, Chongqing (2009)
6. Yang, M.: Comparative Study on Mainstream Dynamic Web Page Technology PHP, JSP, and ASP.NET. Journal of Huaibei Professional and Technical College 1, 9–10 (2011)
7. Sulai, J.: Using PHP and MySQL to Develop Dynamic Website. The Public Science and Technology 3, 14–15 (2011)
8. Zhang, J.: The Advantages of PHP in the Website Background Construction. Development and Application of the Computer 12, 39–40 (2012)
9. Yang, Z.: Several Effective SQL Server Security Configuration. Computer Knowledge and Technology 14, 42–44 (2005)
10. Liao, J.: Design and Implementation of Virtual Machine-based Storage Process. Huazhong University of Science and Technology, Wuhan (2004)
11. Wu, J., Tian, R., Li, Y., et al.: PHP and MySQL Authority Guidelines. Mechanical Industry Press, Beijing (2011) (in Chinese)
12. Meng, F.: Design of Enterprise Office Automation System Based on ASP.NET Technology. Computer and Information Technology (5), 19–21 (2007)

E-Loyalty Building in Competitive E-Service Market of SNS: Resources, Habit, Satisfaction and Switching Costs

Yong Liu[1,*], Shengli Deng[2], and Feng Hu[3]

[1] Chair of Marketing and Innovation, Universität Hamburg, Hamburg, Germany
Yong.Liu@wiso.uni-hamburg.de
[2] Center for study of Information Resources, Wuhan University, Wuhan, China
victorydc@sina.com
[3] Xingzhi College, Zhejiang Normal University, Jinhua, China
hufeng@zjnu.cn

Abstract. Despite considerable efforts have been devoted to study consumer loyalty, there is a lack of knowledge concerning how online service loyalty is or can be established in a competitive e-service market, in which several major service providers coexist to compete for customers. In this study, we attempt to explore the industry environment of Chinese social networking service (SNS), and examine the association between consumer satisfaction and switching costs in building service loyalty. From a resource-based view, unique service resources of SNS (critical mass and supplemental entertainment) are examined regarding their potentials in enhancing consumer satisfaction, habit and switching costs. The results show that habit and the interaction effect of satisfaction and switching cost are the key determinants of SNS loyalty. Critical mass and supplemental entertainment directly or indirectly affect habit and switching costs. This study attempts to bring the thought of competitive environment into e-service loyalty research while new insights for e-service loyalty building in different market environments are discussed.

Keywords: Social networking, service loyalty, habit, critical mass, switching cost, satisfaction, supplemental entertainment, SNS.

1 Introduction

"Even though the results of customer-satisfaction surveys are an important indicator of the health of the business, relying solely on them can be fatal [17, p. 4]."

Today's enterprises are continually looking for ways to enhance customer loyalty mainly through improving consumer satisfaction. Loyal customers are vital to the long-run profitability of any business as acquiring a new customer may cost as much as five times more than retaining an extant customer [23]. Consequently, it is not astonishing that building customer loyalty has been acknowledged as an integral part of doing business or the 'business back-bone' [14]. In currently increasingly service-

---

* Corresponding author.

H. Li et al. (Eds.): I3E 2014, IFIP AICT 445, pp. 79–90, 2014.

based economies [11], a growing interest has been dedicated to consumer loyalty building in an e-service or Internet environment.

A considerable number of studies on e-service loyalty (e-loyalty) have been conducted in recent years, most of which suggest that consumer loyalty, as a result of consumer satisfaction, is mainly gained through improving service quality. However, previous works have claimed that the effect of satisfaction on consumer loyalty varies greatly under the condition of different levels of switching costs in, i.e. various human-mediated physical service environments [18]. There is lack of relevant knowledge and researches in the contexts of e-service community. For instance, little knowledge is available concerning how e-loyalty is altered along with the evolution of business environment, in particular when the market becomes highly competitive with several competent service providers.

In addition, website design tools are under rapid development with the advances in information and communication technology (ICT), which has made it an easy task to develop or even 'duplicate' new website functionality or applications. As a result, it becomes more difficulty today for e-service providers to gain advantages in competition simply relying on a well-working service website, in particular when competitors start to 'learn' fast from each other. In an Internet environment, where competition may be only one click away [36], a rising concern is how to make their service distinguishing, i.e. through offering unique resources.

From a resource-based view, the study sought to investigate the interdependencies between unique e-service resource, habit, consumer satisfaction and switching costs in building consumer loyalty in a highly competitive service market—the social networking service (SNS) market in China. In this market, several competent SNS suppliers coexist to compete for consumers while offering similar web functionalities and presentation. Two unique resources underlying SNS—critical mass and supplemental entertainment, are discussed and examined regarding their roles in facilitating consumer habit, satisfaction and switching costs.

2   Theoretical Background

2.1   Determinants of E-Loyalty: Service Quality and Satisfaction

Customer loyalty has long been a hot research topic for both business and marketing. The last several decades have witnessed a paradigm shift from tangible products (i.e. brand loyalty) to human-mediated service (i.e. hotel and restaurant) and to recent computer-mediated service (or e-service) in this field, along with the advance of information technologies. According to various researches on customer loyalty, the framework of service quality—satisfaction—consumer loyalty or its extension has been widely employed in tangible products and human-mediated service contexts [6, 14]. Apparently, e-service loyalty has been mainly assumed to be a result of satisfaction, which is enhanced by a diversity of service quality.

Despite the widespread emphasis of satisfaction on e-loyalty establishment, recent research on switching behavior of e-service seems to offer a different picture. "New IT choices are only a click away on the Internet, and switching to a competing IT is almost as easy as downloading and installing it, or completing an online registration form to sign up for a different service" [5]. As a result, it becomes increasingly easier for online consumers to switch to other service providers who offer similar services[36]. In certain e-service market (i.e. online gaming industry), it seems to be especially difficult to build customer loyalty, while satisfied consumers may easily switch to other service providers [c.f. 25]. A recent study on massively multiplayer online role playing games (MMORPGs) found that high attractiveness of alternatives and low switching cost significantly motivate game players' switching intention [16]. Steenkamp and Baumgartner (1992) noted that switching behaviors may occur among satisfied customers due to their attempt to try something novel or different for fun or thrill [30]. Customers may be satisfied with their choices, but may still switch to alternatives owing to a desire for novelty or complexity in brand consumption or curiosity, or getting bored, fed up, on repetitively doing the same thing [12, 15, 28]. Balabanis et al. (2006) found that satisfaction is a significant predictor of e-loyalty only for those consumers who have a low level of satisfaction [3]. Bhattacherjee et al. (2012) reported that satisfaction with prior e-service provider only has a small and marginal negative influence on switching behavior [5].

2.2   Business Environment and Consumer Categories

The inconsistence in previous research findings between e-service loyalty and switching behavior may be interpreted from the perspective of business environment. By an investigation of a diversity of tangible products and physical services, a study of Jones and Sasser (1995) noted that a consumer's perceived satisfaction affects loyalty in a different way in different industry settings [18], as shown in Figure 1. An consumer may be loyal to a specific service with a low degree of satisfaction, but also betray a service provider in spite of a high degree of satisfaction [c.f. 17]. As a result, based on the interaction patterns between satisfaction and loyalty, Jones & Sasser (1995, p. 11) suggested that "consumers behave in one of four basic ways: as loyalists, as defectors, as mercenaries, or as hostages", as shown in Table 1.

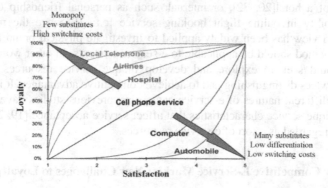

Fig. 1. How the competitive environment affects the satisfaction-loyalty relationship (Source: adapted from [18, 22])

In a highly competitive and low differentiation market, consumers are more likely to be low loyal whilst having a high level of satisfaction, therefore being mercenary. Based on the work of Jones & Sasser (1995), more mercenary consumers will appear in e-service markets as a result of increased competition and decreased differentiation [18]. For instance, Hsiao & Yang (2011) noted that a critical issue in online retailing research is how to identify, attract, and retain customers since online shoppers are typically regarded as less loyal [17].

Table 1. Individual customer satisfaction, loyalty, and behavior (adapted from: Jones & Sasser, 1995)

	Satisfaction	Loyalty	Behavior
Loyalist	High	High	Staying and supportive
Defector	Low to medium	Low to medium	Leaving or having left and unhayppy
Mercenary	High	Low to medium	Coming and going; low commitment
Hostage	Low to medium	High	unable to switch; trapped

2.3   A Resource-based View of E-Loyal Building

From a resource-based view, a firm's performance is founded on a collection of physical and intangible resources (assets or capabilities) that enable it to compete with other firms [34]. Different industry may feature different sorts of resources. For instance, Barney (1991, p. 101) defined resources as "all assets, capabilities, organizational processes, firm attributes, information, knowledge, etc. controlled by a firm that enable the firm to conceive of and implement strategies that improve is efficiency and effectiveness"[4]. A good firm resource should be valuable, heterogeneous, immobile, and non-substitutable, and therefore lead to sustainable competitive advantage of the firm [24]. Previous works suggest that unique service resources may contribute to important dimensions of service quality, which further enhances consumer satisfaction and therefore loyalty. These resources may be external in nature, such as good location and transportation convenience of a hotel[20, 32], or internal, such as personal friendship to customer, routes availability in online flight booking service [c.f. 7]. Since the method of a resource-based view has been widely applied to investigate physical brand and tangible service, this method should be applicable to e-service contexts. In other words, e-service providers should seek to explore and develop unique service resources in order to make their services distinguishing and to achieve competitive advantages. It is noted that, pertaining to different natures of e-service categories, previous studies have identified a number of unique service characteristics that affect service acceptance [19, 29], many of which can be regarded as a sort of e-service resources.

2.4   Highly Competitive E-Service Market: New Challenges to Loyalty Research

Of various Internet services in China, Tencent QQ (QQ), a free instant messaging computer program, is obviously one of the most successful applications. In September 2011, the active QQ user accounts for QQ IM amounted to 711.7 million while its peak concurrent users reached 145.4 million (Tencent, 2011). Note the fact that there were 513 million Internet users in China in 2011 [10][1]. Whilst QQ is initially developed for instant communication, the company seeks to integrate a diversity of value-added services to the QQ and develops it to be a customized "one stop for all" application. An increasing amount of Internet services have been integrated to the simple interface of QQ (like MSN). As a result, through simply clicking on icons in its interface, QQ users can access a diversity of various Internet services, such as email, online music, games, web TV, online shopping, Internet disk, SNS and many others. One stop for all application strategy offers QQ unique advantages compared to other service providers, as there is no more need for users to make additional registration and to remember additional account and password information towards a new service supplier. In addition, friends list of instant messaging service can be easily migrated to SNS, which saves QQ users' lots of efforts to establish a new network for SNS. Also as an integrated function, new updates from SNS (i.e. comments, new posts) will be presented to users through the simple interface of QQ. Considering various inherent advantages of SNS at QQ, it seems that it should be easy to convert loyal QQ instant messaging users to be loyal SNS users alike.

However, as one of the earliest SNS providers, market share of QQ's SNS (or Qzone) has been eaten away by more recent entrants to the industry. Many users shift to other SNS providers despite their loyalty to QQ IM and give up various benefits offered. They are willing to use other SNS services, even if they have to spend additional efforts to open the browser, to fill in account and password information and to search people for reconstructing their new network. Currently, there are 190 million active SNS users for Qzone, 96 million for Renren.com and 40.1 million for Kaixin001.com [31]. The SNS market in China is dominated by several suppliers that provide similar services. After several years of development and competition, SNS suppliers share considerable similarities in their website presentation and functionalities despite limited differences. In this regard, it would be interesting to scrutinize what are unique service features (or resources) to distinguish a SNS supplier from a user's perspective.

3 Research Framework and Hypotheses Development

To evaluate SNS service loyalty, a research framework is proposed. Five hypothesized determinants of loyalty are included, which are habit, satisfaction, switching costs, supplementary entertainments and critical mass.

Satisfaction can be defined as a personal's feeling of pleasure or disappointment resulting from comparing a service's perceived performance (or outcome) in relation to his or her expectations [21]. Previous literature provides unambiguous supports for the positive influence of satisfaction on customer loyalty. Consistent with previous studies, it is expected that the same influence works in SNS alike. Hence, we hypothesized:

H1: Satisfaction positively relates to consumer loyalty.

In service loyalty research, switching costs measure a consumer's perceived difficulty in switching to a new service supplier, which represents anything that

---

[1] Some Internet user may actively use two or more QQ accounts.

makes it more difficult or costly for consumer to change providers [8]. Switching costs are not only economic in nature, but also can be emotional or psychological, such as interpersonal relationships and special treatment [8, 36]. Many studies suggest switching costs exert a direct influence on service loyalty, arguing that if consumers encounter a high switching costs, they are more likely to stay with the service supplier [2, 8, 35, 36]. Therefore, we hypothesized:

H2: Switching costs positively relate to consumer loyalty.

Jones and Sasser (1995) pointed out that the degree of switching costs differ a lot in different service sectors, which may alter the effects of satisfaction on service loyalty [18]. Balabanis et al. (2006) classified satisfaction to be high, moderate and low levels, and hypothesized that influence of perceived switching barriers on e-store loyalty is greater when satisfaction is low [3]. Yang and Peterson (2004) divided e-service users to be two subgroups of unsatisfied and satisfied users, and reported a significant interaction effects of switching costs and satisfaction on customer loyalty in satisfied user group [36]. Lee et al. (2001) studied mobile phone service loyalty and reported a positive interaction effect of switching costs and satisfaction on loyalty [22]. Jones & Sasser (1995) suggested that, in a high competitive and low differentiation market, people become mercenary and have a low commitment. Therefore, it is important to build switching costs to retain customers [18]. Therefore, we hypothesized:

H3: Interaction effects between satisfaction and switching costs positively relate to consumer loyalty.

Habit can be defined as the "learned sequences of acts that become automatic responses to specific situations, which may be functional in obtaining certain goals or end states" [33]. The concept and function of habit has been broadly investigated across a diversity of disciplines, like social psychology, health sciences, marketing/consumer behavior, IT user behavior and organizational behavior [9]. Frequently performed behaviors tend to become habitual and hence automatic over time [26]. Habitual behavior is difficult to resist and substantial conscious efforts are needed for individuals in order to alter their habits. Therefore, altering habit in order to switch to another service provide may exert to be a sort of switching cost. Furthermore, consumers may express their loyalty to an e-service provider simply as a result of habit. In other words, they are 'lazy' to make a change or feel difficult to change their habit of being 'loyal' to existent service provider. Accordingly, the following hypothesis is made:

H4a: Habit positively relates to switching cost.
H4b: Habit positively relates to loyalty.

From a resource-based view, unique resources or capability of a company means the unique service attributes help to discriminate a service provider from others to enhance competitiveness. Unique resources can be internal, accumulated with times and co-created with customers, i.e. hotel stuff's personal relationship or friendship to customers [7]. Concerning SNS, supplementary entertainments can be regarded as a sort of internal resources. Specifically, it refers to the entertainments/games offered by SNS; these entertainments/games are not specifically designed for information sharing or communication purpose, but to offer entertainment that a user can play to compete with peers even when peers are offline. The strategy of providing supplementary entertainments has been dominantly applied by today's SNS providers. Taking 'Happy Farm' social game for instance, it is one of the most popular social networking games in China. At the height of its popularity, there were 23 million daily active users who log in to the game at least every 24 hours[2]. As a value-added element for SNS, a provision of supplementary entertainments potentially brings about more pleasure to SNS user, making consumers more likely to feel satisfied. Therefore, it is proposed:

H5a: Supplementary entertainments positively relate to satisfaction.

Furthermore, frequently interacting with the supplementary entertainments may contribute to a motivator of forming habitual behavior to access SNS. In other words, supplementary entertainments may contribute to the establishment of habitual use of SNS. Therefore, it is proposed:

H5b: Supplementary entertainments positively relate to habit.

External resources refer to the service attributes, which are not generated by the service provider, but contribute to be an important aspect of service, such as the transportation convenience to a hotel customer. Concerning SNS, critical mass is proposed to affect consumer service experience as a kind of external resources. It refers to an important segment of population that chooses to take part in the SNS [1, 25]. Critical mass makes major contributions to the collective action as well as to later subscribers [1, 25]. A new network by its nature requires a group of subscribers if it is to startup; the network becomes mature to move beyond that point in its development where a critical mass has initially assembled [1]. As a result, new subscribers to a mature SNS can join one after another instead of as a group [1]. For instance, a user subscribes to a SNS, for that his/her friends are there already; as a result, s/he subscribes to the SNS by join the already existed circle of his/her friends with no need to establish a new network by his/her own efforts. Hence, users may decide to be loyal to a specific SNS so that they can connect to the people who are important to them, such as friends and relatives. The possibility to communicate with important others inside the network should be an important purpose of using SNS, which enhance their satisfaction to the service. Furthermore, when users have lots of friends using the SNS, it is more likely for them to communicate with their friends via SNS and therefore make the SNS use to be habitual. Therefore, we proposed:

H6a: Critical mass positively relates to habit.

H6b: Critical mass positively relates to switching costs.

4   Research Methodology

4.1   Survey and Questionnaires

As there are many SNS operators in China, Qzone, the biggest social networking service provider in China, is selected as the subject for survey. A five-point Likert-scale ranging from disagree (1) to agree (5) was used to measure each item of the model. The questionnaire survey was conducted online. We provided a hyperlink to

---

[2] See. http://game.dayoo.com/200911/20/68602_100420554.htm

the survey web page and posted it to different forums, while the respondents would be directed to the online version of the questionnaire by clicking on the hyperlink. Finally, 228 complete samples were collected, seven of which have no prior use experience and were therefore omitted from the analysis. The final samples consist of 99 males (44.8%) and 122 females (55.2%). Most respondents are under 25 years old (n = 175) and have used computers for over 3 years (n = 194). 73.3 percent (n = 162) of participants have an experience of using Qzone for over 3 years while 44 respondents (19.9%) has 1-3 years use experience. Only 15 participants (6.8%) have a use experience of less than 1 year.

4.2   Reliability and Validity of Measurement

Partial Least Squares (PLS) path modeling technologies were utilized to analyze the data using SmartPLS 2.0. Convergent validity indicates the degrees to which the items of a scale that are assumed to be theoretically associated are also related in reality. As shown in Table 3 and 4, all values of Cronbach's alpha, factor loadings and CR are over their thresholds of 0.5 and 0.7 respectively.

Discriminant validity refers to whether the items of a scale reflect the construct in question or reflect another related construct. It can be verified if the variances of the average variance extracted for each construct are higher than any squared correlation between this construct and any other construct, the discriminant validity is supported [13].

Table 2. Reliability and convergent validity statistics

Construct	α	CR	Minim. factor loading
Satisfaction	0.941	0.957	0.907
Switching costs	0.871	0.912	0.817
Supplementary entertainments	0.820	0.917	0.917
Critical mass	0.883	0.944	0.944
Habit	0.969	0.980	0.961
Loyalty	0.923	0.951	0.904

Table 3. Discriminant validity
(The bold diagonals are the AVEs of the individual constructs; off diagonal values are the squared correlations between constructs)

Construct	Mean	Sa	Sc	Se	Cm	Ha	Lo
Satisfaction (Sa)	3.26	0.850					
Switching costs (Sc)	2.60	0.149	0.722				
Supplementary entertainments (Se)	3.19	0.247	0.199	0.847			
Critical mass (Cm)	3.45	0.168	0.127	0.098	0.895		
Habit (Ha)	2.96	0.565	0.596	0.383	0.561	0.942	
Loyalty (Lo)	2.45	0.304	0.544	0.165	0.202	0.750	0.868

4.3   Data Analysis and Model Assessment

Figure 2 depicts the results of model testing. As shown in the Figure, both satisfaction and switching cost alone have no significant effect on loyalty. Instead, the interaction effect of satisfaction and switching cost has significant influence on loyalty. Furthermore, habit exerts a substantial and strong influence on both loyalty and switching cost. Critical mass exerts a significant influence on habit, but not on switching cost. Supplementary entertainments significantly affect both habit and satisfaction. Critical mass and supplemental materials have indirect influences on switching cost, which are mediated by habit. The model interprets 35.6 percent of variance of switching cost, 36.3 percent of habit, 24.9 percent of satisfaction and 72 percent of loyalty.

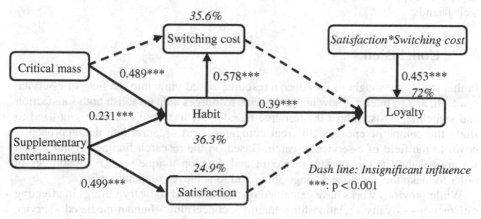

Fig. 2. Results

5   Discussion and Implications

Figure 2 shows that, in this competitive SNS market, consumers tend to be mercenary, whereas a high satisfaction of SNS users doesn't guarantee a high loyalty toward service provider. In addition, satisfaction alone cannot bring about loyalty as well. Consistent with the work of Jones & Sasser (1995), we propose that consumer satisfaction may not be such a reliable indicator of loyalty for e-service market, in particular when the market becomes more competitive with less differentiation. Consumers behave differently in response to different industry environments. In other words, the change of industry environment itself may effectively alter the way that consumers choose to be loyal to a company, i.e. mercenary versus hostage consumers. The results show a strong interaction effect between satisfaction and switching costs on SNS loyalty. In other words, it is really difficult to establish user loyalty in SNS: only when consumers are satisfied with service provider and are feeling difficulty to switch, they become loyal. This also suggests that satisfaction is a necessary condition of service loyalty. Satisfied consumers can be either loyal or not loyal to service provider based on the level of switching cost, but dissatisfied consumers definitely exert no loyalty. Habit has in particular strong and direct influence on SNS loyalty. This implies that, in Chinese SNS market, consumer loyalty is more like a sort of habitual behavior. In other words, there is a high risk of consumer switch, if they are strongly motivated to change their habit, e.g. by their peers (critical mass).

Furthermore, critical mass and supplementary entertainments are found to be important resources that significantly enhance consumer habit, satisfaction and switching costs on SNS. This indicates that, if a SNS maintained an individual's network by hosting the people who are important to him/her, the user is less likely to switch because it will cause a loss of connections to those important people. Also the time and efforts spent on supplementary entertainment, such as social networking games, appears to be valuable to the consumers. They feel more satisfied if they engage in supplementary entertainment, in particular when they play together with their friends.

6   Conclusion

In this study, we sought to introduce a resource-based view into the field of e-service research, and to interpret how unique service resources help establish both satisfaction and switching costs. Whilst the resource-based view has been dominantly utilized to study the competitiveness of different companies and organization, this approach is novel to the field of e-service research. Based on the research findings of this study, we argued that it is important to recruit and develop unique service capabilities in order to make the e-service distinguishing and to achieve competitiveness.

While previous works have examined the influence of industry settings in affecting satisfaction—loyalty relationship mainly concerning human-mediated service industries [18], our study investigates the loyalty building in a competitive e-service market with several competent suppliers. The results of this study illustrate that, in Chinese SNS market, despite of a high satisfaction, consumer loyalty cannot be established without a proper level of switching costs. In Internet environment, the choice of websites is already huge and increasing. Therefore, it is essential for websites to focus more on increasing switching costs than eliminating alternatives. In this regard, our results facilitates a more complete understanding on e-service loyalty building in context and seeks to raise the research interests in unique service resources for developing distinguishing e-service. Previous studies show that switching costs vary in different industry settings while the effects of satisfaction on customer retention will be deteriorated in case of high switching costs [22]. Simply considering e-service loyalty from the view of satisfaction, important information might be underexplored, in particular if industry environment and switching costs become a matter of importance. There is a dearth of relevant wisdom in the e-service field concerning some new and emerging business phenomena, such as the predicament faced by Qzone.

Acknowledgements. This research is partly supported by Wuhan University Academic Development Plan for Scholars after 1970s for the project Research on Internet User Behavior and the National Funds of Social Science (No. 14BTQ044), the National Planning Office of Philosophy and Social Science in China (No. 13CTQ029) and the Zhejiang Province Planning Project of Philosophy and Social Sciences (No. 11JCGL04YB).

References

1. Allen, D.: New telecommunications services: Network externalities and critical mass. Telecomm. Policy. 12(3), 257–271 (1988)
2. Amin, S.M., et al.: Factors Contributing to Customer Loyalty Towards Telecommunication Service Provider. Procedia - Soc. Behav. Sci. 40(6), 282–286 (2012)
3. Balabanis, G., et al.: Bases of e-store loyalty: Perceived switching barriers and satisfaction. J. Bus. Res. 59(2), 214–224 (2006)
4. Barney, J.: Firm resources and sustained competitive advantage. J. Manage. 17(1), 99–120 (1991)
5. Bhattacherjee, A., et al.: User switching of information technology: A theoretical synthesis and empirical test. Inf. Manag. 49(7-8), 327–333 (2012)
6. Bloemer, J., et al.: Linking perceived service quality and service loyalty: a multi-dimensional perspective. Eur. J. Mark. 33(11/12), 1082–1106 (1999)
7. Butcher, K., et al.: Evaluative and relational influences on service loyalty. Int. J. Serv. Ind. Manag. 12(4), 310–327 (2001)
8. Chang, Y.-H., Chen, F.-Y.: Relational benefits, switching barriers and loyalty: A study of airline customers in Taiwan. J. Air Transp. Manag. 13(2), 104–109 (2007)
9. Chou, C., et al.: Understanding Mobile Apps Continuance Usage Behavior and Habit: An Expectance-Confirmation Theory. In: PACIS 2013 Proceedings. p. Paper 132 (2013)
10. CNNIC: The 29th Statistical Report on Internet Development in China (2012)
11. Dai, H., et al.: Antecedents of online service quality, commitment and loyalty. J. Comput. Inf. Syst. Winter, 1–11 (2011)
12. Fiske, D.W., Maddi, S.R.: Functions of varied experience. Dorsey Press, Homewood (1961)
13. Fornell, C., Larcker, D.: Evaluating structural equation models with unobservable variables and measurement error. J. Mark. Res. 18(1), 39–50 (1981)
14. Gremler, D.D., Brown, S.W.: Service loyalty: its nature, importance, and implications. In: Edvardsson, D.D., et al. (eds.) QUIS V: Advancing Service Quality: A Global Perspective, NY, New York, pp. 171–181 (1996)
15. Herrnstein, R.J., Prelec, D.: A theory of addiction. In: Loewenstein, G. and Elster, J. (eds.) Choice Over Time. Russell Sage Press, New York (1992)
16. Hou, A.C.Y., et al.: Migrating to a new virtual world: Exploring MMORPG switching through human migration theory. Comput. Human Behav. 27(5), 1892–1903 (2011)
17. Hsiao, C.H., Yang, C.: The intellectual development of the technology acceptance model: A co-citation analysis. Int. J. Inf. Manage. 31(2), 128–136 (2011)
18. Jones, T., Sasser, W.: Why satisfied customers defect. Harv. Bus. Rev. 76(3) (1995)
19. Kim, J., et al.: The role of etail quality, e-satisfaction and e-trust in online loyalty development process. J. Retail. Consum. Serv. 16(4), 239–247 (2009)
20. Knutson, B.: Frequent travelers: making them happy and bringing them back. Cornell Hotel Restaur. Adm. 29(1), 82–87 (1988)
21. Kotler, P.: Marketing management. Prentice-Hall, Englewood Cliffs (2003)
22. Lee, J., et al.: The impact of switching costs on the customer satisfaction-loyalty link: Mobile phone service in France. J. Serv. Mark. 15(1), 35–48 (2001)

23. Madhavan Parthasarathy, A.B.: Understanding post-adoption behavior in the context of online services. Inf. Syst. Res. 9(4), 362–379 (1998)
24. Mata, F., et al.: Information technology and sustained competitive advantage: a resource-based analysis. MIS Q. 19(4), 487–505 (1995)
25. Oliver, P., et al.: A theory of the critical mass. Am. J. Sociol. 91(3), 522 (1985)
26. Ouellette, J.A., Wood, W.: Habit and intention in everyday life: The multiple processes by which past behavior predicts future behavior. Psychol. Bull. 124(1), 54–74 (1998)
27. Playnomics: Playnomics Quarterly US Player Engagement Study Q3 2012, http://ww1.prweb.com/prfiles/2012/10/18/10028632/Q3reportv4.pdf
28. Raju, P.S.: Optimum Stimulation Level: Its Relationship to Personality, Demographics, and Exploratory Behavior. J. Consum. Res. 7(3), 272–282 (1980)
29. Srinivasan, S.S., et al.: Customer loyalty in e-commerce: an exploration of its antecedents and consequences. J. Retail. 78(1), 41–50 (2002)
30. Steenkamp, J., Baumgartner, H.: The role of optimum stimulation level in exploratory consumer behavior. J. Consum. Res. 19(3), 434–448 (1992)
31. TechRice: Recruiting via China's Social Networks (2012)
32. Tsaur, S.-H., et al.: Determinants of guest loyalty to international tourist hotels—a neural network approach. Tour. Manag. 23(4), 397–405 (2002)
33. Verplanken, B., et al.: Habit, information acquisition, and the process of making travel mode choices. Eur. J. Soc. Psychol. 27, 539–561 (1997)
34. Vinekar, V.: Towards a Unified Model of Information Technology Business Value. In: AMCIS 2006 Proc. (2006)
35. Wang, Y.-S., et al.: The relationship of service failure severity, service recovery justice and perceived switching costs with customer loyalty in the context of e-tailing. Int. J. Inf. Manage. 31(4), 350–359 (2011)
36. Yang, Z., Peterson, R.T.: Customer perceived value, satisfaction, and loyalty: The role of switching costs. Psychol. Mark. 21(10), 799–822 (2004)

Rays of Uniqueness – Cloud in Multinational Organisations: Challenge to Traditional Business System Alignments

Amit Mitra*, Nicholas O'Regan, and Ximing Ruan

Bristol Business School, University of the West of England, Bristol, UK
Amit.Mitra@uwe.ac.uk

Abstract. Cloud computing initiatives support multinational corporations in optimizing resource utilisation and at the same time provide ubiquitous capacity to satisfy expectations of users within it. However, research carried out in the early 1990s demonstrated consequences when there were mismatches between business and IT strategies. The advent of utility computing through cloud based resource development seems to have altered both the perception of IT resources as well as the expectations of their use. In this paper by using a resource based view of cloud computing we examine the nature of unique capacity development in multinational organisations. Evidence for the research was gathered through interviews conducted in two well-known multinational companies in the oil and natural gas and the car manufacturing sectors respectively. The lessons drawn in this research is likely to be beneficial for organizations implementing Cloud based solutions.

Keywords: IS strategy, Business strategy, IS assets, Cloud Computing, Resource sharing.

1 Introduction

Although cloud computing is a term associated with a new innovative development in the world of computing yet the idea of sharing resources originated in the 1960s and 70s [1]. At the same time the contexts of the two developments couldn't be more different. While computing used to be a scarce resource in the 1960s and 1970s it is not so now with the advent of PCs for end users as well as the ubiquitous enabling role played by the Internet. Cloud computing and time sharing are connected to the notion that computing resources can be used as a utility that is distributed as a service like electricity and water. Evidently, cloud computing characterises a fundamentally different way of invention, development, deployment, scaling, updating, paying for information and communication services [2]. Many organisations that are considering moving onto the cloud are initially challenged by the notion of billing at point of use and subsequently by the whole changes to decision making that this sort of service orientation creates. Adoption of core technologies like cloud can also have

---

* Corresponding author.

H. Li et al. (Eds.): I3E 2014, IFIP AICT 445, pp. 91–103, 2014.

influence on organisational adaptations as reported in the context of management innovations [3]. Mitra [4] found duality of impact to be a key feature of technology adaptation processes within British local government. As surplus capacity and re-use or sharing of resources becomes the prevalent obtaining reality, the sharing features of cloud computing will be increasingly an attractive arrangement of choice.

In the quest to estimate cloud use effectiveness it seems obvious that the resource based view (RBV) of the firm would be a credible analytical approach. RBV is based on the notion that organisations succeed in gaining competitive advantage by using its assets as well as sustained competitive advantage is gained by using its existing capabilities. According to Wade and Hulland [5], Information system resources of firms may be visualised through a couple of categories that include IS assets (technology based) and IS capabilities (systems based). IS assets have an inherent disadvantage in that they can be easily copied by competitors and therefore is fragile with regard to sustainability of competitive advantage [cf. 6, 7]. An important issue for this paper is the fact that there is growing evidence to show that competitive advantage often depends on the firm's superior deployment capabilities[8, 9]. It is clear that certain key criticisms of the RBV [10] have stemmed from the static assumptions of the nature of resource, value and sustainable competitive advantage, for the purposes of this paper we'd like to presume both dynamic settings and unique capabilities enable organisations to develop sustainable competitive advantages.

Multinational companies constantly seek to develop unique inimitable capacity so that they can create competitive advantage over their competitors, despite the fact that initiatives for changes are difficult to implement due to resistance.. MNCs that succeed in implementing change by overcoming such resistance usually are supported by leadership that has foresight, is able to handle risk at the same time become considerably unique in the way their business systems operate[11]. Such an orientation as for instance in the study conducted within Nestlé by Mitra and Neale [11] is in contrast to findings which showed that co-ordinating IS plans with business plans impedes effective IS planning[12]. Despite many successful IS implementations yet instances of IS failure has lingered that has enabled the issue of alignment to be pre-eminent in the context of large MNCs. Just like Mitra [4] identified maturity to be a key parameter in Geographic Information Systems implementation within British local government similarly, Luftman [13] categorises alignment to be dependent on six categories of maturity. It would be interesting to see if Hayward's Rigs and Nihon Motors are able to look more closely at how cloud computing can support individual processes [14] rather than how it can support an entire strategy.

2   Nature of Cloud Based Resource Capabilities

In a context where overwhelming evidence shows [15, 11] that IS capacity development leads to inimitable resource advantages, it is clear that both Hayward's Rigs and Nihon Motors did not doubt the possibility of developing cloud based capacities. At the same time maturity of using competencies can make a difference to the way organisations eventually acquire competitive advantages. In the study [4] on British local government based implementations of geographic information systems,

Mitra found that levels of maturity in IS use usually leads to a couple of distinctly different formats in which organisations develop IS capacity. According to Mitra [4] *adoption* and *adaptation* of IS may be manifestly linked to maturity levels of the organisation's implementations of specific IS. Indeed organisations that strive to adapt also seek alignment between businesses and IS strategies for acquiring competitive advantages. Successful alignment of business and IS objectives is a 'process of continuous adaptation and change'[16]. In this study, adoption of cloud computing was never in question. However, Hayward's Rigs and Nihon Motors went through varied adaptations of cloud use to garner specific capacities.

Following Armbrust et al [17], cloud computing in this paper would refer to both the applications delivered as services over the Internet and the hardware and systems software within data centres that provide those services. Alignment of business strategy and information systems strategy has been a longstanding research pivot around which various organisations and their ISs have been explored. It must be borne in mind that the notion of introducing cloud computing is probably somewhat different in contrast to traditional systems. Whilst Bharadwaj [15] implies that IT capability is a rent generating resource that is not easily imitated or substituted in large companies yet, Armbrust et al [17] have clarified that pay as you go as used in cloud computing is clearly tied to usage. Renting usually involves paying a negotiated amount over a fixed period of time irrespective of use. Pay as you go involves metering usage and charging based on actual usage, independently of the time period over which the usage occurs. With the advent of cloud computing, this is perhaps a key difference that has come about in the estimation of IT resources. Dwelling on scale and simplicity as the new dimensions that cloud brings to the context of multinational companies, Grossman [19] concurs, that pay as you go to use cloud capacity is a facet that has hitherto remained unknown. In the context of MNCs as operations scale to international contexts using multiple proprietary providers could lead to challenges in application of security policy [20]. Reductions in budgets and higher space requirements as computing becomes more web based, there is a compulsion that is driving large organisations to rethink their current capacity provisioning [21]. Cloud obviously provides a veritable option that is increasingly being taken seriously by large organisations.

3 Methodology Adopted

Data were collected in two multinational companies that have significant operations within the UK. The first is a British multinational that is well known in the oil and natural gas sector. The second is a Japanese car manufacturer that has significant UK based manufacturing capacities. Prevalent non-disclosure conditions do not permit us to use either the names of the organisations or the staff who have been interviewed for the study. For the purposes of this paper the companies will be referred to using pseudonyms as Hayward's Rigs (HR) and Nihon Motors (NM) respectively. Both companies have assets and personnel spread across various local and global locations that need to seamlessly interact for efficient delivery of business. It is clear that further to their experiences, both companies have realised that merely transferring all of their data across to cloud based repositories is unlikely to work [22]. There were also documents that were used to enrich the analysis that is referred to in this paper. Whilst both companies have been considering using some cloud services ever since 2007 yet faced with various local and global challenges both companies had embarked around 2010 to seriously integrate cloud computing into their operational needs.

A case study approach [23] was used to collate evidence on HR and NM.

Table 1. Antecedents of cloud use

Dimension	Hayward's Rigs	Nihon Motors
Industry expectations	Increased demand from business managers to deploy functionality quickly. HR executives, like many of their counterparts in other organisations, are increasingly "tech savvy"; that is to say they are becoming more alert to the possibilities that developments in technology offer and are keen to utilise these technologies to improve organisation performance. Cloud services are attractive because they allow organisations to seize opportunities quickly and "strike whilst the iron is hot".	Although Nihon Motors began exploring cloud based solutions ever since 2008 yet the need for websites to load swiftly and address customer expectations was a key driver. Personnel within Nihon Motors viewed reliance on cloud services would provide a use and dispose advantage as their envisaged web services could be located in external clouds that needn't be integrated with the rest of the company's operations.
Process standardisation	Moving to cloud-based solutions encourages the adoption of standard configurations and discourages the tendency towards excessive customisation of services for individual users. It focuses attention on the costs of providing customised solutions and encourages organisational members to examine practices and procedures that they previously took as givens.	In a business where there is always a possibility of disruptions to the supply chain (as, for example, the earthquake in Japan highlighted) or the need for parts recall (as, for example, in the recent recalls of cars by Toyota due to airbag faults) providing customers and collaborators with up-to-date information is critical. Cloud technologies have the potential to provide more flexible, efficient and effective communication channels.

Table 1. *(Continued)*

Dimension	Hayward's Rigs	Nihon Motors
Scalability	In the oil and gas industry, as the recent Gulf of Mexico incident vividly demonstrated, it is important to be able to scale up capacity at very short notice to deal with unforeseen events. In addition one-off activities, for example HR's involvement with the Olympic games or its 'commitment to America' advertising campaign, require temporary increases in information and communications capacity. Evidence suggests that flexibility is achieved more easily and cost-effectively through the Cloud.	Host providers have far bigger Internet pipelines than individual firms and flexible charging mechanisms make this provision particularly attractive to firms facing intermittent surges in activity. To quote, *"We could have put in a much bigger pipe to the Internet but that would have been a big pipe that was only used one percent of the time and was unlikely to be cost effective."*
Investment optimisation/ Green credentials	Maintaining "evergreen" in-house capability and capacity requires on-going capital investment that is often difficult to justify from a business perspective. Capital expenditures on upgrades to back office and IT systems add value in indirect ways and, in a climate of resource constraint, are often difficult to defend. Moving to the Cloud alters the structure of costs - in particular it reduces the need to commit large amounts of capital to continuous upgrade and renewal of infrastructure and systems.	The car industry places importance on green factors within its procurement processes. For NM, in particular, it is important that it produces its cars in as environmentally friendly a way as possible. Whilst the main emphasis within the company is on 'greening' it's manufacturing plant, NM as a whole strives to be greener where ever it can and this includes its provision of computing services. Cloud computing delivers environmental benefits to NM in a number of ways. First, the economies of scale available to specialist cloud service providers implies that it is in these vendors' interest to incorporate sophisticated, eco-friendly features into the design and operation and their data centres.

Table 1. *(Continued)*

Dimension	Hayward's Rigs	Nihon Motors
Focus on core capacities	As the pace of development in information and communication technology speeds up so it becomes more difficult for IT departments to keep up with latest developments in all fields. Attempting to maintain leading edge knowledge in areas that support rather than constitute the core business can be a distraction to managers' attention. Handing over responsibility for non-core activities to specialists is likely to produce better results.	Whilst it is possible to develop equivalent capacity and capability in-house, the experience NM gained through its tendering processes suggests that specialist providers offer much cheaper solutions because of they can exploit economies of scale, scope and learning through the re-use of knowledge. Further, cloud vendors offer enhanced search capabilities that can be bought off-the-shelf. This enables firms like NM to improve the performance of web-sites from the end-users perspective quickly and relatively cheaply.

4   Scenario at Hayward's Rigs

4.1   Moving Email Services to the Cloud

One of the key decision parameters for HR was the extent to which the email services it was seeking to procure needed to be customised to HR's specific organisational context. On the one hand the company sought to get the economic and commercial benefits associated with standardised Cloud offers, but on the other hand, recognise that the nature and complexity of its business required some significant elements of custom-build. As one of our respondents explained, "it became a conversation about where we wanted to get Cloud economics and Cloud commercials but actually the reality was that it always looked like more of a custom-built environment."

Following a number of detailed rounds of discussion and negotiation, the contract was eventually awarded to T Systems, the corporate customer division of Deutsche Telekom in July 2012. Under the terms of the contract T Systems will provide a secure private cloud which will enable HR's 830,000 plus employees around the world to access email services from a range of mobile and computer devices. The contract is for a five-year period and is based on a "pay-per-use" model.

4.2   HR's Exploration of Moving an Information Management Platform to the Cloud

HR's exploration of this option has followed a similar path to that of email services in terms of the procurement process. It issued a request for information (RFI) to test the market and to identify potential suppliers and entered into discussion with some of the dominant players like Amazon Web Services (AWS).

These large providers of public Cloud services have presented HR with a new and rather unexpected set of challenges. Providers like AWS provide a standardised service that takes little account of the size or nature of the purchasing organisation. As one of our respondents explained:

" you can forget trying to have a conversation with Amazon in the way we used to with Hewlett Packard (HP) where they (HP) will take on certain service levels and undertake to do special things for you as a customer. No chance [with Amazon]. HR as an organisation has as much firepower with Amazon as I have as an individual customer. It makes no difference whatsoever to them so you need a different set of levers to manage your risk exposure around that and that leads you actually to a different … architecture and a completely different dynamic."

As this quote demonstrates the balance of power between provider and purchaser is currently very different from that which large multinationals have come to expect. HR, like most other global organisations, is used to having a degree of bargaining strength in its negotiation with suppliers but the dynamics of public Cloud services are unusual. A few large players dominate segments of the newly emerging Cloud industry and these first movers have been able to exploit the economies of scale and experience to bring costs down, tipping the balance of advantage in their favour.

4.3   Strategic Implementation at HR

Moves to Cloud-based provision can seem to be relatively inconsequential from the end-users perspectives but from a broader organisational perspective, they can have far reaching and unexpected effects. HR managers, for example, have found that the exploration of Cloud-based solutions has required them to rethink many of their existing practices and processes. Executives involved in the decision-making process report that there is a potential impact on everything from legal frameworks, through billing and charging for IT services to the way performance is measured.

From the end-users point of view a change in the provision of email services or an information management platform can go almost unnoticed but the adoption of cloud services means that "every decision has a new complexion to it". For example in terms of legal contracts, HR's legal teams are used to negotiating specific clauses in contracts but in a "multi-tenant" environment that doesn't make sense and Amazon, for example, requires its customers to adopt standard terms and conditions. The adoption of standard terms and conditions has knock-on implications for the way risks are managed. For instance if AWS went down, how would disaster recovery be managed in this new environment? Similarly, the fact that charges for cloud services are consumption-based means that the way IT services are billed for internally, needs to be altered to reflect this and incentives need to be in place to create economies on the "consumption" of chargeable services. As a respondent put it:

"it's like a prism – take the example of how we charge for IT internally – here we have a specific way of charging the business for services which is not hour by hour/consumption based, it is pretty much year by year consumption. So you can have a situation where, by the very nature of the Cloud, you can peak, move in, see lots of different things and your internal charging models encourage that level of use and variability but that isn't an advantage anymore, that's a problem. So it's how you begin to start to chip away at a series of financial mechanisms of governance that might have been in place for twenty years and … there are lots of dead bodies in stuff like that. You have to work your way through (myriads of issues like this) to begin to leverage the advantage that this other thing (the Cloud) gives."

The same sentiment is expressed in the following comments and illustrates how difficult it is to implement changes that on the surface can appear quite modest but have wide-scale implications.

"When you are trying to create a business case for this (Cloud), You're having to force fit a new world model into an old world model so you try to explain the new world model in old world terms and those things are not natural bedfellows. So even trying to – apples for apples- make financial comparisons can be difficult."

"It took me a long time to actually get to grips with the change. You can intellectually understand it quite quickly but to sort of emotionally buy what you are being told and really begin to deeply understand how to might actually do that [implement a transition to Cloud provision] takes a bit more time."

4.4   Cloud Led Competitive Advantage at HR

Cloud solutions, for example, allow HR to do the things it has always done more speedily. For instance a successful national marketing campaign could be scaled up globally very rapidly or ERP systems could be deployed in ways that achieve further cost savings but the "commodified" nature of the cloud means that adoption of cloud solutions is unlikely to be allowed to encroach into areas of core expertise. HR's has distinctive capabilities in finding new oil and gas reserves, in geo-space analysis and in many other areas that are supported by high-performance computing environments. Strategically it is important that HR retains and develops its knowledge of crucial technologies. Cloud-based computing will allow HR to cut costs, be quicker to market and stay at the leading edge of support technologies so it is perceived to be an important tool that HR needs to deploy but not something that will allows HR to differentiate itself from competitors. In other words from HR's perspective the ability to deploy cloud computing solutions is a very important threshold capability but not a distinctive one.

5   Scenario at Nihon Motors

5.1   First Move to the Cloud

Having made the decision, in principle, to move to a cloud-based solution for its web-sites, NM selected its provider through a traditional tendering process. It approached around five vendors who were known leaders in the web-hosting marketplace and put out a Request for Information (RFI). The RFIs paved the way for formal tenders that were evaluated using NM's usual internal protocols and scoring systems. The contract was awarded to a vendor who had the advantage of being located in the Thames Valley close to NM's own offices and the stand-by site was in London's Docklands.

5.2   NM's Subsequent Cloud Ventures

NM's next major cloud venture accompanied its launch of its sporty, hybrid car, the BS-Y. The marketing team responsible for the launch was keen to utilize the power of social media and came up with the concept of Mode Art. Mode Art was described as a web-site and Facebook application which turned the user's life into art. User information was pulled from Facebook and then merged into a unique art composition based on one of the NM BS-Y's driving modes, namely Sport, Urban or Economy. The resulting artwork could be shared on Facebook, sent to friends or downloaded on to a mobile phone and was designed to act as a catalyst for viral marketing. The challenge for IT was to provide sufficient capacity for storing and processing users' images, particular given there was considerable uncertainty about possible uptake.

The ICT team supporting this launch decided to buy cloud-based storage capacity from Amazon, paying for it by gigabyte per month depending on utilization. This enabled the company to put storage capacity in place in a matter of days and also had the advantage of allowing the company to specify where its data was to be stored, hence avoiding some of the legal issues concerning data protection that can be very complex when data is stored off-shore. Purchasing data storage from Amazon was, however, a commodity transaction and, at the time, Amazon was not geared to corporate customers as our respondent's experience vividly illustrated:

"Amazon weren't geared to corporate buying. It actually ended up with me paying [for data storage] on my own corporate credit card. This was the only way we could do it because they weren't in a position to corporate purchase orders or to invoice the company."

Whilst NM as the purchaser had to accept Amazon's standard terms and conditions and had to navigate a payment system designed for individuals rather than corporates, the big benefit was in terms of price. As it turned out the take-up for this campaign was much lower than expected but the low sunk costs meant that the failure of this particular marketing experiment contained as the following quote illustrates:

"… it does illustrate one of the great benefits of the Cloud in that this particular marketing campaign was not successful in terms of attracting people… But the great advantage was that I was paying something like 5 pence per month for the storage we got from Amazon whereas if we tried to provision it internally, we would have put, I don't know maybe a hundred gigs of storage or so behind it and we would have had to pay for it, provision it etc. --- you would probably have been talking at least five to ten thousand pounds of infrastructure. … The flexibility of Cloud storage and Cloud computing can give you some substantial cost advantages."

Whilst in NM's case its marketing teams had always been urged to take a creative approach to new product launches and were encouraged to take calculated risks, the changes in the cost structure associated with cloud-based web provision mitigates against downside risks and facilitates experimentation.

The choice between private and public cloud solutions required NM to engage with some difficult trade-offs. On the one hand buying off –the-shelf public cloud services provided by firms like Amazon offered significant cost savings and flexible capacity but it also meant that NM still had to do a large amount of work in-house because they were "just buying the infrastructure rather than the solution." On the other hand the specialist providers like Rackspace offered high-end services and "get web-sites up and running quickly with little effort [on the buyer's part]" but highly customized solutions are expensive. In the end NM went for a middle of the road solution that involved some degree of customization.

"The company that runs the IT helpdesk for NM (Europe) has its systems based in India. That involved getting data protection agreements signed with all the NM companies in Europe to say we approve employee data being held in systems in India. It's just a headache."

In terms of cost savings, the move to a quasi-public cloud was estimated by our respondent to

"result in a 30% reduction in annual operating costs and … to deliver a better solution. … They call it the virtual team but if you look at the people that they [Phoenix] have supporting the web-site added to the people we had internally supporting the web-site there is definitely a higher level of support."

5.3   Strategic Implementation at NM

One of the concerns commonly expressed about moves to cloud-based solutions is that there may be resistance to these kinds of developments from in-house IT staff, in part because the move to the cloud has the potential to reduce employment opportunities. This does not appear to have been an issue at NM because there were more than enough new projects continuously coming on stream to fully deploy the existing staff's time and expertise. In addition, many of the capabilities required did not exist in-house.

"Computing is an area where there are always new technologies and new projects and you've got to decide where you are going to put your people. If you look at the skills required we have never had those in-house. Yes, we ran virtual servers in-house but running virtual servers between two sites that requires a level of N-ware expertise that NM never really had in-house and would struggle to afford having in-house. We get 24/7 monitoring from the provider but if that was provisioned in-house we'd have to put our people on to shift systems which we can't do and we want the system to be scalable on demand so that the website automatically adds capacity if there are peaks in usage."

5.4   Cloud Led Competitive Advantage at NM

Our respondent drew parallels between cloud computing technologies and outsourcing. In just the same way as it is not sensible for a firm to outsource activities that were the basis of its competitive advantage so it is not prudent for a firm to move computing activities to the Cloud if those activities formed part of the organisation's distinctive capabilities. To quote our respondent:

"The only way Cloud computing helps in delivering competitive advantage is in a secondary way. Cloud computing can give you cost and speed market advantages so if part of your competitive advantage is getting to market quickly then Cloud computing can help. But, if you view computing in its own right as your competitive advantage then you don't outsource it because, by definition, you're using public things that people can easily copy, easily reproduce and so it very quickly doesn't become your competitive advantage anymore."

Looking to the future, however, it is likely that Cloud computing solutions will take on increasing significance for the car industry. There is a trend towards cars becoming network nodes in their own right – that is to say more information and communication technology being incorporated into vehicles so that cars are permanently online from the manufacturer's perspective. Whilst all car producers would like to gain an advantage by exploiting the opportunities that 'always online' cars potentially offer, it seems unlikely that a single car manufacturer could afford to invest the sums of money necessary to build unique systems and distinctive advantages.

"We'd very much like to have a competitive advantage but, realistically, we probably can't because the cost of provisioning an 'always online' car and having the nationwide networks to do that is well outside the scope of a single car producer. We will probably have to collaborate with mobile network providers. We're going to be using other people to help us and we will need to tap into publicly available services so that is never going to be our competitive advantage. Our competitive advantage will have to be closer to home, for example by designing the interfaces, helping people to use the features of the car and so on."

The fact that Cloud solutions are ubiquitous and easily replicable means that careful consideration does need to be given to which activities are transferred to the Cloud. In the case of CRM and data mining, for example, our respondent was of the opinion that the interrogation of customer data is best done in-house.

"[Interrogating customer data] and data mining is difficult is some ways. You are handling large volumes of data and, yes, that could be a candidate for the Cloud but then you look at the tools you need to handle those large volumes and to what extent are they Cloud-based? You can't shift large data over the network or the Internet. It needs to be closer to home. Lots of the data-mining people are now doing in-memory computing, holding databases in memory [to undertake their analysis]. If you're not careful you will erode your performance advantage by hosting remotely."

NM sees its core capabilities as located in its design of cars and in its manufacturing capabilities so it needs its computing capacity next to the production line. Whilst cloud solutions and external hosting are helping the company to improve the efficiency and effectiveness of its overall operation, its production-related computing activities are likely to remain firmly in-house.

6 Conclusion

It is clear that the motivations and expectations of both companies vary somewhat. At the same time both realise that there are specific advantages that they could garner by using cloud solutions. For instance, HR realises that they'd be able to cut costs, be quicker to market and stay at the leading edge of support technologies. However, HR doesn't consider cloud to be able to provide it with a capability that would allow it to differentiate itself from competitors. In contrast to the oil and natural gas sector, the car industry is quite heavily customer orientated. In such a context as was evident through the facebook exercise initiated by NM for its BS-Y model there are serious limitations on what can be achieved in-house. Further as hybrid cars become more network reliant, cloud based capacity may become an imperative. Probably the most fascinating outcome of the research for this study is embedded in the challenges being experienced by both MNCs in implementing cloud solutions. Both HR and NM have reported that decision making is getting affected as provisioning of computing resources becomes more commodified. Here the duality dimensions of influence as found by Mitra [4] seems to become gradually evident as the companies move towards fully cloud orientated organisations.

References

1. Waschke, M.: Cloud Standards: Agreements that hold together clouds. CA technologies (2012)
2. Marston, S., Li, Z., Bandopadhyay, S., Zhang, J., Ghalsasi, A.: Cloud computing – The business perspective. Decision Support Systems 51, 176–189 (2011)
3. Khanagha, S., Volberda, H., Sidhu, J., Oshri, I.: Management innovation and adoption of emerging technologies: The case of cloud computing. European Management Review 10(1), 51–67 (2013)
4. Mitra, A.: An Interpretation of the Organisational Context of Geographic Information System use in British Local Government. Unpublished PhD dissertation, University of Birmingham (2001)
5. Wade, M., Hulland, J.: The resource based view and IS research: Review, extension, and suggestions for future research. MIS Quarterly 28(1), 107–142 (2004)
6. Leonard-Barton, D.: Core capabilities and core rigidities: A paradox in managing new product development. Strategic Management Journal 13, 111–125 (1992)
7. Teece, D.J., Pisano, G., Shuen, A.: Dynamic Capabilities and Strategic Management. Strategic Management Journal 18(7), 509–533 (1997)
8. Christensen, C.M., Overdorf, M.: Meeting the challenge of disruptive change. Harvard Business Review 78(2), 66–77 (2000)
9. Day, G.S.: The capabilities of market driven organisations. Journal of Marketing 58(4), 37–52 (1994)
10. Kraaijenbrink, J., Spender, J.-C., Groen, A.J.: The resource based view: A review and assessment of its critiques. Journal of Management 36(1), 349–372 (2010)
11. Mitra, A., Neale, P.: Visions of a pole position: Developing inimitable resource capacity through enterprise systems implementation in Nestlé. Strategic Change 23(3-4), 225–235 (2014)
12. Lederer, A.L., Mendelow, A.L.: Co-ordination of information systems plans with business plans. Journal of Management Information Systems 6(2), 5–19 (1989)
13. Luftman, J.: Assessing IT/Business alignment. Information Systems Management 20(4), 9–15 (2003)
14. Tallon, P.P.: A process-oriented perspective on the alignment of information technology and business strategy. Journal of Management Information Systems 24(3), 227–268 (2007)
15. Bharadwaj, A.S.: A resource based perspective on information technology capability and firm performance: an empirical investigation. MIS Quarterly 24(1), 169–196 (2000)
16. Hirschheim, R., Sabherwal, R.: Detours in the path toward strategic information systems alignment. California Management Review 44(1), 87–108 (2001)

17. Armbrust, M., Fox, A., Griffith, R., Joseph, A.D., Katz, R., Konwinski, A., Lee, G., Patterson, D., Rabkin, A., Stoica, I., Zaharia, M.: A view of cloud computing. Communications of the ACM 53(4), 50–58 (2010)

18. Ba, S., Stalleart, J., Whinston, A.B.: Research commentary: Introducing a third dimension in information systems design – The case for incentive alignment. Information Systems Research 12(3), 225–239 (2001)

19. Grossman, R.L.: The case for cloud computing, pp. 23–27. IEEE Computer Society (2009)

20. Jaeger, P.T., Lin, J., Grimes, J.M.: Cloud computing and information policy: Computing in a policy cloud? Journal of Information Technology & Politics 5(3), 269–283 (2008)

21. Sarkar, P., Young, L.: Sailing the cloud: A case study of perceptions and changing roles in an Australian University. In: Proceedings of the European Conference on Information Systems. Paper 124, Aalto University, Helsinki, Finland (2011)

22. Runciman, B.: The IT Linguist, ITNOW. Journal of the British Computer Society, 56–57 (June 2014)

23. Yin, R.K.: Case study research: Design and methods. Sage (2003)

An Improved Signal Subspace Algorithm for Speech Enhancement

Xuzheng Dai, Baoxian Yu, and Xianhua Dai[*]

School of Information Science and Technology, Sun Yat-sen University, China
463418661@qq.com
yubx@mail2.sysu.edu.cn
issdxh@yubx@mail.sysu.edu.cn

Abstract. Most of the algorithms for speech enhancement are designed to improve the speech listening comfort. However the frequency spectrum character is destroyed seriously after the speech enhancement. To achieve better speech listening comfort with less frequency spectral damages, we present an improved signal subspace algorithm for speech enhancement. Compared with the traditional signal space method, the improved algorithm can decrease the Mel-frequency Cepstral Coefficients (MFCC) distance, an evaluation measure which means less frequency spectral damages to the voice and keep the voices' intelligence at the same time. Besides, the method can enlarge the distance of the easily confused voices, which means the improvement of the voice recognition ratio. Thus we get the purpose of the speech enhancement. The improved algorithm is used in a speech recognition program and has a good performance.

Keywords: Speech enhancement, signal subspace method, wiener filtering, prior SNR, Mel-frequency Cepstral Coefficients.

1 Introduction

Speech enhancement and voice recognition have been widely used in recent years. In some occasions, the noisy environment will destroy the frequency spectrum character and lead to an erroneous recognition. Thus the speech enhancement algorithm needs to reduce the noise and keep integrality of the frequency spectrum character at the same time. Ephraim (1995) proposed the signal subspace approach to minimize the speech distortion and keep the residual noise below a preset threshold [1]. Hu (2003) proposed a generalized subspace approach for speech enhancement in both white noise and colored noise environment, and derived a time-domain estimator constraint and a spectral domain constraint[2]. The well-known decision-directed technique for speech enhancement limits the musical noise well[3], but the estimated priori signal-to-noise ratio (SNR) is biased since it depends on the speech spectrum estimation of the previous frame which degrades the noise reduction performance. Plapous (2004)

---

[*] Corresponding author.

H. Li et al. (Eds.): I3E 2014, IFIP AICT 445, pp. 104–114, 2014.

proposed a two-step noise reduction (TSNR) technique to solve this problem while maintaining the effect of the decision-directed approach [4]. Plapous (2005) also proposed a harmonic regeneration noise reduction method (HRNR) for solving the harmonic distortion in enhanced speech by regenerating the degraded harmonics of the distorted signal in an efficient way [5]. Objective and subjective measures prove the improvement than TRNR approach. However, the TSNR method destroyed the frequency spectrum character of the speech more seriously than TRNR. So TSNR is the algorithm we need in this paper.

The key of TRNR is to estimate the SNR, while a deviation of SNR in signal subspace approach significantly affects the speech performance. To solve the problem, we present an improved signal subspace algorithm that combines the signal subspace approach and TRNR that can get a better performance in speech enhancement.

Mel-frequency Cepstral Coefficients (MFCC) is one of the best approaches for voice recognition[6][7]. We propose the distance of the Mel-frequency Cepstral Coefficients as a measure to evaluate the effect of speech enhancement methods. The less the distance means the less damage to the frequency of the voices and better effect of the speech enhancement. The evaluation measure is based on the spectral features so as to evaluate the result of the speech enhancement from the perspective of speech recognition. The experiments verify that the new algorithm has a better performance than the others.

2   Speech Enhancment Approaches

In this section, we briefly review the signal subspace approach and TSNR algorithm. Then we discuss the shortcoming of each algorithm.

A. Signal Subspcace Approach

The signal subspace approach is based on the theory of projecting the signal onto two subspaces: the signal-plus-noise subspace and the noise subspace. Thus we can remove the noise part through the decomposition of the signal. The decomposition can be either the singular value decomposition (EVD) or the eigenvalue decomposition (SVD), and in fact the two-decomposition method can be mutual transformed.

A linear clean signal $\mathbf{x}$ can be described as:

$$\hat{\mathbf{x}} = \mathbf{H}\mathbf{y} \tag{1}$$

Where $\mathbf{y}$ is the noisy signal and $\mathbf{H}$ is a $K \times K$ matrix whose rank is M , and $M < K$. Thus the error of the signal can be obtained by:

$$\varepsilon = \hat{\mathbf{x}} - \mathbf{x} = \mathbf{H}\mathbf{y} - \mathbf{x} = (\mathbf{H} - \mathbf{I})\mathbf{x} + \mathbf{H}\mathbf{d}$$
$$= \varepsilon_x + \varepsilon_d \tag{2}$$

Where ε_x represents the speech distortion and ε_d represents the residual noise. So the time-domain constrained optimization is given by making:

$$\min_H \overline{\varepsilon}_x^2 \tag{3}$$

Subject to:

$$\frac{1}{K}\overline{\varepsilon}_d^2 \leq \alpha\sigma^2 \tag{4}$$

Where σ^2 is a positive constant and $0 < \alpha < 1$ for scaling. We can get the answer by constructing a Lagrange multiplier, and we can get the optimization of $\mathbf{H}$ in white noise environment:

$$\begin{aligned} \mathbf{H}_{opt} &= \mathbf{R}_d \mathbf{U}\Lambda_\Sigma (\Lambda_\Sigma + \mu\mathbf{I})^{-1}\mathbf{U}^T \\ &= \mathbf{U}^{-T}\Lambda_\Sigma (\Lambda_\Sigma + \mu\mathbf{I})^{-1}\mathbf{U}^T \\ &= \mathbf{U}^{-T}\mathbf{G}\mathbf{U}^T \end{aligned} \tag{5}$$

Where $\mathbf{G} = \Lambda_\Sigma (\Lambda_\Sigma + \mu\mathbf{I})^{-1}$. μ is the multiplier factor, $\mathbf{R}_x$ is the covariance matrix of clean signal, and $\mathbf{R}_d$ is the covariance matrix of noise. Λ_Σ and $\mathbf{U}$ are the eigenvalue matrix and eigenvector matrix of Σ, where $\Sigma = \mathbf{R}_d^{-1}\mathbf{R}_x$.

Searle in his book tells the theory that a matrix $\mathbf{U}$ which can simultaneously diagonalize $\mathbf{R}_x$ and $\mathbf{R}_d$ [8]. Thus we can get:

$$\begin{aligned} \mathbf{U}^T \mathbf{R}_x \mathbf{U} &= \Lambda_\Sigma \\ \mathbf{U}^T \mathbf{R}_d \mathbf{U} &= \mathbf{I} \end{aligned} \tag{6}$$

The matrix $\mathbf{G}$ is a diagonal matrix, thus the k th diagonal element g_{kk} is given by:

$$g_{kk} = \begin{cases} \dfrac{\lambda_x^k}{\lambda_x^k + \mu}, & k = 1, 2, \cdots, M \\ 0, & k = M+1, \cdots K \end{cases} \tag{7}$$

The value of μ affects the quality of the enhanced speech directly. According to Dendrinos's theory, μ depends on the short time of SNR [9]:

$$\mu = \begin{cases} \mu_0 - (SNR_{dB})/s, & -5 \leq SNR_{dB} \leq 20 \\ \mu_{\min}, & SNR_{dB} > 20 \\ \mu_{\max}, & SNR_{dB} < 5 \end{cases} \tag{8}$$

Where:

$$SNR_{dB} = 10\log_{10} SNR$$
$$\mu_0 = \mu_{\min} + 20*s \tag{9}$$
$$s = \frac{(\mu_{\max} - \mu_{\min})}{25}$$

The value of $\mu_{\max}$ and $\mu_{\min}$ represent the maximum and minimum of μ, and they are chosen experimentally. SNR can be given by:

$$SNR = \frac{tr(\mathbf{V}^T\mathbf{R}_x\mathbf{V})}{tr(\mathbf{V}^T\mathbf{R}_d\mathbf{V})} = \frac{\sum_{k=1}^{M}\lambda_x^{(k)}}{K} \tag{10}$$

Thus we can get the time-domain constrained optimization of the signal subspace approach. Form (10) we can find that the SNR in each frame is a value rather than a vector, which means that the transmission function we got based on SNR is not that accurate, so we need to find some other ways to get more accurate SNR.

B. Two Step Noise Reduction approach

In some speech enhancement algorithms based on the SNR, two parameters are needed: the posteriori SNR and the priori SNR, which are computed by:

$$SNR_{post}^{local}(p,k) = \frac{|Y(p,k)|^2}{|N(p,k)|^2} \tag{11}$$

and

$$SNR_{pri}^{local}(p,k) = \frac{|X(p,k)|^2}{|N(p,k)|^2} \tag{12}$$

Where $Y(p,k)$, $X(p,k)$ and $N(p,k)$ represent the frequency spectral of the noisy speech, the clean speech and the noise. The directed-decision algorithm says that the posterior SNR can be given by[10]:

$$S\hat{N}R_{pri}^{DD}(p,k) = \beta\frac{|\hat{X}(p-1,k)|^2}{\hat{\gamma}_n(p,k)} + (1-\beta)P[S\hat{N}R_{post}(p,k)-1] \tag{13}$$

Where $S\hat{N}R_{pri}^{DD}(p,k)$ represents the priori SNR got from the directed-decision algorithm, $\hat{\gamma}_n(p,k)$ represents the estimated noise as we can't know the exact noise, and β is a constant to balance the result, usually, $\beta=0.97$.

According to the Weiner filtering theory, the transmission function $H_{DDopt}(p,k)$ is given by:

$$H_{DDopt}(p,k) = \frac{S\hat{N}R_{pri}^{DD}(p,k)}{1+S\hat{N}R_{pri}^{DD}(p,k)} \qquad (14)$$

The experiment demonstrates that the directed-decision algorithm can reduce the "music noise" well, however the SNR got from (13) has a frame delay compares with the speech, especially at the speech onset and offset moment, which will limit the noise reduction performance and bring in some new reverberation effect.

The TSNR approach computes the SNR of the next frame using the transmission function got by directed-decision approach as:

$$S\hat{N}R_{pri}^{TSNR}(p,k) = S\hat{N}R_{pri}^{DD}(p+1,k)$$

$$= \beta' \frac{\left|H_{DDopt}(p,k)X(p,k)\right|^2}{\hat{\gamma}_n(p,k)} + (1-\beta')P[S\hat{N}R_{post}(p+1,k)-1] \qquad (15)$$

Where β' has the same effect as β, and we make $\beta'=1$ because we can't know the information in the $p+1$ th frame. The experiment suggests that the TSNR approach has a good performance on estimating the SNR. However, the TSNR has a large attenuation of the signal energy, which leads to a low sound of the voice signal. The low energy of the voice is disadvantage for the speech recognition. So even though the TSNR has a good performance in SNR and listening intelligence, it is not a good algorithm for speech recognition.

3   The Improved Signal Subspace Algorithm

In signal subspace approach, the SNR calculated in (10) is a value rather than a vector, which means that the SNR is not accurate. Thus we instead the (10) by (15), which is much more accurate than (10). Thus the whole process of the new algorithm is shown in Figure 1.

The proposed algorithm can be formulated in the following ten steps. For each frame of the voice signal:

Step 1: Compute the covariance matrix $\mathbf{R}_y$ of the noisy signal, and compute $\Sigma = \mathbf{R}_d^{-1}\mathbf{R}_y - \mathbf{I}$. Then update the matrix of the noise $\mathbf{R}_d$.

Step 2: Compute the decomposition of matrix Σ by $\Sigma\mathbf{U} = \mathbf{U}\Lambda_\Sigma$.

Step 3: Sorting the eigenvalue of the matrix Σ as $\lambda_{\Sigma,1} \geq \lambda_{\Sigma,2} \geq \cdots \geq \lambda_{\Sigma,P}$, and we can estimate the rank of the speech signal subspace as $M = \max\limits_{1\leq k\leq P} \arg\{\lambda_{\Sigma,k} > 0\}$.

Step 4: Get the frequency spectrum of the noisy speech signal by FFT, and estimate the frequency spectrum of the noise signal at the same time.

Step 5: Compute the posteriori SNR by (11), and then compute the priori SNR by (13).

Step 6: Compute the TSNR SNR by (15).

Step 7: Compute the multiplier factor μ by (8) and (9) using the TSNR SNR we get in the previous step.

Step 8: Compute the diagonal elements g_{kk} of the matrix $\mathbf{G}$, and get the matrix $\mathbf{G}$ by:

$$G = diag\{g_{11}, g_{22}, ..., g_{MM}\} \tag{16}$$

Step 9: Compute the optimization $\mathbf{H}$ by (5).

Step 10: Estimate the enhanced speech signal by $\hat{\mathbf{x}} = \mathbf{H}\mathbf{y}$.

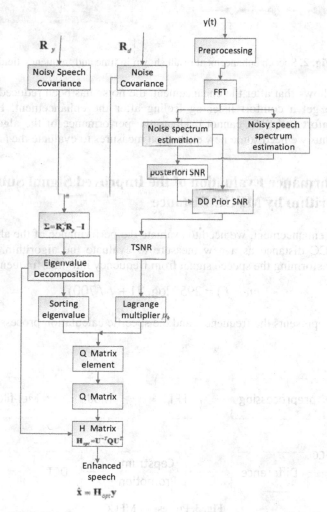

Fig. 1. The process of the new speech enhancement algorithm

To illustrate the better performance of the improved algorithm, we take the speech enhancement result of a 10s length voice which is polluted by the factory noise from database NOISEX.92 as the example. The SNR after speech enhancement improves about 2.5dB in heavy noise environment. Figure 2 shows the enhancement result in time and frequency field.

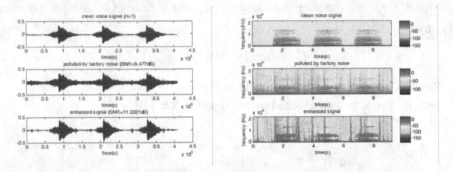

Fig. 2. Speech enhancement result shown in time and frequency field

Figure 2 shows that after the enhancement, the noise has been reduced pretty well. We can also get a comfort listening feeling after the enhancement. However the listening comfort and SNR cannot evaluate the performance of the algorithm in all directions. Thus we need some new evaluation measures to evaluate the algorithm.

4   Performance Evaluation of the Improved Signal Subspece Algorithm by MFCC Distance

After speech enhancement, we need to evaluate the performance of the algorithm. We propose MFCC distance as a new measure to evaluate the algorithm. The key of MFCC is transforming the speech signal from frequency into mel-frequency by:

$$mel(f) = 295 * \log_{10}(1 + f / 700) \tag{17}$$

Where f represents the frequency, and the specific calculation process of MFCC is in Figure 3:

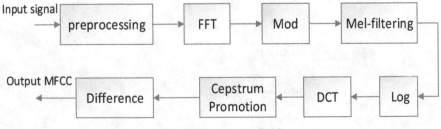

Fig. 3. Process of MFCC

Figure 3 shows the process of getting the MFCC of the speech. The preprocessing includes Frame Blocking, Pre-emphasis and Hamming-windowing. Then get the frequency spectrum by FFT (Fast Fourier Transform Algorithm), and get the mode of the spectrum. Mel-filtering is as shown as below:

$$H_m(k) = \begin{cases} 0 & ,k < f(m-1) \\ \dfrac{k-f(m-1)}{f(m)-f(m-1)}, & f(m-1) \leq k \leq f(m) \\ \dfrac{f(m+1)-k}{f(m)-f(m-1)}, & f(m) \leq k \leq f(m+1) \\ 0 & ,k > f(m-1) \end{cases} \tag{18}$$

Where $H_m(k)$ represents the coefficients of the Mel-filtering, k represents the frequency, and $f(m)$ represents the center frequency of the Mel-filtering, where $m = 1, 2, ..., 20$.

And we can get the logarithm of the frequency spectrum by:

$$s(p,m) = \ln(\sum_{k=0}^{N-1} |X_a(p,k)|^2 H_m(k)), \quad 0 \leq m \leq M \tag{19}$$

Where $X_a(p,k)$ is the frequency spectrum through FFT of the signal, and $s(p,m)$ is the logarithm of the frequency spectrum in p th frame, the m th order of the Mel-filtering.

The DCT means Discrete Cosine Transformation:

$$C(p,n) = \sum_{m=0}^{N-1} s(p,m)\cos(\frac{\pi n(m-0.5)}{M}), \quad n = 1, 2, ..., L \tag{20}$$

Where $C(p,n)$ is the MFCC coefficient, and L is the MFCC order number, usually we make $L = 12$.

Usually the MFCC coefficients value in the low frequency is more easily interfered by the channel than in the high part, while the high part has a too high influence on speech recognition, thus the center part of the coefficients is the most useful and important. Thus we need a Cepstrum Promotion for the signal to promote the center part of the coefficients by:

$$W(n) = 1 + \frac{L}{2}\sin\frac{\pi n}{L} \tag{21}$$

$$mf(p,n) = C(p,n)W(n), \quad n = 1, 2, ..., L$$

Where $W(n)$ is the Cepstrum promotion transmission function, and $mf(p,n)$ is the MFCC coefficients after Cepstrum Promotion. However the MFCC coefficients now can only reflect the voice parameters of the current frame without the change of the front and rear frame. So we take the difference of the MFCC coefficients in the front and rear frame into account by:

$$dmf(p,n) = \frac{2(mf(n+2)-mf(n-2))}{3}$$
$$+\frac{+(mf(n+1)-mf(n-1))}{3}$$

(22)

$dmf(p,n)$ represents the difference coefficients. Then we combine the MFCC coefficients $mf(p,n)$ and difference coefficients $dmf(p,n)$ as the whole MFCC coefficients of the speech. We calculate the distance of the coefficients between the noisy speech and the clean speech. The less the distance is, the better performance of the enhancement algorithm is. And if the distance between the easily confused speech signals gets larger after enhancement, it means the enhancement algorithm has a better performance.

Table 1. Coefficients distance

algorithm / Noise kind	White Noise	Factory Noise	Average of both Noise
DD approach	2.803	2.759	2.713
TSNR	3.125	3.012	2.930
Signal Subspace	2.860	2.658	2.780
New algorithm	**2.000**	**2.236**	**2.144**

Table 1 shows the MFCC distance between the noisy speech and the clean speech. The speech signal is a toy voice of 10s length, and the speech signal is chosen from NOISEX.92 database. We can see that the distance of the new algorithm is less than TSNR and signal subspace approach, which concludes that the new algorithm has a better performance than TSNR and signal subspace approach.

Almost every speech enhancement algorithm will damage the frequency feature of the speech, then why do we still need enhancement algorithm? The reason is that some speeches' feature polluted by noise are easily got confused. Thus after the speech enhancement, the easily confused speech should be judged correctly. There are two German alphabets 'e' and 'i' which sound very similar, and are easily confused in severe noisy environment. After the enhancement, the misjudged signal can be judged correctly as seen in below:

Table 2. The new algorithm for confused voice acting

speech Clean speech	i	e	Distance difference
Polluted i	1.8651	1.9104	0.0453
Polluted e	2.1306	2.1709	**-0.0403**
Enhanced i	3.2937	3.3849	0.0912
Enhanced e	3.9398	3.7787	**0.1611**

From the table II we can get the conclusion that after the enhancement, the distance difference of alphabet 'i' is enlarged from 0.0453 to 0.0912 which means that the new algorithm makes the signal easier to be identified. Also, the distance difference of alphabet 'e' is -0.0403, which means that the alphabet 'e' can't be identified correctly because of the noisy pollution. But after the enhancement, the distance becomes 0.1611, means that 'e' can be identified correctly, which demonstrates the role of the new method for speech recognition.

5   The Application of the Algorithm

In cooperation of our laboratory with a toy company, we are asked to design software which is used at the production line in the factory to recognize if the voices of the toys are right or not. The toys can sing many kinds of voices, and many of them are usually confused, such as the German alphabets 'e' and 'i' we test in the fourth part. The alphabets are more difficult to recognize in the noisy factory because of the loud noise. So we need some algorithms to remove the noise. We used MFCC algorithm for the voice recognition, but most of the speech enhancement algorithms would enlarge the MFCC distance, which will lead voice recognition to a failure. The algorithm can decrease the back noise, improve the feeling of our hearing and do not destroy the spectral features of the voice at the same time as far as possible, which is useful for the voice recognition. The use of our algorithm is shown in Figure 4:

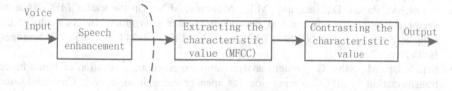

Fig. 4. The use of the algorithm

Figure 4 shows the use of the improved algorithm. The left pat of the dotted line is the place where the algorithm is used in this paper. Because of the loud noise in the factory, the voices of the toys are polluted seriously, which will influence the recognition rate. We tried many algorithms include the traditional signal subspace algorithm, but none of them can meet their demands of the recognition rate. So we tried many improvements until we found the algorithm we propose in this paper. The improved algorithm meets the demand they want. As is shown in the fourth part, the improved algorithm can decrease the MFCC distance, keep the frequency correlation and voice intelligence at the same time. Also, the method can make the distance of the easily confused words and letters larger, which make it not that easy to be erroneously judged. After we use the improved algorithm the average recognition rate of the toys' voices in the factory production line is from 73% to 91%, thus we get the purpose of the speech enhancement, and that is the value of the improved algorithm.

However, the complexity of the algorithm in this paper becomes larger, which is a big drawback of the algorithm. In the following research, the main focus is to reduce the complexity of the algorithm, and continue to improve the speech recognition rate toys.

6   Summary and Conclusions

In this paper we propose a new algorithm for speech enhancement that combine the TSNR and signal subspace approach. And we propose MFCC coefficients distance to evaluate the performance of the speech enhancement algorithms. The experiments verify that the new algorithm has a better performance than the single.

References

1. Ephraim, Y., Van Trees, H.L.: A signal subspace approach for speech enhancement. IEEE Transactions on Speech and Audio Processing 3(4), 251–266 (1995)
2. Hu, Y., Loizou, P.C.: A generalized subspace approach for enhancing speech corrupted by colored noise. IEEE Transactions on Speech and Audio Processing 11(4), 334–341 (2003)
3. Ephraim, Y., Malah, D.: Speech enhancement using a minimum-mean square error short-time spectral amplitude estimator. IEEE Transactions on Acoustics, Speech and Signal Processing 32(6), 1109–1121 (1984)
4. Plapous, C., Marro, C., Mauuary, L., et al.: A two-step noise reduction technique. In: Proceedings of IEEE International Conference on Acoustics, Speech, and Signal Processing(ICASSP 2004), vol. 1, pp. I-289–I-292. IEEE (2004)
5. Plapous, C., Marro, C., Scalart, P.: Speech enhancement using harmonic regeneration. In: IEEE International Conference on Acoustics, Speech and Signal Processing (ICASSP 2005) (2005)
6. Samal, A., Parida, D., Satapathy, M.R., Mohanty, M.N.: On the use of MFCC feature vector clustering for efficient text dependent speaker recognition. In: Satapathy, S.C., Udgata, S.K., Biswal, B.N. (eds.) FICTA 2013. AISC, vol. 247, pp. 305–312. Springer, Heidelberg (2014)
7. Sahidullah, M., Saha, G.: Design, analysis and experimental evaluation of block based transformation in MFCC computation for speaker recognition. Speech Communication 54(4), 543–565 (2012), doi:10.1016/j.specom.2011.11.004
8. Searle, S.R.: Matrix algebra useful for statistics, New York (1982)
9. Dendrinos, M., Bakamidis, S., Carayannis, G.: Speech enhancement from noise: A regenerative approach. Speech Communication 10(1), 45–57 (1991)
10. Cohen, I.: On the decision-directed estimation approach of Ephraim and Malah. In: Proceedings of IEEE International Conference on Acoustics, Speech, and Signal Processing (ICASSP 2004), vol. 1, pp. I-293–I-296. IEEE (2004)

The Key Algorithm of the Sterilization Effectiveness of Pulsating Vacuum Sterilizer

Xicheng Fu[1,*], Mingcai Lin[2], Xingdong Ma[1]

[1] School of Information Engineering, Hainan Institute of Science and Technology,
China 571126
davidxm99@126.com
[2] School of Information and Technology, Hainan Normal University, China 571158

Abstract. The relationship between the physical parameters of pulsating vacuum sterilizer and the piping system states is analyzed based on cluster analysis and interpolation approximation algorithm. The cluster analysis is performed first, followed by the interpolation approximation of the cluster data. The algorithm provides a comprehensive sterilization efficacy evaluation, as well a feasible method to analyze the states of the system.

Keywords: data mining, cluster analysis, the average quadratic interpolation.

1 Introduction

Pulsating vacuum sterilizer is mainly used in hospitals, chemical industry, food industry, and scientific research. It is also widely used in the sterilization of the routine items in medicinal product testing, vaccination, and bioengineering industries. Having replaced the exhaust sterilizer, pulsating vacuum autoclaves are most widely used in hospitals. However, the solenoid valves, the check valves and other valves in the exhaust piping system are prone to failure and need repeated maintenance due to the complexity of the structure of the pulsating vacuum sterilizer piping system, The safety of the pulsating vacuum sterilizer also is a key problem.

2 B-D Test

B-D tests are performed to assess the residual air in the pulsating vacuum sterilizer and steam penetration (Figure 1); it is specially designed to test the air exhaust effect of the pulsating vacuum pressure steam sterilizer. The batch challenge test (Process Challenge Device, PCD) is used to determine whether the entire load steam sterilization cycle has achieved the sterilization conditions, often need to release early using the biological indicators or the fifth class of chemical sterilization indicator. The study of sterilization parameters is the prerequisite of a small probability of surviving microbes.

---

* Corresponding author.

H. Li et al. (Eds.): I3E 2014, IFIP AICT 445, pp. 115–122, 2014.

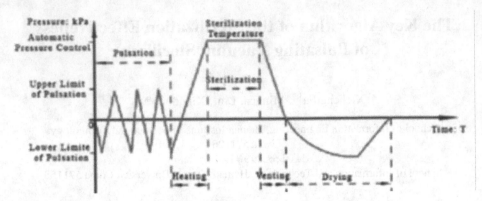

Fig. 1. B-D Test Program

It is often difficult to detect the microbes using the current sterility test method. Therefore, it is necessary to verify the reliability of the method of sterilization. The values of F and F0 can be used as the indicators to verify the reliability of the parameters of sterilization process.

Sterilization parameters are mainly:

1) Value of D

At a certain temperature, the sterilization time required to kill 90% of the microorganism (or 10% residual):

Microorganism death rate: $\lg N0 - \lg Nt = kt / 2.303$

$$D = t = 2.303 / k (\lg 100 - \lg 10) \qquad (2\text{-}1)$$

D is the time required to reduce the number of microbes to the one-tenth of the original number, or to reduce to one logarithmic units or decrease ($\lg 100$ reduce to $\lg 10$). The value of D depends on the different types of sterilization methods and different types of microbes.

2) Value of Z

Z is the value of temperature increment that is required to lower the value of $\lg D$, i.e. sterilization time required to reduce to one tenth of the original number of microbes, or the temperature increment required to kill 99% of the microbes in the same time.

$$Z = T2 - T1 / \log D2 - \log D1 \qquad (2\text{-}1)$$

3) Value of F

The time required at the sterilization temperature (T) to achieve the same sterilization effect as with the reference temperature (T0), under a certain value of Z. It is commonly used in dry heat sterilization with the unit of min.

$$F = \Delta t \sum 10 \qquad (2\text{-}2)$$

4) Value of F0

The time that is required to sterilize at a temperature (T) with Z $=10\ ^\circ$C to achieve the same sterilization effect at 121 $^\circ$C with Z $= 10\ ^\circ$C. F0 value is limited to autoclaving only. The biological F0 value is equivalent to the time needed to kill all the microbes at 121 °C during hot sterilization. F0 reflects the unity of sterilization temperature and time on the sterilizing effect. This value is more accurate and practical. A safety factor should be increased to ensure the sterilization effect, typically 50% more than the theoretical value in order to achieve the expected remaining number of microbes, i.e., the probability of contamination.

The value of physical F0 is defined as: $F0 = \Delta t \sum 10T - 121 / Z$ (2-3)

The value of biological F0 is defined as: $F0 = D121°C \times (\lg N0 - \lg Nt)$ (2-4)

How to correctly and timely asses of the relationships between the operating conditions, such as solenoid valves, check valves in the piping system and the physical parameters that are associated with the sterilization effectiveness will be able to provide a feasible analysis algorithm for the equipment states. This paper constructs the correlation algorithm between the sterilization parameters and the state of the piping system based on the cluster analysis and numerical interpolation approximation.

3   Cluster Analysis

The concept of data mining (DM, Data Mining) was first proposed at the ACM annual meeting in 1995. It is the process of extracting implicit, undiscovered, and potentially valuable information from the database. Data mining is the product that IT has reached to a certain stage of development. It requires large-scale databases, efficient computing power, and effective calculation algorithms. Data mining is the process of extracting a useful knowledge from the large amount of data that stored in the database, data warehouse or other libraries.

According to the large number of historical data, cluster analysis is performed to determine the distribution of device parameters under different operating conditions. In this paper, the improved K-Means clustering algorithm is used.

The improved K-means clustering algorithm is discussed as the following:

Method of selecting the initial cluster centers:

(1). Select the object that is the farthest from the mean point as the initial cluster centers first cluster (seed) O_1;

(2). Select object O_2, which is the farthest from object O_1, as the initial center of the second cluster;

(3). Calculate of the minimum distance between of the remaining unallocated objects and the selected the seeds, and find the maximum value among the minimum distances. The object corresponds to this maximum value is taken as the initial seed of the next cluster;

(4). Repeat the process K times to achieve K points. These K points are the initial the initial cluster seeds.

In this paper, K=3 is selected. The reason is that the status of the devices is always one of the following: qualified, sub-qualified and disqualified. The corresponding data will be presented in three different distributions. After digging out the three clustering results, the quadratic interpolation is performed with respect to sterilization temperature, sterilization time, etc., to determine the distribution of the three clusters of data which are representative of the state apparatus. Results of data mining are closely related to the types of the data selected. The result of the analysis and the actual experiment may vary greatly if the sample data is not comprehensive. Then we need to select new sample data for data mining.

4   Lagrange Interpolation

Problem 1.1: Let $y = f(x)$ has the values of y_0, y_1 at the different nodes of x_0, x_1, construct a polynomial with the power of n, with $n \leq 1$:

$$p_1(x) = a_0 + a_1 x \tag{4-1}$$

so that it satisfies $p_1(x_0) = y_0; p_1(x_1) = y_1$. This interpolation is a linear interpolation, also known as Lagrange interpolation.

Linear interpolation calculation is easy; it replaces the curve by a straight line. Therefore it generally requires small interpolation interval $[x_0, x_1]$, and changes over this interval should be stable, and otherwise the errors may be significant. To overcome this drawback, we consider using a simple curve to approximate the complex curve. [2]

Problem 1.2: Let $y = f(x)$ has the values of y_0, y_1, y_2 at the different nodes of x_0, x_1, x_2, construct a polynomial with the power of n, with $n \leq 2$:

$$p_2(x) = a_0 + a_1 x + a_2 x^2 \tag{4-2}$$

So that it satisfies $p_2(x_0) = y_0; p_2(x_1) = y_1; p_2(x_2) = y_2$

This interpolation problem is defined by parabolic interpolation, also known as Lagrange quadratic interpolation [3]. Its geometric meaning is to seek a parabolic curve of $p_2(x)$ to approximate $f(x)$ via the three nodes $(x_0, y_0), (x_1, y_1), (x_2, y_2)$ that are on $f(x)$.

5   The Construction of the Key Algorithm of Equipment Status Analysis by Piecewise Average Quadratic Interpolation Based on Cluster Analysis

5.1   Improved K-Means Algorithm

K-Means clustering algorithm refers to the grouping the collection of physical or abstract objects into different clusters according to the similarities the objects.

Input: A database with N objects, the number of clusters K, with $\beta \in (0,1)$

Output: Clustering result, i.e., K clusters.

Method:

(1). Select K initial cluster seeds according to the initial cluster centers shown previously;

(2). Repeat;

(3). Assign each object the most similar class according to the principle of the most similarity within the class (the closest distance between the clusters and the cluster seeds);

(4). Calculate the minimum similarity $MinSim_{t(k-1)}$ between with the cluster seed $O_{t(k-1)}$ and the cluster data, with the threshold value of $1 - \beta * \left(1 - MinSim_{t(k-1)}\right)$;

(5). Select the data with the similarity greater than the threshold in the cluster $C_{t(k-1)}$ and the cluster seeds $O_{t(k-1)}$, to get the set of $CN_{t(k-1)}$;

(6). Calculate the mean value of $CN_{t(k-1)}$, thus being the cluster seed;

(7). Repeat until there is no further change between the two adjacent cluster seeds.

5.2   Piecewise Average Quadratic Interpolation

The piecewise average quadratic interpolation will be constructed in this paper. The procedure of building the average quadratic interpolation is shown as the following:

(1) Partition the interval $[a,b]$ as: $\Delta : a = x_0 < x_1 < x_2 < \cdots < x_k = b$

(2) Construct the Lagrange quadratic interpolation polynomial $p_{i2}(x)$ over the interval of $\left[x_{i-1}, x_i, x_{i+1}\right]$;

$$p_{i2}(x) = y_{i-1} \frac{(x-x_i)(x-x_{i+1})}{(x_{i-1}-x_i)(x_{i-1}-x_{i+1})} + y_i \frac{(x-x_{i-1})(x-x_{i+1})}{(x_i-x_{i-1})(x_i-x_{i+1})} + y_{i+1} \frac{(x-x_{i-1})(x-x_i)}{(x_{i+1}-x_{i-1})(x_{i+1}-x_i)}$$

$$(5-1)$$

Likewise, construct the Lagrange quadratic interpolation polynomial $p_{(i+1)2}(x)$ over the interval $\left[x_i, x_{i+1}, x_{i+2}\right]$:

$$p_{(i+1)2}(x) = y_i \frac{(x-x_{i+1})(x-x_{i+2})}{(x_i-x_{i+1})(x_i-x_{i+2})} + y_{i+1} \frac{(x-x_i)(x-x_{i+2})}{(x_{i+1}-x_i)(x_{i+1}-x_{i+2})} + y_{i+2} \frac{(x-x_i)(x-x_{i+1})}{(x_{i+2}-x_i)(x_{i+2}-x_{i+1})}$$

$$(5-2)$$

(3)   Construct the interpolation polynomial over every interval of $\left[x_i, x_{i+1}\right]$:

$$p_i(x) = \frac{p_{i2}(x) + p_{(i+1)_2}(x)}{2} \tag{5-3}$$

(4)   Stitch together all the $p_i(x)$ as $p(x)$, take $p(x)$ as the interpolation function of $f(x)$ over $[a,b]$, i.e.:

$$p(x) = p_i(x) \qquad x \in [x_i, x_{i+1}] \tag{5-4}$$

6   Example of Relationship between the Sterilization Parameters and the Pipeline Equipment States

The data of 1000 pot experiments are collected using the XG1-DWED0.8 pulsating vacuum sterilizer from Shandong Xinhua Medical Instrument Corporation. The data is show as Figure 2. It is shown that the clustering analysis of the equipment operating status with respect to the F0 value using the improved K-means algorithm. The values of F0 for the three clusters, qualified $F0_M$, sub-qualified $F0_N$, and disqualified $F0_L$, are calculated by:

$$F0 = D121°C \times (\lg N0 - \lg Nt) \tag{6-1}$$

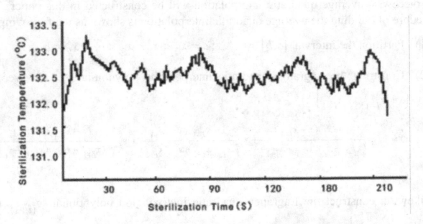

Fig. 2. The curve of sterilization temperature vs. sterilization time

The Lagrange quadratic interpolation function can be constructed using multiple discrete data points according to equation (5-4). The values of F0 for the three clusters are taken as: $F0_M$ for qualified state, $F0_N$ for sub-qualified state, and $F0_L$ disqualified state. Different values of F0 correspond to different operating states. The interpolation curve of disqualified $F0_L$ according to the above algorithm is shown in

Figure 3. It shows the interpolation of $F0_L$ when the sterilization steam trap vale malfunctions. The temperature drops below 132°C because of the malfunction. The interpolation curve reflects the equipment state very well. In the mean time, the interpolation function is able to eliminate the interferences of different operating states so that the accuracy of the analysis is improved.

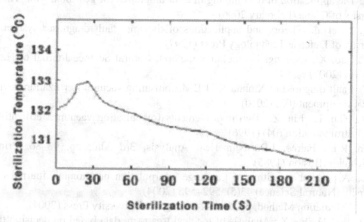

Fig. 3. The Interpolation Curve of $F0_L$

The working states of the sterilization device can be automatically informed according to the clustering interpolation data curve, with pre-set threshold values of the piping system operating states.

In this example, the equipment operating status is analyzed using the parametric statistical method using the collected data, and good results are achieved. The automatic recording and the automatic notification before reaching the specified number of times are realized.

7   Conclusion

In summary, the operating states of the sterilization system and the status of the units can be analyzed using the algorithm proposed in this paper. Thus improve the quality of sterilization. It has ensured the supply of the sterile items in clinical and hospital operating rooms. It can also provide a practical algorithm to analyze system states for other equipments.

Acknowledgements. The work of this paper is supported by the Hainan Natural Science Foundation under grant number 614249, and the Scientific Research Project of Hainan Education Department under grant number HNKY2014-111.

References

1. Association for the Advancement of Medical Instrumentation. ANSI/AAMI ST 79:2006 Comprehensive guide to steam sterilization and sterility assurance in health care facilities (2006)
2. Xu, Z.: The application of data mining in fault diagnoses for gear box. MSc Thesis, North University of China (February 2006)
3. Zhang, Y., et al.: Theory and applications of dynamic fault diagnoses system. National University of Defense Technology Press (1997)
4. Fu, X., Guo, X.: Lagrange interpolation method. Natural Science Journal of Hainan University (6) (2007)
5. Mo, T.: Fault diagnose of Xinhua XG1.DME pulsating vacuum sterilization device. China Medical Equipment (03) (2010)
6. Liu, Y., Zhu, L., Liu, Z.: Two improvements for pulsating vacuum sterilization device. China Instrumentation (04) (1996)
7. Burden, R.L., Faires, J.D.: Numerical Analysis, 3rd edn., pp. 98–86. Prindle, Weber&Schmidt, Boston (1985)
8. Lin, L.: A parallel algorithm of Lagrange interpolation polynomial. Journal of Xiamen University (Natural Science) 43(5), 592–595 (2004)
9. Wu, Z.: Calculation Method, pp. 61–84. Tshinghua University Press (2004)
10. Wu, Q., Wu, D., Fu, X.: Analysis of medical treatment data based on decision tree Computer CD Software and Applications (January 2014)

Study on Driving Forces of UGC Adoption Behavior in Service Industry: A Platform Feature Based Model

Ying Hua[1,*], Jin Chen[1], and Yonggui Wang[2]

[1] Information School, University of International Business and Economics, Beijing, China
cissyhy@gmail.com, chenjin008@hotmail.com
[2] Business School, University of International Business and Economics, Beijing, China
jcewang@gmail.com

Abstract. Nowadays, User Generated Content (UGC) influences consumers profoundly in their decision-making. UGC is more credible than advertising as common users generate it. The influencing power traditionally held by enterprises has shift to consumers dramatically. Current research mostly focus on the changes in consumer behavior after adoption of UGC, ignoring the source---factors that influence customers to choose and adopt certain UGC information, especially those related to social media platform features, emotion and attitude. Based on Information Adoption Model (IAM), this study introduces two types of trust and platform feature related factors to construct a new theoretical model, to comprehensively interpret information adoption behavior in service industry, aiming to provide theoretical contributions by extending IAM and according theory in new research settings.

Keywords: User Generated Content, User Created Content, UGC, UCC, information adoption, social media.

1 Introduction

Internet and Information Technology have led to technology fusion which triggers the explosive growth of various social media platforms as Blog, Micro-blog, social network, and etc. Social media is a group of internet-based applications that build on the ideological and technological foundations of Web 2.0, allowing the creation and exchange of User Generated Content (UGC), which takes the form of online review, online post, blog/micro-blog content and etc. The explosive growth of UGC has been one of the most important developments in the areas of media and information systems over the past decade for its significant impact on consumer behavior as well as enterprise marketing strategies. UGC is increasingly becoming a major source of information for many consumers in decision-making. More than 90% consumers are expected to seek suggestions and opinions from UGC (eMarketer 2010). Meanwhile UGC provides first-hand user data to companies for service innovation. Social media platforms have become the popular marketing channel.

---

* Corresponding author.

H. Li et al. (Eds.): I3E 2014, IFIP AICT 445, pp. 123–131, 2014.

Existing research and practices recognize UGC has higher persuasiveness over advertising in affecting consumer behavior. Consumer adoption of UGC represents how it changes their attitude toward products or services, and may help them make purchase decisions. Thus, UGC adoption is an effective signal of consumers' future performance. While prior studies focus more on the result of UGC adoption such as impact on purchase decision, perceived value, loyalty and etc., ignoring the source—the factors that determine adoption or acceptance of the content, especially those related to social media platform feature, emotion and attitude. Some studies have also attempted to investigate the factors contributing to UGC adoption, most of which simply consider usefulness as the only antecedent of adoption behavior that may not be sufficient in the new era. Therefore, a thorough understanding of the determinants of users' UGC adoption behavior is critical if firms are to maximize use of users' power in their social marketing strategy. Therefore, there are both theoretical and managerial needs for a more in-depth understanding of precursors of UGC adoption.

Based on Information Adoption Model and current research, this paper proposes a theoretical model of influencing factors of UGC adoption behavior by introducing new elements. The rest of the paper is structured as follows. First, a literature review of related theories and variables is provided. We then develop a new theoretical model with explanation of newly introduced constructs. The implications and conclusions are provided in the final session.

2   Literature Review

2.1   Social Media and UGC

User Generated Content (UGC), also known as User Create Content (UCC) or Consumer Generated Content (CGC) has been defined by different scholars from different angles [1,2]. In general, UGC refers to any content created and uploaded to the Internet outside of professional routines and practices, which is the aggregation and leveraging of users' contributions on the web [3], that can be seen as the sum of all ways in which people make use of social media [3]. While there is a lack of a formal definition, social media can be generally understood as a wide range of Internet-based applications that build on the ideological and technological foundations of Web 2.0 and that allow the creation and exchange of User Generated Content [3]. UGC represents the essence of social media. UGC is related to, but not identical with eWOM (electronic word of mouth). The difference lies in the originality of the content [4,5]. Online reviews may be the dominant form of UGC and has attracted most research attention.

There are abundant researches concentrating on the result of UGC adoption that is the impact on various marketing outcomes as purchase decision [6], sales [7], customer loyalty [8], knowledge sharing [9], trust, and etc. There are not sufficient studies on the source problem, such as what determine customers' choose and adoption of certain UGC information when facing numerous information on social media platforms. The influencing factors especially those related with platform features remain unclear.

2.2   Information Adoption Behavior Research

2.2.1   Information Adoption Model (IAM)

The nature of UGC is information, so IAM (Information Adoption Model) rather than TAM (Technology Adoption Model) is selected as the theoretic basis of this study. In IAM, argument quality, also called information quality and source (UGC creator) credibility positively affect perceived usefulness, which further affects information adoption behavior. IAM is widely applied in the research of online community [11], social network, eWOM [11]. Perceived usefulness is the only predictor of information adoption in IAM. Current research has approved that perceived usefulness has significant impact on users' self-evaluation system and behavior intention such as information adoption [12]. However, Cheung and Lee (2008) found that usefulness can only account for 46% of the variance of UGC adoption. Chen et al. (2011) got similar results. The authors proposed that some important variables may be missing. Oum and Han (2011) also indicated only usefulness is not enough to interpret and influence consumer behavior in new media context. Therefore, it's necessary to explore other important precursors based on prior research and the features of social media and service industry.

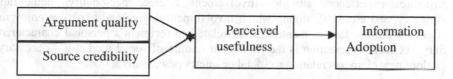

Fig. 1. IAM: Information Adoption Model, Sussman & Siegal 2003[10]

2.2.2   Influencing Factors of Information Persuasiveness and Information Adoption

According to Communication Theory, the persuasiveness of information and according adoption behavior are influenced by three categories of factors as information quality, information source and characteristics of recipient. Current research has covered all the three categories while ignoring factors related to social media features.

(1) Argument quality (information quality)

Argument quality, also information quality refers to the persuasive strength of arguments embedded in an informational message [13]. Abundant researches have been conducted with multi-dimensions [11]. Accuracy, relevance, currency, variety, completeness, understandability, dynamism, and personalization are the measures used in recent commerce studies [14].

A prominent feature of UGC information on social media platform is the diversification of information presentation, which reflects the complexity and all involved costs in UGC creation and publication. Thus, the diversification of information presentation will affect user's perceived usefulness of information. Further study is needed to explore the impact of this information quality related factor.

(2) Source credibility (UGC creator)

Source credibility is defined as the extent to which an information source is perceived to be believable, competent and trustworthy by information recipients, affecting the persuasiveness of information [15]. Expertise and trustworthiness of UGC creator are the key measures widely accepted in existing research [11].

As for expertise, there is a debate. Only when recipient himself (herself) has enough expertise can he or she have the ability to evaluate creator's expertise, which is a demanding task. Meanwhile, the anonymity and weak tie between information creator and recipient make it difficult to distinguish the identity of UGC creator. Many researches show that credibility, trust and information persuasiveness will increase when more personal information of the creator is disclosed [16]. While Lee et al.'s [17] research achieved the opposite conclusion. Therefore, self-disclosure can promote user interaction, while its effect on UGC adoption behavior remains unclear, which requires further analysis.

(3) Characteristics of information recipient

Information may exert different influences to different information recipients due to different perceptions and experiences of individuals, especially in service decision for its higher subjectivity and potential risk. So personal characteristics of information recipient is included as another category in current research, measured by knowledge structure, experience, attitude, involvement degree, personality, demographic variables, and etc. [10], among which involvement degree and expertise enjoy most attention. All the factors mentioned are related to recipient's personal characteristics only. As social interaction is the prominent feature of social media, it's necessary to explore new elements relating to social or interpersonal interaction.

2.2.3   Platform Feature Related Factors

This is the category ignored in current research. Due to the anonymity and weak tie connection of social media users, platforms are perceived as the major participants in UGC interaction, which indicates platform level factors are far more important. Existing website feature related researches mainly focus on technology aspects, such as rating system [18], layout and interface design [19], without consideration of social interaction related elements that can improve the attractiveness of websites. On the other hand, evidence shows that where social media content is posted is critical as different platforms are perceived differently in credibility and trustworthiness [20]. Therefore, platform features play a vital role in predicting users' adoption behavior.

2.2.4   Trust

Numerous studies have indicated that trust is indispensable for predicting the activity of online consumers [21].The literature suggests that trust is essential when exchanging opinions in online communities. The higher the level of trust between individuals is, the higher the possibility of engaging in information seeking. Trust is considered a vital factor affecting online consumer activities, such as the acceptance of others' advices. In the context of social media, trust can be regarded as users' evaluation of information and its source, which help to promote the dissemination and adoption of UGC.

While most of prior researches focus on interpersonal trust from a more cognitive aspect, there are no discussions on different mechanisms and impact of different types of trust classified by interaction subjects. And no consensus has been achieved as for the impact of dimensions of trust on user behavioral intention [22]. As mentioned before, social media platforms play an important role in UGC interaction. Therefore, user trust may stem both from UGC creator and from the platform. Hence, a thorough exploration of the classification of trust is necessary to fully uncover the different construction mechanism and impact of different types of trust.

2.3   Summary

In general, existing UGC adoption behavior researches mostly focus on the results, also are the marketing outcomes after adoption of UGC, with insufficient research on the source what influence users to choose and adopt certain information. And most of the closely related studies take usefulness as the only antecedent of adoption behavior, ignoring the typical emotional or attitudinal factors, which are important in service decision. Prior researches has covered three categories of influencing factors of information adoption behavior, without consideration of non-technological platform level elements that represent the social interaction feature of social media. As for the three categories of influencing factors, few researches have discussed the impact of diversification of UGC representation, and there exist no consensus on the impact of UGC creator's self-disclosure degree. Therefore, it's necessary to construct a new theoretical framework by introducing new elements, to comprehensively understand the driving forces of users' UGC adoption behavior on social media platforms in service context.

3   Research Model Development

On the basis of an intensive review of the literature, we propose a theoretical model of the influencing factors of UGC adoption behavior as shown in Fig.2, followed with the explanation of newly introduced elements.

3.1   Two Types of Trust

As discussed in the literature review, trust is a crucial aspect to determine the intention to follow advice of others that helps to promote the dissemination and adoption of UGC on social media platforms. In UGC interaction, platform is perceived as the other party involved. Therefore, trust may stem both from information creator and from the social media platform, with distinct mechanisms and effects. According to the interaction subjects, two types of trust are introduced.

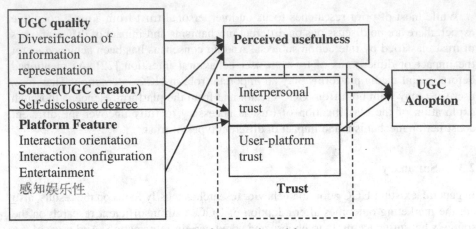

Fig. 2. Research model

3.1.1   Interpersonal Trust

As shown in the literature review, most of prior researches on trust concentrate on interpersonal trust, the trust between UGC creator and recipient. The consumer will have the intention to accept the other party's advice if they feel he or she honest, competent and benevolent. In online environments, some cues i.e. the perceived credibility of information creator lead to the emergence of interpersonal trust, which will further the interaction to influence consumer behavior and purchase intention [23]. That is, interpersonal trust can increase the persuasiveness of information, thus influence online consumers' attitude and behavior [6], as UGC adoption behavior.

3.1.2   User-Platform Trust

As face-to-face interaction is replaced by screen to face interaction, platforms are the intuitive party perceived by users. In practice, interaction platform is trusted as a whole since it's hard for consumers to evaluate source credibility [24]. The same information shown on different platforms receives different degrees of trustworthiness, indicating there is a type of trust related to platform feature. The researches of Senecal and Nantel (2004) and Burgess et al. (2009) support this argument with empirical study in online travel community, while only focusing on the host of the website not social interaction related classification. Therefore, interpersonal trust and user-platform trust are distinct concepts with different developing mechanisms and impact, which is of great value to study thoroughly and separately.

3.2   Platform Features

Social media are characterized by social interaction, user integration, personalization and exchange of content, among which the last three aspects are implied in the construction of the research model. Here we concentrate on platform level elements related to social interaction, which are indispensable to induce trust and information adoption behavior, as discussed above. In the context of social media, the desired platform features with attractiveness can be understood as a shared aggregated positive valence of users [24], which covers rational and affective aspects of usage scenarios, From the attitudinal level, it would affect individual perceptions and attitudes as trust. From the action-based level, it would also positively affect user behavior as information adoption.

According to the literature review and successive analysis, three dimensions are included. ①Interaction orientation. Interaction is a crucial element of social media and the corresponding properties offered can be summarized as interaction orientation, which covers users' need for interactive content and the corresponding expectations regarding the provider of these offers [24]. Such an interaction-related strategy is referred to as interaction orientation. ②Interaction configuration, which refers to the perceived structure of interaction process on a certain platform [25]. Companies continuously process integrated data from user interaction to serve as the orientation and configuration basis of added value.③Entertainment. The attractive platform features should include both hedonic and/or cognitive aspects. Most of the services like travel purchases belong to the category of hedonic goods, in the decision process of which involved requirements for socio-emotional benefits and experiences such as fantasy, fun and pleasure. Hence perceived entertainment is a crucial element that can affect user decision, especially of greater importance in service decisions [26]. Langlois et al. (2000) indicated attractive platform features can affect users' self-perception i.e. trust and usefulness. Therefore, there exist correlations between different dimensions of platform feature and the proposed mediators.

3.2.1   Diversification of Information Representation
The presentation of UGC information on social media platforms takes various forms as text, images, audios, videos and etc., which is defined as the diversification of information presentation. Since it reflects the complexity and all involved costs in UGC creation and publication, the diversification of information presentation will affect user's perceived usefulness of information. So this construct is especially included to present the measurement of information quality.

3.2.2   Self-Disclosure Degree
Self-disclosure refers to any personal information one shares with others, which can promote user interaction. As discussed before, high expertise is required for UGC recipients to accurately access the creator's credibility, which is always a problem. Thus self-disclosure of personal information becomes a vital indicator of credibility of the creator. While some researches all recognized its correlations with both cognitive and attitudinal elements as perceived usefulness and interpersonal trust, there is no consensus on whether more or less personal information disclosed by UGC creator can positively affect source credibility, trust and information persuasiveness [16]. Therefore, the construct of self-disclosure degree is introduced to further test how it affects perceived usefulness and trust.

4   Conclusion

With the prosperity of UGC and social media, the power of persuasiveness shits from companies to users. Both academia and industries recognize the significant impacts of UGC on consumer behavior and business marketing strategies. Most prior studies focus on the results rather than the source--driving forces of UGC adoption behavior, which requires more in-depth research especially when confronted with over-loaded information on social media platforms. Based on an intensive review of the literature and understanding of research context, this paper proposes a theoretical model by extending Information Adoption Model (IAM) with new elements as follows: the attitudinal elements two types of trust as the mediators, platform feature related elements with three dimensions as antecedents, diversification of information representation and creator's self-disclosure degree as the measurements of information quality and source credibility respectively. The theoretical contribution of this study lies in the extension of information adoption theory in the research context of social media platform and service industry, which hopefully could provide some lights to the emerging research in UGC behavior field.

References

1. Wang, Y., et al.: Customers'Perceived Benefits of Interacting in a Virtual Brand Community in China. Journal of Electronic Commerce Research 14(1), 46–69 (2013)
2. Cooke, M., Buckley, N.: Web 2.0, social networks and the future of market research. International Journal of Market Research 50(2), 267–292 (2008)
3. Kapla, A.M., Haenlein, M.: Social media: back to the roots and back to the future. Journal of Systems and Information Technology 14(2), 101–104 (2012)
4. Hennig-Thurau, T., Gwinner, K.P., Walsh, G., Gremler, D.D.: Electronic Word-of-Mouth via Consumer-Opinion Platforms: What Motivates Consumers To Articulate Themselves on the Internet? Journal of Interactive Marketing 18(1), 38–52 (2004)
5. Cheong, H.J., Morrison, M.A.: Consumers' Reliance on Product Information and Recommendations Found in UGC. Journal of Interactive Advertising 8(2), 38–49 (2008)
6. Lu, W., Stepchenkova, S.: Ecotourism experiences reported online: Classification of satisfaction attributes. Tourism Management 33(3), 702–712 (2012)
7. Ye, Q., et al.: The influence of user-generated content on traveler behavior: An empiricalinvestigation on the effects of e-word-of-mouth to hotel online bookings. Computers in Human Behavior 27(2), 634–639 (2011)
8. Rondan-Cataluna, F.J., Sanchez-Franco, M.J., Villarejo-Ramos, A.F.: Searching for latent class segments in technological services. Service Industries Journal 30(6), 831–849 (2010)
9. Qu, H., Lee, H.: Travelers' social identification and membership behaviors in online travel community. Tourism Management 32(6), 1262–1270 (2011)
10. Sussman, S.W., Siegal, W.S.: Informational Influence in Organizations: An Integrated Approach to Knowledge Adoption. Information Systems Research 14(1), 47–65 (2003)
11. Cheung, C.M.K., Lee, M.K.O., Rabjohn, N.: The impact of electronic word-of-mouth:The adoption of online opinions in online customer communities. Internet Research 18(3), 229–247 (2008)

12. Morosan, C., Jeong, M.: The Role of the Internet in the Process of Travel Information Search. Information Technology in Hospitality 5(1), 13–23 (2008)
13. Bhattacherjee, A., Sanford, C.: Influence Processes for Information Technology Acceptance: An Elaboration Likelihood Model. Management Information Systems Quarterly 30(4), 805–825 (2006)
14. DeLone, W.H., McLean, E.R.: The DeLone and McLean Model of Information Systems Success: A Ten-Year Update. Journal of Management Information Systems 19(4), 9–30 (2003)
15. Westerman, D., Spence, P.R., VanDer Heide, B.: A social network as information: The effect of system generated reports of connectedness on credibility on Twitter. Computers in Human Behavior 28(1), 199–206 (2012)
16. Head, M.M., Hassanein, K.: TRUST IN E-COMMERCE. Quarterly Journal of Electronic Commerce 3(3), 307 (2002)
17. Ip, C., Law, R., 'Andy' Lee, H.: A review of website evaluation studies in the tourism and hospitality fields from 1996 to 2009 13(3), 234–265 (2011)
18. Lee, D., Yoon, S.N., Choi, H., Park, Y.: Electronic word of mouth systems and acceptance of user-generated contents. Int. J. Information and Decision Sciences 3(1), 54–69 (2011)
19. Fan, Y.-W., Miao, Y.-F., Fang, Y.-H., Lin, R.-Y.: Establishing the Adoption of Electronic Word-of-Mouth through Consumers' Perceived Credibility. International Business Research 6(3), 58–65 (2013)
20. Zheng, X., Ulrike, G.: Role of social media in online travel information search. Tourism Management 31(2), 179–188 (2010)
21. Awad, N.F., Ragowsky, A.: Establishing Trust in Electronic Commerce Through Online Word of Mouth: An Examination Across Genders. Journal of Management Information Systems 24(4), 101–121 (2008)
22. Hwang, Y., Lee, K.C.: Investigating the moderating role of uncertainty avoidance cultural values on multidimensional online trust. Information & Management 49(3-4), 171 176 (2012)
23. Sparks, B.A., Perkins, H.E., Buckley, R.: Online travel reviews as persuasive communication: The effects of content type, source, and certification logos on consumer behavior. Tourism Management 39, 1–9 (2013)
24. Wirtz, B.W., Piehler, R., Ullrich, S.: Determinants of Social Media Website Attractiveness. Journal of Electronic Commerce Research 14(1), 1611 (2013)
25. Chung, C., Austria, K.P.: Attitudes Toward Product Messages on Social Media: An Examination of Online Shopping Perspectives Among Young Consumers. International Journal of E-Services and Mobile Applications 4(4), 1–14 (2012)
26. Ayeh, J.K., Au, N., Law, R.: Predicting the intention to use consumer-generated media for travel planning. Tourism Management 35, 132–143 (2013)

Quantum-Behaved Particle Swarm Optimization Based on Diversity-Controlled

HaiXia Long[*], Haiyan Fu, and Chun Shi

School of Information Science Technology, Hainan Normal University, Haikou 571158,
Hainan, China
haixia_long@163.com
{48318159,605515770}@qq.com

Abstract. Quantum-behaved particle swarm optimization (QPSO) algorithm is a global convergence guaranteed algorithms, which outperforms original PSO in search ability but has fewer parameters to control. But QPSO algorithm is to be easily trapped into local optima as a result of the rapid decline in diversity. So this paper describes diversity-controlled into QPSO (QPSO-DC) to enhance the diversity of particle swarm, and then improve the search ability of QPSO. The experiment results on benchmark functions show that QPSO-DC has stronger global search ability than QPSO and standard PSO.

Keywords: global convergence, quantum-behaved particle swarm optimization, diversity-controlled, benchmark function.

1 Introduction

Particle swarm optimization (PSO) is a kind of stochastic optimization algorithms proposed by Kennedy and Eberhart [1] that can be easily implemented and is computationally inexpensive. The core of PSO is based on an analogy of the social behavior of flocks of birds when they search for food. PSO has been proved to be an efficient approach for many continuous global optimization problems. However, as demonstrated by F. Van Den Bergh [2], PSO is not a global convergence guaranteed algorithm because the particle is restricted to a finite sampling space for each of the iterations. This restriction weakens the global search ability of the algorithm and may lead to premature convergence in many cases.

Several authors developed strategies to improve on PSO. Clerc [3] suggested a PSO variant in which the velocity to the best point found by the swarm is replaced by the velocity to the current best point of the swarm, although he does not test this variant. Clerc [4] and Zhang et al. [5] dynamically change the size of the swarm according to the performance of the algorithm. Eberhart and Shi [6], He et al. [7] adopted strategies based on dynamically modifying the value of the PSO parameter called inertia weight. Various other solutions have been proposed for preventing premature convergence:

---

[*] Corresponding author.

H. Li et al. (Eds.): I3E 2014, IFIP AICT 445, pp. 132–143, 2014.

objective functions that change over time [8]; noisy evaluation of the function objective [9]; repulsion to keep particles away from the optimum [10]; dispersion between particles that are too close to one another [11]; reduction of the attraction of the swarm center to prevent the particles clustering too tightly in one region of the search space [12]; hybrids with other meta-heuristic such as genetic algorithms [13]; or ant colony optimization [14]; an up-to-date overview of the PSO [15].

Recently, a new variant of PSO, called Quantum-behaved Particle Swarm Optimization (QPSO) [16, 17], which is inspired by quantum mechanics and particle swarm optimization model. QPSO has only the position vector without velocity, so it is simpler than standard particle swarm optimization algorithm. Furthermore, several benchmark test functions show that QPSO performs better than standard particle swarm optimization algorithm. Although the QPSO algorithm is a promising algorithm for the optimization problems, like other evolutionary algorithm, QPSO also confronts the problem of premature convergence, and decrease the diversity in the latter period of the search. Therefore a lot of revised QPSO algorithms have been proposed since the QPSO had emerged. In Sun et al. [18], the mechanism of probability distribution was proposed to make the swarm more efficient in global search. Simulated Annealing is further adopted to effectively employ both the ability to jump out of the local minima in Simulated Annealing and the capability of searching the global optimum in QPSO algorithm [19]. Mutation operator with Gaussian probability distribution was introduced to enhance the performance of QPSO in Coelho [20]. Immune operator based on the immune memory and vaccination was introduced into QPSO to increase the convergent speed by using the characteristic of the problem to guide the search process [21].

In this paper, QPSO with diversity-controlled is introduced. This strategy is to prevent the diversity of particle swarm declining in the search of later stage.

The rest of the paper is organized as follows. In Section 2, the principle of the PSO is introduced. The concept of QPSO is presented in Section 3 and the QPSO with diversity-controlled is proposed in Section 4. Section 5 gives the numerical results on some benchmark functions and discussion. Some concluding remarks and future work are presented in the last section.

2   PSO Algorithm

In the original PSO with M individuals, each individual is treated as an infinitesimal particle in the D-dimensional space, with the position vector and velocity vector of particle i, $X_i(t) = (X_{i1}(t), X_{i2}(t), \cdots, X_{iD}(t))$ and $V_i(t) = (V_{i1}(t), V_{i2}(t), \cdots, V_{iD}(t))$. The particle moves according to the following equations:

$$V_{ij}(t+1) = V_{ij}(t) + c_1 \cdot r_1 \cdot (P_{ij}(t) - X_{ij}(t)) + c_2 \cdot r_2 \cdot (P_{gj}(t) - X_{ij}(t)) \tag{1}$$

$$X_{ij}(t+1) = X_{ij}(t) + V_{ij}(t+1) \tag{2}$$

for $i = 1, 2, \cdots M; j = 1, 2 \cdots, D$. The parameters c_1 and c_2 are called the acceleration coefficients. Vector $P_i = (P_{i1}, P_{i2}, \cdots, P_{iD})$ known as the personal best position, is the best previous position (the position giving the best fitness value so far) of particle i, vector $P_g = (P_{g1}, P_{g2}, \cdots, P_{gD})$ is the position of the best particle among all the particles and is known as the global best position. The parameters r_1 and r_2 are two random numbers distributed uniformly in (0, 1), that is $r_1, r_2 \sim U(0,1)$. Generally, the value of V_{ij} is restricted in the interval $[-V_{max}, V_{max}]$.

Many revised versions of PSO algorithm are proposed to improve the performance since its origin in 1995. Two most important improvements are the version with an Inertia Weight [22], and a Constriction Factor [23]. In the inertia-weighted PSO the velocity is updated by using

$$V_{ij}(t+1) = w \cdot V_{ij}(t) + c_1 \cdot r_1 (P_{ij}(t) - X_{ij}(t)) + c_2 \cdot r_2 \cdot (P_{gj} - X_{ij}(t)) \tag{3}$$

while in the Constriction Factor model the velocity is calculated by using

$$V_{ij}(t+1) = K \cdot [V_{ij}(t) + c_1 \cdot r_2 \cdot (P_{ij}(t) - X_{ij}(t)) + c_2 \cdot r_2 \cdot (P_{gj} - X_{ij}(t))] \tag{4}$$

where

$$k = \frac{2}{\left| 2 - \varphi - \sqrt{\varphi^2 - 4\varphi} \right|} \quad \varphi = c_1 + c_2, \quad \varphi > 4 \tag{5}$$

The inertia-weighted PSO was introduced by Shi and Eberhart [6] and is known as the Standard PSO.

3   QPSO Algorithm

Trajectory analyses in Clerc and Kennedy [24] demonstrated the fact that convergence of PSO algorithm may be achieved if each particle converges to its local attractor $p_i = (p_{i1}, p_{i2}, \cdots p_{iD})$ with coordinates

$$p_{ij}(t) = (c_1 r_1 P_{ij}(t) + c_2 r_2 P_{gj}(t))/(c_1 r_1 + c_2 r_2), \quad \text{or} \quad p_{ij}(t) = \varphi \cdot P_{ij}(t) + (1 - \varphi) \cdot P_{gj}(t) \tag{6}$$

where $\varphi = c_1 r_1 / (c_1 r_1 + c_2 r_2)$. It can be seen that the local attractor is a stochastic attractor of particle i that lies in a hyper-rectangle with P_i and P_g being two ends of its diagonal. We introduce the concepts of QPSO as follows.

Assume that each individual particle move in the search space with a δ potential on each dimension, of which the center is the point p_{ij}. For simplicity, we consider a particle in one-dimensional space, with point p the center of potential. Solving Schrödinger equation of one-dimensional δ potential well, we can get the probability distribution function $D(x) = e^{-2|p-x|/L}$. Using Monte Carlo method, we obtain

$$x = p \pm \frac{L}{2} \ln(1/u) \quad u \sim U(0,1) \tag{7}$$

The above is the fundamental iterative equation of QPSO.

In Sun et al. (2004b) a global point called Mainstream Thought or Mean Best Position of the population is introduced into PSO. The mean best position, denoted as C, is defined as the mean of the personal best positions among all particles. That is

$$C(t) = (C_1(t), C_2(t), \cdots, C_D(t)) = \left(\frac{1}{M} \sum_{i=1}^{M} P_{i1}(t), \quad \frac{1}{M} \sum_{i=1}^{M} P_{i2}(t), \quad \cdots, \quad \frac{1}{M} \sum_{i=1}^{M} P_{iD}(t) \right) \tag{8}$$

where M is the population size and P_i is the personal best position of particle i. Then the value of L is evaluated by $L = 2\alpha \cdot |C_j(t) - X_{ij}(t)|$ and the position are updated by

$$X_{ij}(t+1) = p_{ij}(t) \pm \alpha \cdot |C_j(t) - X_{ij}(t)| \cdot \ln(1/u) \tag{9}$$

where parameter α is called Contraction-Expansion (CE) Coefficient, which can be tuned to control the convergence speed of the algorithms. Generally, we always call the PSO with equation (9) Quantum-behaved Particle Swarm Optimization (QPSO), where parameter α must be set as $\alpha < 1.782$ to guarantee convergence of the particle [16]. In most cases, α decrease linearly from α_0 to α_1 ($\alpha_0 < \alpha_1$).

We outline the procedure of the QPSO algorithm as follows:
Procedure of the QPSO algorithm:
Step1: Initialize the population;
Step2: Computer the personal position and global best position;
Step3: Computer the mean best position C;
Step4: Properly select the value of α;
Step 5: Update the particle position according to Eq. (9);
Step6: While the termination condition is not met, return to step (2);
Step7: Output the results.

4   QPSO with Diversity-Controlled

QPSO is a promising optimization problem solver that outperforms PSO in many real application areas. First of all, the introduced exponential distribution of positions makes QPSO global convergent. The QPSO algorithm in the initial stage of search, as the particle swarm initialization, its diversity is relatively high. In the subsequent search process, due to the gradual convergence of the particle, the diversity of the population continues to decline. As the result, the ability of local search ability is continuously enhanced, and the global convergence ability is continuously weakened. In early and middle search, reducing the diversity of particle swarm optimization for contraction efficiency improvement is necessary, however, to late stage of search, because the particles are gathered in a relatively small range, particles swarm diversity is very low, the global search ability becomes very weak, the ability for a large range of search has been very small, this algorithm will occur the phenomenon of premature.

To overcome this shortcoming, we introduce diversity-controlled into QPSO.

The population diversity of the QPSO-DC is denoted as diversity (pbest) and is measured by average Euclidean distance from the particle's personal best position to the mean best position, namely

$$\text{diversity(pbest)} = \frac{1}{M \cdot |A|} \cdot \Sigma_{i=1}^{M} \sqrt{\Sigma_{j=1}^{D} (\text{pbest}_{i,j} - \overline{\text{pbes}}}$$ (10)

where M is the population of the particle, $|A|$ is the length of longest the diagonal in the search pace, and D is the dimension of the problem. Hence, we may guide the search of the particles with the diversity measures when the algorithm is running.

In the QPSO-DC algorithm, only low bound d_{low} is set for diversity (pbest) to prevent the diversity from constantly decreasing. The procedure of the algorithm is as follows. After initialization, the algorithm is running in convergence mode. In process of convergence, the convergence mode is realized by Contraction-Expansion (CE) Coefficient. On the course of evolution, if the diversity measure diversity (pbest) of the swarm drops to below the low bound d_{low}, the mean best position is reinitialized.

5   Experiment Results and Discussion

To test the performance of the QPSO with diversity-controlled, seven widely known benchmark functions listed in Table 1 are tested for comparison with Standard PSO (SPSO), QPSO. These functions are all minimization problems with minimum objective function values zeros. The initial range of the population listed in Table 2 is asymmetry as used in Shi and Eberhart [25]. Table 2 also lists V_{max} for SPSO. The fitness value is set as function value and the neighborhood of a particle is the whole population.

As in Angeline [22], for each function, three different dimension sizes are tested. They are dimension sizes: 10, 20 and 30. The maximum number of generations is set as 1000, 1500, and 2000 corresponding to the dimensions 10, 20, and 30 for first six functions, respectively. The maximum generation for the last function is 2000.In order to investigate whether the QPSO-RS algorithm well or not, different population sizes are used for each function with different dimension. They are population sizes of 20, 40, and 80.For SPSO, the acceleration coefficients are set to be c1=c2=2 and the inertia weight is decreasing linearly from 0.9 to 0.4 as in Shi and Eberhart [25]. In experiments for QPSO, the value of CE Coefficient varies from 1.0 to 0.5 linearly over the running of the algorithm as in [18], while in QPSO-DC, the value of CE Coefficient is listed in Table 3 and Table 4. From the Table 3 and Table 4, we also obtain the CE COFFICIENT of QPSO-DC decreases from 0.8 to 0.5 linearly. We had 50 trial runs for every instance and recorded mean best fitness and standard deviation

The mean values and standard deviations of best fitness values for 50 runs of each function are recorded in Table 5 to Table 11.

Table 1. Expression of the five tested benchmark functions

	Function Expression	Search Domain
Sphere	$f_1(X) = \sum\limits_{i=1}^{n} x_i^2$	$-100 \le x_i \le 100$
Rosenbrock	$f_2(X) = \sum\limits_{i=1}^{n-1} (100 \cdot (x_{i+1} - x_i^2)^2 + (x_i - 1)^2)$	$-100 \le x_i \le 100$
Rastrigrin	$f_3(X) = \sum\limits_{i=1}^{n} (x_i^2 - 10 \cdot \cos(2\pi x_i) + 10)$	$-10 \le x_i \le 10$
Greiwank	$f_4(X) = \dfrac{1}{4000} \sum\limits_{i=1}^{n} x_i^2 - \prod\limits_{i=1}^{n} \cos\left(\dfrac{x_i}{\sqrt{i}}\right) + 1$	$-600 \le x_i \le 600$
Ackley	$f_5 = 20 + e - 20 e^{-\frac{1}{5}\sqrt{\frac{1}{n}\sum_{i=1}^{n} x_i^2}} - e^{-\frac{1}{n}\sum_{i=1}^{n} \cos(2\pi x_i)}$	$-30 \le x_i \le 30$
Schwefel	$f_6 = 418.9829\, n - \sum\limits_{i=1}^{n} (x_i \sin\sqrt{|x_i|})$	$-500 \le x_i \le 500$
Shaffer's	$f_7(X) = 0.5 + \dfrac{(\sin(\sqrt{x^2 + y^2})^2}{(1.0 + 0.001(x^2 + y^2))^2}$	$-100 \le x_i \le 100$

Table 2. The initial range of population for all the tested algorithms and Vmax for SPSO

	Initial Range	Vmax
f_1	(50, 100)	100
f_2	(15, 30)	100
f_3	(2.56, 5.12)	10
f_4	(300, 600)	600
f_5	(15,30)	30
f_6	(250,500)	500
f_7	(30, 100)	100

Table 3. Parameter value of QPSO-dc

CE Coefficient	Sphere function		Rosenbrock function		Rastrigrin function	
	fitness	St. Dev.	fitness	St. Dev.	fitness	St. Dev.
(1.0,0.5)	1.0050e-013	2.8029e-013	63.6787	112.2467	96.4876	11.2102
(0.9,0.5)	2.1644e-017	3.3232e-017	53.2366	75.9345	94.8874	11.2350
(0.8,0.5)	4.9846e-026	1.7413e-025	39.2303	37.6453	89.7231	9.9548
(0.7,0.5)	9.7934e-022	3.6896e-021	73.4979	100.0602	83.4991	11.6756
(1.0,0.4)	4.6673e-012	2.4765e-011	48.4907	46.6895	81.5502	13.5262
(0.9,0.4)	1.9889e-014	7.9829e-014	78.9340	124.4234	80.8902	12.1062
(0.8,0.4)	6.0362e-015	3.4761e-014	72.1817	102.6235	74.1639	10.4042
(0.7,0.4)	1.6104e-018	4.5714e-018	102.6623	155.5243	70.7840	15.0399
(1.0,0.3)	1.7248e-010	4.4884e-010	59.0231	74.3503	64.9413	19.3783
(0.9,0.3)	1.0957e-011	6.1253e-011	76.6616	103.9478	59.4859	17.9470
(0.8,0.3)	6.2854e-014	1.2280e-013	80.0582	91.4265	56.3751	17.4575
(0.7,0.3)	2.4054e-013	9.3876e-013	106.6513	133.7599	48.8128	20.5664

Table 4. Table 3 (continue)

CE Coefficient	Griewank function		Ackley function		Schwefel function		Shaffer's f6 function	
	fitness	St. Dev.	fitness	St. Dev.	fitness	St. Dev.	fitness	St. Dev.
(1.0,0.5)	0.0219	0.0254	3.0209e-013	4.3354e-013	5.0840e+003	238.6654	7.7802e-004	0.0027
(0.9,0.5)	0.0165	0.0171	3.2649e-014	8.3201e-014	4.9791e+003	218.1045	5.8366e-004	0.0023
(0.8,0.5)	0.0078	0.0096	1.0978e-014	9.2924e-015	3.9491e+003	682.3425	7.8734e-004	0.0027
(0.7,0.5)	0.0141	0.0183	3.9826e-014	5.6935e-014	4.4273e+003	347.6864	0.0019	0.0039
(1.0,0.4)	0.0131	0.0170	3.7380e-011	1.3192e-010	5.0548e+003	251.4751	5.8369e-004	0.0023
(0.9,0.4)	0.0182	0.0191	1.5916e-012	3.8720e-012	5.0319e+003	238.8422	7.9738e-004	0.0027
(0.8,0.4)	0.0096	0.0106	8.0437e-013	1.2495e-012	4.8485e+003	267.2251	0.0016	0.0036
(0.7,0.4)	0.0130	0.0175	0.0231	0.1634	4.3537e+003	513.4628	0.0027	0.0044
(1.0,0.3)	0.0139	0.0143	3.9977e-010	7.1079e-010	4.9716e+003	262.9720	1.0843e-007	2.0595e-007
(0.9,0.3)	0.0129	0.0207	8.3162e-011	1.1640e-010	4.9157e+003	242.6659	7.9009e-004	0.0027
(0.8,0.3)	0.0148	0.0141	1.4957e-010	3.6527e-010	4.7433e+003	334.3465	0.0017	0.0038
(0.7,0.3)	0.0142	0.0123	0.0231	0.1634	4.8584e+003	273.0154	0.0027	0.0044

Table 5. Numerical results on Sphere function

Dim.	Gmax	SPSO		QPSO		QPSO-DC	
		fitness	St. Dev.	fitness	St. Dev.	fitness	St. Dev.
10	1000	4.6119e-021	1.0352e-020	3.1979e-043	2.2057e-042	5.2719e-045	1.5684e-045
20 20	1500	9.0266e-012	3.5473e-011	1.9197e-024	7.1551e-024	2.5427e-026	4.2876e-026
30	2000	3.9672e-008	7.0434e-008	7.3736e-015	2.1796e-014	3.2486e-016	5.3971e-016
10	1000	1.3178e-024	3.7737e-024	2.7600e-076	1.8954e-075	8.4527e-074	2.3460e-074
40 20	1500	2.3057e-015	6.6821e-015	3.9152e-044	1.8438e-043	5.4381e-045	8.5426e-045
30	2000	1.1286e-010	2.2994e-010	2.6424e-031	6.3855e-031	2.5419e-032	1.3645e-032
10	1000	1.6097e-028	7.1089e-028	2.5607e-103	6.4847e-103	1.9824e-104	2.5681e-104
80 20	1500	6.3876e-018	1.4821e-017	1.4113e-068	7.4451e-068	1.5873e-068	4.3876e-068
30	2000	3.2771e-013	7.5971e-013	6.5764e-050	3.6048e-049	2.0254e-050	1.2574e-050

Table 6. Numerical results on Rosenbrock function

M	Dim.	Gmax	SPSO		QPSO		QPSO-DC	
			fitness	St. Dev.	fitness	St. Dev.	fitness	St. Dev.
	10	1000	51.0633	153.7913	9.5657	16.6365	7.5341	0.8634
20	20	1500	100.2386	140.9822	82.4294	138.2429	32.8419	28.3476
	30	2000	160.4400	214.0316	98.7948	122.5744	69.3102	54.3791
	10	1000	24.9641	49.5707	8.9983	17.8202	4.9810	3.2094

Table 6. (*Continued*)

M	Dim.	Gmax						
40	20	1500	59.8256	95.9586	40.7449	41.1751	15.3429	12.3076
	30	2000	124.1786	269.7275	43.5582	38.0533	35.2487	24.8619
	10	1000	19.0259	41.6069	6.8312	0.3355	4.1086	3.5476
80	20	1500	40.2289	46.8491	33.5287	31.6415	15.6271	0.0367
	30	2000	56.8773	57.8794	44.5946	31.6739	22.3706	0.1648

Table 7. Numerical results on Rastrigrin function

M	Dim.	Gmax	SPSO		QPSO		QPSO-DC	
			fitness	St. Dev.	fitness	St. Dev.	fitness	St. Dev.
	10	1000	5.8310	2.5023	4.0032	2.1409	2.6871	1.9684
20	20	1500	23.3922	6.9939	15.0648	6.0725	13.5970	12.3458
	30	2000	51.1831	12.5231	28.3027	12.5612	22.3694	4.2061
	10	1000	3.7812	1.4767	2.6452	1.5397	1.8541	2.1380
40	20	1500	18.5002	5.5980	11.3109	3.5995	12.3471	7.2684
	30	2000	39.5259	10.3430	18.9279	4.8342	14.3796	6.9173
	10	1000	2.3890	1.1020	2.2617	1.4811	1.0367	1.3574
80	20	1500	12.8594	3.6767	8.4121	2.5798	6.3247	4.6873
	30	2000	30.2140	7.0279	14.8574	5.0408	13.6975	5.3671

Table 8. Numerical results on Griewank function

M	Dim.	Gmax	SPSO		QPSO		QPSO-DC	
			fitness	St. Dev.	fitness	St. Dev.	fitness	St. Dev.
	10	1000	0.0920	0.0469	0.0739	0.0559	0.0657	0.0502
20	20	1500	0.0288	0.0285	0.0190	0.0208	0.0262	0.0227
	30	2000	0.0150	0.0145	0.0075	0.0114	0.0002	0.0235
	10	1000	0.0873	0.0430	0.0487	0.0241	0.0587	0.0822
40	20	1500	0.0353	0.0300	0.0206	0.0197	0.0135	0.0431
	30	2000	0.0116	0.0186	0.0079	0.0092	0.0020	0.0123
	10	1000	0.0658	0.0266	0.0416	0.0323	0.0424	0.0682
80	20	1500	0.0304	0.0248	0.0137	0.0135	0.0050	0.0103
	30	2000	0.0161	0.0174	0.0071	0.0109	1.0570e-006	5.3089e-006

Table 9. Numerical results on Ackley function

M	Dim.	Gmax	SPSO		QPSO		QPSO-DC	
			fitness	St. Dev.	fitness	St. Dev.	fitness	St. Dev.
	10	1000	2.0489e-011	3.0775e-011	6.8985e-012	1.2600e-011	2.7356e-015	5.0243e-016
20	20	1500	0.0285	0.2013	1.5270e-008	2.1060e-008	9.7700e-015	5.8751e-015
	30	2000	0.2044	0.4899	4.3113e-007	4.4188e-007	7.1797e-013	1.5299e-012
	10	1000	2.4460e-013	5.2901e-013	2.5935e-015	5.0243e-016	2.6645e-015	0
40	20	1500	2.6078e-008	4.0653e-008	6.3491e-013	1.3305e-012	3.5882e-015	1.5742e-015
	30	2000	3.7506e-006	6.4828e-006	5.5577e-011	8.6059e-011	6.5015e-015	1.8772e-015

Table 9. (*Continued*)

	10	1000	5.5778e-015	7.4138e-015	2.4514e-015	8.5229e-016	2.2382e-015	1.1662e-015
80	20	1500	5.2979e-010	7.2683e-010	5.8620e-015	1.6446e-015	2.7356e-015	5.0243e-016
	30	2000	1.6353e-007	2.1300e-007	6.7253e-014	4.4759e-014	5.4357e-015	1.4866e-015

Table 10. Numerical results on Schwefel function

M	Dim.	Gmax	SPSO		QPSO		QPSO-DC	
			fitness	St. Dev.	fitness	St. Dev.	fitness	St. Dev.
	10	1000	630.6798	169.9164	1.0784e+003	342.1967	740.4763	293.0045
20	20	1500	2.0827e+003	327.0313	4.0026e+003	679.1461	3.4182e+003	1.0382e+003
	30	2000	4.1097e+003	481.8801	6.5075e+003	589.9931	6.8196e+003	1.1614e+003
	10	1000	494.0236	169.5594	1.4359e+003	246.9290	876.2897	382.1407
40	20	1500	1.8516e+003	273.8690	4.5580e+003	242.8220	3.9461e+003	816.6767
	30	2000	3.4042e+003	384.3150	7.3380e+003	288.9463	7.3466e+003	879.8090
	10	1000	426.0796	130.5357	1.6060e+003	132.0009	979.6775	363.0876
80	20	1500	1.5466e+003	270.5587	4.6407e+003	225.9431	4.1601e+003	518.4552
	30	2000	3.0325e+003	419.8580	8.0583e+003	226.3796	7.3655e+003	852.6205

Table 11. Numerical results on Shaffer's f6 function

M	Dim.	Gmax	SPSO		QPSO		QPSO-DC	
			fitness	St. Dev.	fitness	St. Dev.	fitness	St. Dev.
20	2	2000	3.8965e-004	0.0019	0.0016	0.0036	1.3764e-007	7.3542e-007
40	2	2000	1.9432e-004	0.0014	5.8308e-004	0.0023	1.2681e-008	3.6984e-008
80	2	2000	0.0000	0.0000	7.5695e-009	2.5080e-008	5.9540e-009	8.0367e-009

The results show that both QPSO, QPSO-DC are superior to SPSO except on Schwefel and Shaffer's f6 function. On Shpere Function the QPSO works better than QPSO-DC when the warm size is 40 and dimension is 10, and when the warm size is 80 and dimension is 20. Expect for the above two instance, the best result is QPSO-DC. The Rosenbrock function is a mono-modal function, but its optimal solution lies in a narrow area that the particles are always apt to escape. The experiment results on Rosenbrock function show that the QPSO-DC outperforms the QPSO. Rastrigrin function and Griewank function are both multi-modal and usually tested for comparing the global search ability of the algorithm. On Rastrigrin function, it is also shown that the QPSO-DC generated best results than QPSO. On Griewank function, QPSO-DC has better performance than QPSO algorithm. On Ackley function, QPSO-DC has best performance when the dimension is 10, except for above functions, the QPSO-DC has minimal value. On the Schwefel function, SPSO is best performance in any situation. On Shaffer's function, it has sub-optima 0.0097. The QPSO-DC shows its stronger ability to escape the local minima 0.0097 than the QPSO, and even than SPSO. Generally speaking, the QPSO-DC has better global search ability than SPSO and QPSO.

Figure 1 shows the convergence process of the three algorithms on the first four benchmark functions with dimension 30 and swarm size 40 averaged on 50 trail runs. It is shown that, although QPSO-DC converge more slowly than the QPSO during the early stage of search, it may catch up with QPSO at later stage and could generate better solutions at the end of search.

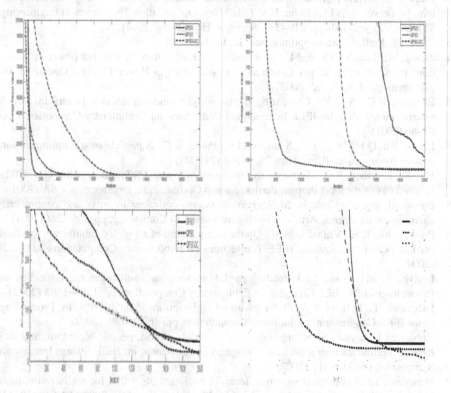

Fig. 1. Convergence process of the three algorithm on the first four benchmark functions with dimension 30 and swarm size 40 averaged on 50 trail runs

From the results above in the tables and figures, it can be concluded that the QPSO-DC has better global search ability than SPSO, QPSO.

Acknowledgements. This work was financially supported by the National Natural Science Fund (No. 71461008, No. 61362016) and the Hainan Province Natural Science Fund (No. 614235), and the Higher School Scientific Research Project of Hainan Province (Hjkj2013-22).

References

1. Kennedy, J., Eberhart, R.: Particle swarm optimization. In: Proc.IEEE Int. Conf. Neural Networks, pp. 1942–1948 (1995); Van den Bergh, F.: An Analysis of Particle Swarm Optimizers. University of Pretoria, South Africa (2001)
2. Clerc, M.: Discrete particle swarm optimization illustrated by the traveling salesman problem. In: Onwubolu, G.C., Babu, B.V. (eds.) New Optimization Techniques in Engineering. STUDFUZZ, vol. 141, pp. 219–239. Springer, Heidelberg (2004)
3. Clerc, M.: Particle swarm optimization. In: ISTE (2006)
4. Zhang, W., Liu, Y., Clerc, M.: An adaptive PSO algorithm for reactive power optimization, In: Sixth International Conference on Advances in Power Control, Operation and Management, Hong Kong (2003)
5. Eberhart, R.C., Shi, Y.: Comparing inertia weights and constriction factors in particle swarm optimization. In: IEEE International Conference on Evolutionary Computation, pp. 81–86 (2001)
6. He, S., Wu, Q.H., Wen, J.Y., Saunders, J.R., Paton, R.C.: A particle swarm optimizer with passive congregation. Biosystems 78, 135–147 (2004)
7. Hu, X., Eberhart, R.C.: Tracking dynamic systems with PSO: where's the cheese? In: Proceedings of the workshop on Particle Swarm Optimization, Indianapolis, USA (2001)
8. Parsopoulos, K.E., Vrahatis, M.N.: Particle swarm optimizer in noisy and continuously changing environments. Artificial Intelligence and Soft Computing, pp. 289–294 (2001)
9. Parsopoulos, K.E., Vrahatis, M.N.: On the computation of all global minimizers thorough particle swarm optimization. IEEE Transactions on Evolutionary Computation 8, 211–224 (2004)
10. LoZvbjerg, M., Krink, T.: Extending particle swarms with self-organized criticality. In: Proceedings of the IEEE Congress on Evolutionary Computation, pp. 1588–1593 (2002)
11. Blackwell, T., Bentley, P.J.: Don't push me! Collision-avoiding swarms. In: Proceedings of the IEEE Congress on Evolutionary Computation, pp. 1691–1696 (2002)
12. Robinson, J., Sinton, S., Rahmat-Samii, Y.: Particle swarm, genetic algorithm, and their hybrids: optimization of a profiled corrugated horn antenna. In: IEEE Swarm Intelligence Symposium, pp. 314–317 (2002)
13. Hendtlass, T.: A combined swarm differential evolution algorithm for optimization problems. In: Monostori, L., Váncza, J., Ali, M. (eds.) IEA/AIE 2001. LNCS (LNAI), vol. 2070, pp. 11–18. Springer, Heidelberg (2001)
14. Poli, R., Kennedy, J., Blackwell, T.: Particle swarm optimization. Swarm Intelligence 1, 33–57 (2007)
15. Sun, J., Xu, W.B., Feng, B.: Particle swarm optimization with particles having quantum behavior. In: Proc. Congress on Evolutionary Computation, pp. 325–331 (2004)
16. Sun, J., Xu, W.B., Feng, B.: A global search strategy of quantum behaved particle swarm optimization. In: Proc. IEEE Conference on Cybernetics and Intelligent Systems, pp. 111–116 (2004)
17. Sun, J., Xu, W.B., Fang, W.: Quantum-behaved particle swarm optimization with a hybrid probability distribution. In: The Proceeding of 9th Pacific Rim International Conference on Artificial Intelligence (2006)
18. Liu, J., Sun, J., Xu, W.: Quantum-Behaved Particle Swarm Optimization with Immune Operator. In: Esposito, F., Raś, Z.W., Malerba, D., Semeraro, G. (eds.) ISMIS 2006. LNCS (LNAI), vol. 4203, pp. 77–83. Springer, Heidelberg (2006)
19. Coelho, L.S.: Novel Gaussian quantum-behaved particle swarm optimizer applied to electromagnetic design. Science, Measurement & Technology 1, 290–294 (2007)

20. Liu, J., Sun, J., Xu, W.B.: Quantum-behaved particle swarm optimization with immune memory and vaccination. In: Proc. IEEE International Conference on Granular Computing, USA, pp. 453–456 (2006)
21. Angeline, P.J.: Using Selection to Improve Particle Swarm Optimization. In: Proc. 1998 IEEE International Conference on Evolutionary Computation, Piscataway, NJ, pp. 84–89 (1998)
22. Shi, Y., Eberhart, R.C.: A Modified Particle Swarm. In: Proc.1998 IEEE International Conference on Evolutionary Computation, Piscataway, NJ, pp. 69–73 (1998)
23. Clerc, M., Kennedy, J.: The Particle Swarm: Explosion, Stability, and Convergence in a Multi-dimensional Complex Space. IEEE Transactions on Evolutionary Computation 6, 58–73 (2002)
24. Shi, Y., Eberhart, R.: Empirical Study of Particle Swarm Optimization. In: Proc. 1999 Congress on Evolutionary Computation, Piscataway, NJ, pp. 1945–1950 (1999)

Modified K-means Algorithm for Clustering Analysis of Hainan Green Tangerine Peel

Ying Luo and Haiyan Fu[*]

School of Information Science Technology, Hainan Normal University, Haikou, China
yanzi78@163.com

Abstract. K-means is a classic, the division of the clustering algorithm, apply to the classification of the globular data. According to the initial clustering center, this paper comprehensive consideration the characteristics of various Hierarchical cluster algorithms and choose the appropriate Hierarchical cluster algorithm to improve K-means, and combined with Hainan Green Tangerine Peel cluster analysis of data which is compared experiments. The results indicate that the improved algorithm have increasing the distance between classes with each others, get a stable of cluster results and better implementation data mining. Finally to summary the two algorithms and the further research direction.

Keywords: clustering, Modified K-means algorithm, initial clustering center, average link.

1   Introduction

The K-means is a classic clustering algorithm which is a data exploration technique that allows samples with similar characteristics to be clustered together in order to facilitate their further processing and usually has applications in identification of patterns hidden in data. Although this algorithm has high flexibility and efficiency, in some case, it has not a good performance for random initialization centre. Owing to trial and error to achieve good representation it spend more memory and CPU time.

To address this issue, some scholar suggested improving algorithm based on other clustering algorithm that has better performance with speed, efficiency and clustering results than previous traditional clustering algorithms when solve some specified problems.

In this paper, a modified K-means algorithm based on two stages strategy is introduced. Modified K-means employed a hierarchical clustering algorithm- average linkage technology- to initialize the centre for the second stage and then uses the traditional K-means for further classification.

2   Related Research

In the traditional K-means, process depends on a near-optimal solution. For the first iteration, the algorithm is initialized randomly. Then, the clustering centers from the

---

[*] Corresponding author.

H. Li et al. (Eds.): I3E 2014, IFIP AICT 445, pp. 144–150, 2014.
© IFIP International Federation for Information Processing 2014

last iteration are used to initialize in subsequent iterations till there is no change in the average value. In worse case, a group unavailable initial clustering center lead to the K-means stops to a local minimum, that it to say, their true distribution in the problem domain will not be reflected by the distribution of objects in the result [1].

Hierarchy clustering is famous because of easy to read, but it not flexible enough due to cannot be operated with samples reversibly. Initially, all genes are considered as individual clusters followed by sequential merging of the two closest clusters in each subsequent step based on their distance, the final step has only one clustering left with all the genes in it [2]. So, this paper argues that, hierarchical clustering is suitable for used to preliminary classification. In lots of representative algorithm, there are three different technology; single linkage, average linkage and complete linkage.

The single linkage technology uses the distance of samples with most similar in different groups. Sometimes, a clustering distribution is non-homogeneous that has a bad impact on the performance of the algorithm. The complete linkage technology, also called furthest neighbor, uses the distance of samples with most dissimilar in two groups and as a complete vision, they have a better performance. However, this technology is susceptible to noise, and unable to get satisfactory results due to few samples which away from the center of group. The third technology is the average linkage technology that uses the average distance between all pairs in different clustering. This algorithm considers the structure of dataset, the most similarity group tend to be merged.

$$d(r,s) = \frac{1}{n_r n_s} \sum_{i=1}^{n_r} \sum_{j=1}^{n_s} dist(x_{ri}, x_{sj}) \tag{1}$$

where r and s are group, there are n_r samples in group r, and n_s samples in group s. Therefore, employed average linkage technology in subset of dataset for achieve initial clustering center to the second stage is available.

3   Algorithm Design

The modified K-means algorithm works in two stages: The first stage makes preliminary classification with the average linkage technology. The second stage uses the clustering centers from the first stage as initial clustering center and employed the traditional K-means algorithm on the further classification.

K is the expected number of clusters, in practice, it is empirically chosen by the user depending on characteristics of the dataset [3].

Algorithm process is as follows:

The First Stage:

Step 1: Choose appropriate function calculate distance metric used for similarity of samples. The details of function choosing are explained in later.

Step 2: The similarity samples x_i and x_j are merged, the first clustering named D1, D1= {x_i, x_j}.

Step 3: Another sample x_p is added to the D1 if the average distance between x_p and samples that in the D1 less than Threshold, and then D1= {x_i, x_j, x_p}.Else, x_p is taken out and a new clustering is made with it.

Step 4: Step 3 are repeated till the number of cluster is equal to k.

The Second Stage:

Step 1: clustering centers from the first stage are initialized.

Step 2: Samples in dataset are assigned to the cluster that the most similar initial centers are.

Step 3: Recalculated clustering average value as centers for the sequential iterative.

Step 4: Step 2 and Step 3 are repeated till there is no change in clusters.

In the first stage, some function are common used for computes the distance metric that defined dissimilarity between samples, such as Euclidean, Hamming distance and so on. The Hamming distance between two samples is the percentage of coordinates that differ.

$$d(x_i, x_j) = \frac{m - \sum_{\alpha=1}^{m} (\sum x_i[v_\alpha] \oplus x_j[v_\alpha])}{m}$$ (2)

where x_i and x_j are two samples in sets $X = \{x_j \mid j = 1, 2, \cdots n\}$, $[v_\alpha]$ is the α-th attribute, and each samples have m attributes and they are discrete.

Clustering Performance Criterion Function

Clustering is aimed to divide a set of samples $X = \{x_1, x_2, \cdots, x_n\}$ into K disjointed subsets X_1; X_2; ...; X_K so that points in the same subset share common properties while points which belong to different subsets do not share these properties. We evaluate the clustering performance using error sum of squares criterion function, it is defined as:

$$ESS = \sum_{i=1}^{K} \sum_{p \in X_i} \|p - m_i\|$$ (3)

where average vector $m_i = \frac{1}{n_i} \sum_{p \in X_i} p \quad i = 1, 2, \cdots, k$.

Algorithm steps are as follows:

input: $X = \{x_1, x_2, \cdots, x_n\}$ samples sets

          K   numbers of clustering

output: $S = \{X_1, X_2, \cdots X_K\}$ the results of clustering

the dissimilarity function: hamming distance

criterion function: error sum of squares criterion function ESS

initial state:

$d = 0$ %initial threshold

$S = \{\{x_1\}, \{x_2\}, \cdots \{x_n\}\}$ %each sample is a group *step*1 : compute similarity of samples, and obtained adjacent matrix C.

*step*2 : examine every samples in X, and emerge the most similar x_i and x_j, the get new group: $D_1 = \{x_i, x_j\}$.

*step*3 : calculate average distance of sample x_p outside of group D_1 and samples in D_1

> *if*   *average* $\leq d$
>
> *then*
> $$D_1 = \{x_i, x_j, x_p\}$$
> *else*
> $$D_2 = \{x_p\}$$
> $$d = d + 1$$

until the number of clustering is equal to K.

*step*4 : obtain the centriod of K group $W = \{\mu_i \mid i = 1, 2, \cdots K\}$.

*step*5 : update centriod, recalculated clustering average value as centriod for the sequential iterative.

*step*6 : *step*5 are repeated until the criterion function - error sum of squares criterion function is constriction.

The first stage is important to the whole process of avoid some worse case which dependent on random initialization of the traditional K-means. While the average linkage technology is completed, lots of samples are in true distribution in the problem domain and reduce the number of iterative in the second stage.

Two parameters are compared during the testing: clustering quality and stability.

4   Comparison for Green Peel Cluster Analysis of Hainan Island

The performance of the modified K-means algorithm is compared with traditional K-means algorithms using our dataset that comes from the experiment of Green Tanger-ine Peel clustering Analysis of Hainan. We select 100 leaves evenly with DNA molecular markers method at five regions (group) in Hainan Island: ShiMeiWan (SMW), SanYa (SY), BaWangLing (BWL), JianFengLing (JFL), and WenChang (WC).

The dataset has 100 arrays, 206 columns (each sample has 206 characteristics) and 15 times are run for each test. All data as follow are the average value.

Two algorithms are run with Inter(R) Core(TM) 2Duo CPU T6500@2.1GHz. and 2.0GB RAM. The operation system is Vista32 and the software is written in MATLAB7.0.

The centroid of the cluster in the second stage of modified K-means is obtained by average linkage technology, and traditional K-means randomly chooses K points to be the initial centroids, we use hamming distance to calculate the dissimilarity.

4.1   Evaluation of Clustering Quality

The goal of clustering is to minimize the intra-cluster distances and to maximize the inter-cluster distances [3]. In order to evaluate the quality of the results of the two algorithms, the distance between centers in two clusters are used to intra-cluster distances while distances between a clustering center and the objects belonging to it are inter-cluster distances. We clustered dataset and get the results of 2-10 numbers clustering.

Table 1. The intra-cluster distances of cluster used in two algorithms

cluster	2	3	4	5	6	7	8	9	10
K-means	0.500	0.599	0.621	0.565	0.606	0.624	0.635	0.635	0.633
AK-means	0.602	0.603	0.605	0.593	0.601	0.609	0.615	0.620	0.633

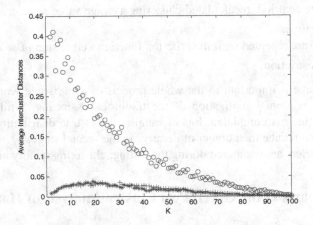

Fig. 1. The "o" is on behalf the average intercluster distances of the K-means algorithm, and the "*" is represent the corresponding of the modified K-means

On the whole, the performance of modified K-means are superior to the traditional K-means algorithm (the average inter-cluster distances are smaller).

4.2   Stability of the Clustering

For compare the stability which is important that reflects the similarity between different runs using the same program of each algorithm, we need a statistic. In the K-means algorithm, based on iterative till there is no change in each cluster, at the same time, the value of the total amount of distortion wills no change more. The distortion of a cluster which describes the dispersion degree in the distribution of the samples is defined as the sum of the squared Hamming distances between its centre and the objects belonging to it.

$$S = \sqrt{\frac{\sum \left(ESS_i - \overline{ESS} \right)^2}{m-1}} \qquad (4)$$

Standard deviation of distortion in test using the same program could distribution the range of distortion. If vary of standard deviation of distortion are wildly, the clustering results are unstable, and vice versa. For comparison, we have run 12 times for each and recorded the result.

Table 2. Comparison of standard deviation of distortion to complete analysis of two clustering algorithms

cluster	2	3	4	5	6	7	8	9	10
K-means	2.499	1.929	2.334	1.606	1.255	0.654	0.358	0.178	0
AK-means	0	0	0	0	0	0	0	0	0

The table 2 shows, as we can see, the first stage of the modified K-means benefited from the average link technique performs (the standard deviation of distortion =0) better than the corresponding (which is>0). Namely, the stability of cluster is improved.

Significantly, the modified K-means clustering algorithm is better than the traditional K-means in time taken, clustering quality, and stability of algorithm.

5   Summary

The modified algorithm employed a hierarchical clustering algorithm- average linkage technology- to initialize the centre avoid a bad clustering results in finally by samples which are in the edge of the group as center in the first iterative in the traditional K-means algorithm. At the same time, thanks to reducing the influence by noise, the modified algorithm having improve the Evaluation of clustering quality in a certain degree. Besides, random selection of the initial clustering center lead to the algorithm stops in a local minimum and achieve different results in many tests and modified K-means is an available strategy in this solution.

However, as the K-means, the modified K-means control the border of clustering with diameter of semi-sphere, if the clustering is not spherical, two algorithms will not work very well.

Acknowledgements. This work is supported by NSF of China (70940007 and 71461008).

References

1. Jin, W., Chen, H.P.: A k-means algorithm based on Hierarchical clustering. Journal of Hohai University-Changzhou 1, 7–10 (2007)
2. Lin, J.B., Liu, M.D., Chen, X.: Data mining and OLAP theory and practice. Tsinghua university press, Beijing (2003)
3. Hu, R.F., Yin, G.F., Tan, Y.: A Hybrid Clustering Algorithm and It's Application. Journal of Sichuan University: Engineering Science Edition 5, 68–73 (2006)
4. Wang, H.R., Zhao, L.M., Pei, J.: Equilibrium Modified K-Means Clustering Method. Journal of Jilin University(Information Science Edition) 2, 41–44 (2006)
5. Velmurugan, T.: Performance based analysis between k-Means and Fuzzy C-Means clustering algorithms for connection oriented telecommunication data. Applied Soft Computing 19, 134–146 (2014)
6. Wan Mohd, W.M.B., Beg, A.H., Herawan, T., Rabbi, K.F.: An Improved Parameter less Data Clustering Technique based on Maximum Distance of Data and Lioyd k-means Algorithm. Procedia Technology 1, 367–371 (2012)

Perceived Significance of Improved Patient Data and Healthcare Services in the Formation of Inter-organizational Healthcare IT Governance

Tomi Dahlberg

Turku School of Economics At University of Turku, Finland
tomi.dahlberg@utu.fi

Abstract. IT is seen as the means to develop healthcare and social welfare services for citizens and to improve the quality of medical and social welfare data. This is deemed to require better IT cooperation between organizations. My research investigates the formation of voluntary inter-organizational IT governance in healthcare and social welfare IT engaging over 100 organizations. Attention is placed especially on the perceived benefits of IT governance. Results suggest that concrete benefits are necessary for the formation of inter-organizational IT governance arrangements. The anticipated benefits of patient data improvement and short-term benefits were evaluated more important than the expected long-term improvements of healthcare and social welfare reforms.

Keywords: Healthcare IT, Social welfare IT, IT Governance.

1 Introduction

Healthcare and social welfare professionals, national governments and international organizations consider questions such as: How to arrange and provide healthcare and social welfare services for citizens? How to improve the integration and quality of patient / customer data? How to govern and manage information technology (IT) and patient / customer data as a part of healthcare and social welfare services?

Societal developments, especially the aging of population, make it necessary to consider reforms to national healthcare and social welfare systems [18]. Data mining, analysis, visualization etc. techniques offer huge potentials to improve healthcare and social welfare services [17]. At the same time the fragmentation and poor quality of data and data storages as well as lack of message and data model standards hamper electronic data transfer and the deployment of these potentials [11]. IT-enabled services are seen as one key element both for reforms and for data quality improvements, at the same time as healthcare and social welfare IT faces several "make ends meet" type challenges that limit the potential of IT [26].

According to an UN 2012 estimate [20], persons older than 60 years accounted for 11 per cent of the world's population. During the next four decades, the proportion of older population is projected to increase to 22 per cent [20]. Within the European Union (EU), including the country of this study, the demographic changes with longer

H. Li et al. (Eds.): I3E 2014, IFIP AICT 445, pp. 151–163, 2014.

life expectancies and lower birth rates have dramatic effects. In EU, dependency ratio is estimated to drop from 1:4 to 1:2, cost of (health and social) care to increase with 4-8 % of GDP and work force to shrink with 20 million by 2025 [26]. Increasing demand for healthcare and social welfare services is deemed to require coordinated activities and better information sharing between social welfare, basic healthcare and specialized medical care services. The same is true for all groups with disabilities and special needs [12]. IT is seen to support reforms in two ways. Automation of routines, integration of information systems (IS), and the usage of data storages and electronic data transfer are expected to produce better services. Interoperability, standards and cooperation between organizations are needed for this. Secondly, new IT-based services are called for so that citizens are able to take more responsibility for their well-being and for living more healthily, are empowered to active aging with new types of work arrangements and longer participation to the activities of a society [26]. Ease of use and new health and social care models are necessary for this. Economic growth, markets and novel business models are needed to support needed innovations [26].

At the moment, fragmentation and inconsistency of healthcare and social welfare data appears to be more the norm than the exception [12], [17]. For example, out of the 194 member states of World Health Organization (WHO) only 34 members were able to provide reliable health data in 2012 [12]. A major reason for this is the way, how healthcare and social welfare service providers run their patient/customer ISs and databases [12]. They register medical measurements, events, radiographs, prescriptions, case summaries, etc., social welfare service actions, decisions etc., and other data into their unique databases. Data sharing remains challenging, even when organizations have purchased the same IS from the same vendor. Organization-specific implementations result in differences in data models, in data coding schemes and in data handling practices, which then prevent electronic data sharing [17]. Data structures and messages need to be standardized together with the related identification and authentication services and data transfer interfaces. Also data collection and processing practices need to be standardized to the extent of producing consistent data.

Organizational silos and limited coordination contribute to the existence of "making end needs" type challenges of healthcare and social welfare IT. Resources are insufficient as organizations develop similar IT solutions independently. Possibilities to recruit highly skilled specialists and to purchase rare IT assets or skills are limited due to financial, availability, size etc. constrains. The power of one organization is little in negotiations with IT service vendors as compared to pooled purchases and negotiation power. The absence of established inter-organizational (IT) cooperation and governance arrangements limits cooperation to experience exchange instead of resource sharing and joint IT service development and operations.

The usage of IT to reform national healthcare and social welfare systems, to improve data quality and interoperability, and to solve the "making ends meet" challenges of IT resourcing lead to the same conclusion. Better inter-organizational IT cooperation and governance is needed. IT governance bodies should define clear objectives for IT cooperation, direct cooperation by agreeing accountabilities for plans, activities etc., and monitor cooperation by reviewing outcomes. National governments typically emphasize reforms' long-term benefits and impacts on data quality. Local professionals probably focus more on short-term service and data interoperability benefits and IT professionals may also wish to solve IT-specific challenges.

The current IT governance research addresses IT governance mainly within one organization [21], [23] with a few case studies as the exceptions. These case studies describe inter-organizational IT governance between related business units or partners in domestic [4, 5] or in global company [7] contexts. My research examines the formation of voluntary inter-organizational IT governance engaging a large number of independent healthcare and social welfare organizations. Research has also examined the benefits of IT governance limitedly. Financial metrics such as cost, revenue and return on assets have been investigated [23]. My study draws on the resource based view (RBV), transaction cost economics (TCE) and social network theories to complement the theoretical basis. They address alliances (=inter-organizational governance) as a form of governance [6], [26], [28], have been used in prior IS research [e.g. [19], [24] and have rich metrics for the benefits of inter-organizational governance.

To summarize, with this research I investigate the formation of voluntary inter-organizational IT governance between legally independent organizations. Special attention is paid to IT cooperation benefits during the formation of an IT governance arrangement. The two research questions are: What is the significance of IT cooperation benefits for the formation of inter-organizational IT governance? Secondly, how important are short-term benefits in healthcare and social welfare patient / customer data and in "making ends meet IT" as compared to long-term service improvements?

Section 2 explains how the objectives of this research were achieved and the research questions answered by theoretically discussing IT and health information governance, RBV, TCE and social network theory. Section 3 describes how that was done through a case study with a related expert survey. Results from the empirical case study and the survey are shown in Section 4. Finally, in Section 5, the theoretical and empirical findings of this study are discussed and conclusions are drawn.

2   Theoretical Background

This study follows the social network theory definition of inter-organizational (IT) governance [14]: "Network governance involves a select, persistent, and structured set of autonomous firms engaged in creating products or services based on implicit and open-ended contracts to adapt to environmental contingencies and to coordinate and safeguard exchanges. These contracts are socially—not legally—binding." The case I investigated has all characteristics of the definition although "firms" are nonprofit public sector organizations. The establishment of (IT) governance is one phase in its lifecycle, also called the formation of an alliance [6]. Literature is reviewed to understand, why do organizations decide to form alliances, such as formal IT governance arrangements? What benefits do participating organizations expect from that? Since IT governance research on inter-organizational governance is scarce, constructs from RBV, TCE and social network theory are reviewed to answer these questions.

IT Governance and Health Information Governance research: As explained above, prior studies have examined mainly intra-organizational IT governance. However, [4] and [7] investigated IT governance arrangements within single-organization multi-unit and [5] within value network contexts. Value networks included single buyer – multi-vendor (contractual) and single service organizer – multi-partner (consensual) arrangements. The mentioned articles suggest that the formation of both intra-organizational and inter-organizational IT governance follow rather similar practices. De Haes et al. [7] explain how IT governance structures, processes and cooperation mechanisms could be extended from intra-organizational to inter-organizational settings. The generic purpose of IT governance is to "enable both business and IT to execute their responsibilities in support of business and in the creation of business value from IT investments". [7] It is noteworthy that IT governance benefits are classified into business (e.g. healthcare and social welfare services and better patient / customer data) and into IT performance benefits. Business and IT performance benefits are also built in into IT governacne best practice methods, especially COBIT.

The benefit proposition is that "IT governance effectiveness is positively associated with organizational performance" [16]. Mohamed et al. [16] note that the benefits of IT governance consist of financial, customer, learning and growth, and of internal and business process benefits (=the 4 dimensions of the balanced scorecard). Weill et al. investigated the business performance impacts of alternative IT governance arrangements with three types of measures; cost-efficient use of IT, business profitability and revenue growth [23]. Huang et al. [13] classified the IT performance impacts of IT governance into the efficiency of IT use, the breath of current IT use and the breath of potential IT use. Bradley et al [2] studied IT governance outcomes in US hospitals by dividing them into social risk management, technical risk management, market responsiveness, external relationship management and operational IT effectiveness. In summary, the current IT governance literature considers rather thinly, why organizations cooperate and what are the benefits of inter-organizational IT governance.

The definitions of health information governance [17], [11] differ slightly from those used in IT governance research and show how elusive the IT governance concept is. Definitions emphasize "standardizing the management, policy design and risk management associated with information". Hovenga [12] investigated the versatile beneficial impacts of health information governance on one nation's (Australia) healthcare system. Potential benefits include improved evaluating and comparing of health outcomes from transitioning to digital environment, and better sustainability of heath systems. Other potential benefits are: improved leadership and governance (national strategic policy framework, health process management and strategic health policy frameworks); financing of healthcare (health insurance versus funding, terminology and coding systems etc.); health workforce (planning of roles, occupational categories, workforce registration, licensing etc.); medical products, devices and technologies (device and technology types, medical device regulations, medical device nomenclature, etc.); health service delivery types and regulations (harmonization of service delivery data, accessing healthcare services, planning and managing resources and health service delivery, epidemiology issues etc.); and information and research (for leaders and governors, other key decision makers, professionals, researchers etc.)

RBV, TCE and Social Network Theory research: RBV, TCE and social network theory [25] regard single organizations (vertical governance in TCE), markets (market governance in TCE) and networks / alliances (hierarchical or relational governance in TCE) as alternative governance models. (Strategic) alliances are voluntary cooperative inter-organizational agreements that aim to produce competitive advantages for alliance partners [6]. Within the context of this research, voluntary inter-organizational IT governance is seen as an arrangement to agree how to govern (the management of) IT as an alliance with the motive to provide better healthcare / social welfare services for citizens through the means of IT and digital data deployment.

Barney [1] introduced the work of Penrose into IS research. RBV sees each organization unique [1], [24], as an equivalent to the broad set of tangible and intangible resources that it owns (semi-)permanently [9], [24]. Intangible resources include IT and information, and the usage arrangements of resources. The resources of the organization as a whole, and especially imperfectly mobile, non-imitable and non-substitutable resources determine the organization's value creation potential. RBV proposes that inter-organizational cooperation happens, if cooperation provides win-win value for the participants by pooling, aggregating, sharing and exchanging their valuable unique resources, and if such value cannot be better achieved in other ways.

Barney proposed that four properties of (IT) resources make them unique, valuable, rare, imperfectly imitable and non-substitutable [1]. Das et al. [6] treat mobility, imitability and substitutability as the basic unique properties of resources. Resources have also other attributes. They can be property based (human, intellectual or physical) or knowledge based (organizational, technological or managerial). Resources can be supplementary, complementary, surplus or wasteful. Melville et al. [15] classify IT resources into human and technical. According to Wiengarten et al. [24] IT resources are typically complementary to an organization's business resources. They can be complementary to other assets, capabilities, organizational processes, firm attributes, information, knowledge, etc. controlled by the organization. IT resources can thus be complementary both to non-IT resources and to other IT resources.

Wiengarten et al. [24] propose that IT resources provide value to an organization / alliance: "(1) When IT resources are aligned, integrated or coupled with organizational processes such as knowledge processes, or management processes, synergies and higher-order capabilities can be created. (2) Moreover, these higher-order capabilities may result in long-term performance improvement. (3) To develop these long-term performance enhancement capabilities, a firm may need to redesign its processes in terms of its employed resources (i.e. human resources and facilities)." Park et al. [19] classify the outcomes of resource properties as: (1) conserve resources, (2) share risks, (3) obtain information, (4) access complementary resources, (5) reduce product development costs, (6) improve technological capabilities and (7) enhance reliability.

TCE introduced by Coase was developed further and operationalized by Williamson [25]. He introduced asset specificity, uncertainty and transaction frequency as the determinants by which firms (=organizations) choose to make, buy or ally while they execute transactions. Geyskens et al. [9] define relational governance (alliance) as:

"Governance modes characterized by the parties to a transaction jointly developing policies directed toward the achievement of certain goals." The voluntary inter-organizational IT governance arrangement investigated here is relational in its nature.

Geyskens et al. [9] provide operationalized construct definitions and metrics for asset specificity, volume uncertainty, technological uncertainty and behavioral uncertainty. The metrics of asset specificity include organization nature, (healthcare / social welfare) product and services, customer (patient) complexity, need of training to use assets etc. Volume uncertainty metrics contain expected volume fluctuation and uncertainty of volume estimates, and the stability of industry (e.g. healthcare and social welfare) volumes, ability to predict trends, etc. Technology uncertainly is captured with such metrics as technology improvement and specification changes. The metrics of behavioral uncertainty include buyer expectations, (healthcare and social welfare IT services) seller competition, price quality in relation to seller competition and the importance on non-selling activities (e.g. support to IT services). Ability to codify transactions (for example, patient/customer data) and task complexity (for example, the design of enterprise architecture) are other metrics of asset specificity. Similarly, intention to conduct R&D (for example, the development of novel IT-enabled self-diagnostic healthcare services to citizens) is another technology uncertainty metric.

Geyskens et al. [9] divided the performance impacts (benefits) of TCE into cost inclusive and into cost exclusive categories. Levels and changes of revenues and costs, abnormal returns on assets (e.g. IT investments) are cost inclusive metrics. Levels and changes of activities are cost exclusive. Other metrics used so far include speed of development (e.g., time needed to execute healthcare and social welfare IT projects), and level and changes in quality (for example, in the quality of patient / customer data) [8]. Wang et al. [22] investigated the impacts of information visibility (assets specificity), inter-organizational supply chain flexibility (behavioral uncertainty) and integration (technological uncertainty) on supply chain performance. They used three measures for outcomes; fit of work practices between organizations, fit of work outcomes between individuals and fit in routines between organizations [22, appendix].

The social network theory started to investigate the social embeddedness of institutions with the so-called structural macro-level studies [10]. Markets are seen to consist of social rather than economic structures [10], and to contain strong or weak ties between people and organizations [28]. Social network theory proposes that an organization perceives the world through a network with a structural rather than an individual lens. Such perception gives the organization a more complete, but not necessary only view to reality. A network, such as healthcare and social welfare IT cooperation, offers opportunities and constraints to participants, which then guide actions taken.

Embeddedness of organizations is the key construct of social network theory. Embeddedness impacts actions that organizations take in their responses to markets. The longer and the tighter the relationships between organizations, the more embedded they are [28]. Zaheer et al. [28] discovered that organizations in tight networks are likely to take different actions than traditional un-networked organizations. Social capital and structural holes are other key constructs. Social capital describes the value of connections between organizations. Structural hole means the lack of a straight connection between two organizations. Social network governance uses social mechanisms – cooperation mechanisms in traditional IT governance terminology - to guide actions and to solve network problems. Jones et al. [14] identified four social mechanisms; restrictions to access the network, collective sanctions, use of social memory and cultural processes. Restricted access and cultural processes are used to coordinate network operations. Collective sanctions and social memory are used to safeguard network operations. Brass et al. [3] classified networks on the basis of following characteristics: strategic alliance and collaboration, flow of information, affect or friendship, good and services work flow, influence or advice, and overlapping group memberships. Organizations form and join networks motivated by their needs, such as responding to market challenges (e.g. reforms) and need to development services.

Antecedents to collaborate vary at personal, inter-unit (internal units of an organization) and inter-organizational (between organizations) levels. Actor similarities, personalities, proximity and organizational structure, and environmental factors are personal level antecedents. Interpersonal ties, functional ties and organizational processes, and control mechanisms are inter-unit antecedents. Personal and inter-unit antecedents are valid also on inter-organizational level. Motives (new resources, uncertainty reduction, bigger legitimacy, common goals), learning about an industry or about networking, relational trust, common norms and monitoring, equity between partners and context affecting all participating organizations constitute inter-organizational antecedents to collaborate [3]. Networks transmit information to equalize common knowledge and share innovations, mediate issues between organizations and individuals and share resources and power among network participants.

3   Case and Expert Survey

The researcher acted as a consultant during the formation of the investigated IT governance arrangement. Thus, reflective observation approach in a single case context is used. Guidelines for case studies [27] and for the building of research constructs are followed. For research, I have access to the full range of the case material. Volumewise workshop materials represent the largest proportion of empirical data. Workshop materials range from RACI charts to discussion notes; from agendas to thick pre-reading documents; and from draft revisions of the Law on Arranging Social Welfare and Health Care Services to the drafts and final memorandums of the project. Empirical data includes also memos, emails, presentation slides, and answers to Q/A.

The case covers the formation of voluntary inter-organizational IT governance for healthcare and social welfare IT in one of Finland's five Special Catchment Areas. The arrangement engages over 100 organizations; 68 cities, towns or municipalities, 5 healthcare districts, 9 hospitals including one university central hospital, 33 healthcare centers and 5 social welfare development districts. The idea was to use this case as the pilot for the other four areas and to provide input to law reform work. No IT governance arrangement had existed earlier between the organizations. The IT governance arrangement also consolidated the IT services of specialized medical care, basic health care and social welfare services for the first time.

The project group that accomplished the task during 9 months consisted of 17 persons. They included the chief medical officer of one healthcare district, three healthcare district CIOs, the CIO of a major city, two enterprise architects, three specialist from social welfare development centers and the National Institute for Health and Welfare, one IT expert from another Special Catchment Area, two senior advisors from the Ministry of Social Affairs and Health, two senior advisors from the Association of Local and Regional Authorities plus the researcher as a consultant. The proposal of the project was accepted in February 2014 and its implementation started immediately. In March 2014, the Government of Finland decided to give the responsibility for the arrangement of healthcare and social welfare services to five "Special Catchment Areas". In June 2014, all political parties agreed the same. The proposal accepted in February is in compliance with these national level decisions.

The definition of concrete IT cooperation benefits was one task accomplished by the project. Nine concrete benefits were defined. Need to improve patient / customer data interoperability and data quality was considered the most burning challenge in the current status. Four additional specific data related benefits were defined for this reason. The 13 statements are shown in appendix 1. Also the connections between each statement and the theoretical constructs reviewed in Section 2 are shown.

A web-enabled self-administered survey was conducted to offer leading healthcare and social welfare experts an opportunity to evaluate the defined benefits and other aspects of the proposed IT governance arrangement prior its acceptance. This study uses only that part of the available data that concentrates on IT governance benefits. In addition to the IT governance benefit statements, the survey included demographic, situational and behavioral control variables also shown in Appendix 1. The survey items of IT cooperation benefits and the behavioral control variables were formulated into statements. Respondents were asked to evaluate each statement on a 7-point Likert scale from totally disagree (=1) to totally agree (=7).

An invitation to participate to the survey was emailed to 260 healthcare and social welfare experts throughout the country. After one reminder 68 responses (26 % response rate) were received. Survey respondents had the following characteristics:

- Slightly over half (37) of the 68 responses were from the geographic area of the case.
- Seven respondents participated to the formation of the inter-organizational IT governance arrangement. Statistical comparisons to other respondents were not made.
- Slightly over half (53) of the respondents worked in healthcare districts, 37 % in cities, towns or municipalities, and 10 % in other organizations.
- 66 % were executives and managers and 34 % were experts.
- 43 % were healthcare or social welfare managers or executives and 12 % specialists. CIOs and IT managers accounted for 23 % and IT specialists for 22 %.
- Healthcare and social welfare executives and managers had worked in those positions on average for 15.9 years, experts for 12.7 years. Healthcare and social welfare CIOs and IT managers had worked in those positions on average for 7.2 years and IT experts for 8.8 years.

The expectation drawn from the theoretical background was that expert evaluations would be high. On the other hand, no expectations were made regarding which benefits would be evaluated most important. Due to the limited number of responses multivariate statistical methods, including structured equation models, could not be used. Statistical analysis on the differences in the means of survey items was done with the two-tailed Stu dent t-test. For statistical analyses responses to behavioral control variables were classified into agree (values 5-7) and non-agree (values 1-3) classes.

4   Results

The need for IT cooperation by pooling, sharing and exchanging IT resources with envisioned concrete benefits was decisive for the willingness to participate to the formation of the IT governance arrangement. The project started with an effort to rely solely on IT governance knowledge. That approach proved insufficient in terms of participants' and organizations' commitment. One of the CIOs coined the problem: "Although I understand the governance matrix with RACI roles that we crafted..., that has been totally useless in my organization. Doctors do not understand from the matrix the benefits of IT cooperation. They ask me, why should we use time to consider inter-organizational IT governance when there are so many burning IT issues in our own organization." Improved quality of patient / customer data was also regarded important. At the dissemination presentations of the drafted IT governance model one medical director stated: "Actually, I only want to know, when will I get reliable, consistent and holistic patient and medical data from all the data sources we have."

As expected the evaluations of IT governance benefits were very high on the scale from 1 (low value) to 7 (high value) as Table 1 shows. More specific results are:

1. Patient / customer information benefits were considered more important than generic IT governance benefits.
2. Making ends meet benefits were considered more important than the long-term improvements of reforms to healthcare and social welfare services. The background documents used in the project discuss the necessities to improve the national healthcare and social welfare systems. Those appear to remain abstract in comparison to concrete needs to improve services and related IT. Alternatively it is possible that the solving of concrete challenges is seen to preside reforms.
3. None of the demographic, situational or behavioral control variables were related to variations in the means of IT governance benefits in statistically significant way (t-test). Thus, in this survey, IT governance benefits were evaluated similarly independent of geographic location, profession, professional status, length of experience, attitudes toward IT or attitudes toward the use of IT within respondent's organization. Table 1 shows averages by professional orientation.

Table 1. Results of the expert survey

Survey Item	Mean	Median	Proportion of strongly agree	Mean, Biz experts	Mean, IT experts
Generic IT Governance Benefits: Improved Healthcare and Social Welfare Services (Resulting from Reforms) - - Inter-Organizational Well-Organized Cooperation in Healthcare and Social Welfare IT Is Needed in Order to ...					
Increase the interoperability of patient/customer information systems and data storages	6.3	6.5	86.8%	6.3	6.3
Ensure ability to participate to the national level development of healthcare and social welfare services	5.8	6.0	72.1%	5.8	5.7
Ensure the availability of equal healthcare and social welfare services everywhere in the country	5.7	6.0	67.6%	5.8	5.5
Generic IT Governance Benefits: Bnefits from Better Usage of Resources (Making Ends Meet) - Inter-Organizational Well-Organized Cooperation in Healthcare and Social Welfare IT Is Needed in Order to ...					
Avoid the development of overlapping and difficult to integrate IT services	6.4	7.0	86.8%	6.4	6.4
Create enterprise architectures	6.1	6.0	79.4%	6.2	6.0
(Co-)source IT-services cost-efficiently and effectively	6.0	6.0	75.0%	6.2	5.7
Implement national level healthcare and social welfare IT services efficiently and effectively in the area	6.0	6.0	75.0%	6.2	5.9
Use IT resources and assets efficiently and effectively	5.7	6.0	70.6%	5.7	5.8
Ensure access to specialized capabilities and competencies everywhere in the area	5.8	6.0	69.1%	5.9	5.7
IT Governance Benefits from Improved Patient Data - The development of patient / customer information systems and data storages requires ...					
Tighter cooperation on national level	6.4	7.0	89.7%	6.4	6.3
Tighter cooperation on regional level (this Special Catchment Area)	6.3	6.5	85.3%	6.4	6.1
The creation of jointly agreed data models and sticking to them	6.3	7.0	85.3%	6.4	6.3
Tighter cooperation between healthcare and social welfare	6.3	7.0	79.4%	6.3	6.4

5   Discussion and Conclusions

This research shows that IT governance knowledge alone was insufficient for the formation of inter-organizational IT governance arrangement in the investigated case. Results suggest that need to cooperate with concrete benefits is critical to the formation of inter-organizational IT governance. This is the answer to the first research question presented in section one. Improvements to patient / customer data were perceived more important than long-term improvements to healthcare and social welfare services. The solving of making ends meet issues were also evaluated more important than long-term improvements to healthcare and social welfare services. Additional studies are necessary to verify these results, as this research is a single case study.

This research opens up news venues for IT governance and health information governacne research, for which RBV, TCE and social network theories offer rich and theoretically proven basis. The mentioned research traditions address three governance arrangements, single organization, markets and alliances / networks. So far IT and health information governance research has investigated mainly single organizations. IT governance arrangements could be compared in various contexts such as IT or data outsourcing including cloud services, supplier-customer networks, value chain orchestration or virtual organizations. Health information governance is by nature largely inter-organizational. Linking of inter-organizational governance research approaches to this context is another possible research direction.

The advice to practitioners is to pay serious attention to concrete level IT governance benefits when inter-organizational IT governance arrangements are designed. In national reforms of healthcare and social welfare sectors, concrete advancements in the quality of patient / customer data have a key role. It is especially important to make the long-term improvement objectives of reforms operational and concrete in order to entice experts to take necessary development efforts. This probably requires that long-term healthcare and social welfare reform objectives, such as the empowerment to active aging, be turned into measurable short-term milestones.

References

1. Barney, J.B.: Firm Resources and Sustained Competitive Advantage. J. Manage. 17(1), 99–120 (1991)
2. Bradley, R.V., Byrd, T.A., Pridmore, J.L., Thrasher, E., Pratt, R.M., Mbarika, V.W.: An Empirical Examination of Antecedents and Consequences of IT Governance in US Hospitals. J. Inf. Technol. 27(3), 156–177 (2012)
3. Brass, D.J., Galaskiewicz, J., Greve, H.R., Tsai, W.: Taking Stock of Networks and Organizations: a Multilevel Perspective. Acad. Manage. J. 47(6), 795–817 (2004)
4. Croteau, A.-M., Bergeron, F.: Interorganizational Governance of Information Technology. In: Proceedings of the 42nd Annual Hawaii International Conference on System Sciences (HICSS 2009), 8 pages. IEEE Press, New York (2009)
5. Croteau, A.-M., Bergeron, F., Dubsky, J.: Contractual and Consensual Profiles for an Inte rorganizational Governance of Information Technology. International Business Research 6(9), 30 43 (2013)
6. Das, T.K., Teng, B.-S.: A Resource-Based Theory of Strategic Alliances. J. Manage. 26(1), 31–61 (2000)
7. De Haes, S., Van Grembergen, W., Gemke, D., Thorp, J.: Inter-Organizational Governance of Information Technology: Learning from a Global Multi-Business-Unit Environment. International Journal of IT/Business Alignment and Governance 3(1), 27–46 (2013)
8. Gereffi, G., Humphrey, J., Sturgeon, T.: The governance of global value chains. Rev. Int. Polit. Econ. 12(1), 78–104 (2005)
9. Geyskens, I., Steenkamp, J.-B.E.M., Kumar, N.: Make, Buy, or Ally: A Transaction Cost Theory Meta-Analysis. Acad. Manage. J. 49(3), 519–543 (2006)
10. Granovetter, M.S.: Economic Action and Social Structure: The Problem of Embeddedness. Am. J. Sociol. 91, 481–510 (1985)
11. Hovenga, E.J.S.: National Healthcare Systems and the Need for Health Information Governance. In: Hovenga, E.J.S., Grain, H. (eds.) Health Information Governance in a Digital Environment. Studies in Health Technology and Informatics, vol. 193, pp. 3–23. IOS Press, Amsterdam (2013)
12. Hovenga, E.J.S.: Impact of Data Governance on a Nation's Healthcare System Building Blocks. In: Hovenga, E.J.S., Grain, H. (eds.) Health Information Governance in a Digital Environment. Studies in Health Technology and Informatics, vol. 193, pp. 24–66. IOS Press, Amsterdam (2013)

13. Huang, R., Zmud, R.W., Price, L.R.: Influencing the Effectiveness of IT Governance Practices through Steering Committees and Communication Policies. Eur. J. Inform. Syst. 19(16), 288–302 (2010)
14. Jones, C., Hesterly, W.S., Borgatti, S.P.: A General Theory of Network Governance: Exchange Conditions and Social Mechanisms. Acad. Manage. Rev. 22(4), 911–945 (1997)
15. Melville, N., Kraemer, K., Gurbaxani, V.: Information Technology and Organizational Performance: An Integrative Model of IT Business Value. MIS Quart. 28(2), 283–322 (2004)
16. Mohamed, N., Singh, G.: A Conceptual Framework for Information Technology Governance Effectiveness in Private Organizations. Information Management & Computer Security 20(2), 88–106 (2010)
17. Nahar, J., Imam, T., Tickle, K.S.: Issues of Data Governance Associated with Data Mining in Medical Research: Experiences from an Empirical Study. In: Hovenga, E.J.S., Grain, H. (eds.) Health Information Governance in a Digital Environment. Studies in Health Technology and Informatics, vol. 193, pp. 332–361. IOS Press, Amsterdam (2013)
18. Obi, T., Auffret, J.-P., Iwasaki, N.: Aging Society and ICT: Global Silver Innovation. IOS Press, Amsterdam (2013)
19. Park, N.K., Mezias, J.M., Song, J.: A Resource-based View of Strategic Alliances and Firm Value in the Electronic Marketplace. J. Manage. 30(1), 7–27 (2004)
20. United Nations: World Population Prospects: The 2012 Revision. United Nations, Department of Economic and Social Affairs (2014) Downloadable open data excel spreadsheet, http://esa.un.org/wpp/WPP2012_POP_F13_A_OLD_AGE_DEPENDENCY_RATIO_1564.XLS
21. Van Grembergen, W., De Haes, S.: Implementing Information Technology Governance: Models, Practices and Cases. Idea Group Global, Hershey (2008)
22. Wang, E.T.G., Wei, H.-L.: Interorganizational Governance Value Creation: Coordinating for Information Visibility and Flexibility in Supply Chains. Decision Sci. 38(4), 647–674 (2007)
23. Weill, P., Ross, J.: IT Governance: How Top Performers Manage IT Decision Rights for Superior Results. Harvard Business School Press, Boston (2004)
24. Wiengarten, F., Humphreys, P., Cao, G., McHugh, M.: Exploring the Important Role of Organizational Factors in IT Business Value: Taking a Contingency Perspective on the Resource-Based View. Int J. Manag. Rev. 15(1), 30–46 (2013)
25. Williamson, O.E.: The Economic Institutions of Capitalism. Free Press, New York (1985)
26. Wintley-Jensen, P.: EU Activities on Aging. In: Obi, T., Auffret, J.-P., Iwasaki, N. (eds.) Aging Society and ICT: Global Silver Innovation, pp. 180–187. IOS Press, Amsterdam (2013)
27. Yin, R.K.: Case Study Research, Design and Methods, 4th edn. SAGE publications, Thousand Oaks (2009)
28. Zaheer, A., Gözübüyük, R., Milanov, H.: It's the Connections: The Network Perspective in Interorganizational Research. Acad. Manage. Perspect 24(1), 62–77 (2010)

Appendix 1

Operational definitions of IT governance benefits used also as survey items with their references to theoretical constructs reviewed as the theoretical basis of the research

Construct	SURVEY ITEM (Evaluate the following statements)	Construct in IT Governance Research - Benefits of IT Governance	Construct in RBV - Properties of Resources and Their Outcomes	Construct in TCE - Properties of Transactions and Their Outcomes	Construct in Social Network Theory - Properties of Social Ties and Their Outcomes	REFERENCES
	Inter-organizational well-organized cooperation in healthcare and social welfare IT is needed in order to					
	Avoid the development of overlapping and difficult to integrate IT services	Cooperation mechanism, IT governance matrix	Valuable, surplus, wasteful, conserve resources	Asset specificity-product and services	Good & service workflow, tight relations, control mechanism	(2,13,16,21,23), [1,6,19], (25), (3,28)
	Increase the interoperability of patient/customer information systems and data storages	Health information governance	Non-substitutable, mobile, complementary to non-IT, synergies	Asset specificity-information visibility, technology uncertainty-technology improvement	Tight relations, flow of information, motives-new resources & common goal	(2,11,12), [1,6,19,24], (25), (3,28)
	Create enterprise architectures	Key IT decision areas	Non-substitutable, knowledge based-technological, higher-order capabilities	Asset specificity-task complexity, behavior uncertainty-inter-organizational flexibility	Cultural process, strategic alliance and collaboration, motives-new resources	(2,23), [1,6,15,19,24], (8,9,22), (3,14,28)
	(Co-)source IT-services cost-efficiently and effectively	IT Unit Costs	Non-substitutable, supplementary, complementary to IT, conserve resources	Asset specificity-product and services, behavior uncertainty-seller competition	Strategic alliance and collaboration, motives-bigger legitimacy, opportunity set	(2,23), [1,6,19,24], (9), (3,14,28)
	Implement national level healthcare and social welfare IT services efficiently and effectively	Health information governance	Non-imitable, knowledge based-technological, share risks	Asset specificity-task complexity, technology uncertainty-specification	Social capital, structural holes, access the network, motives-common goal	(2,11,12,13,16), [1,6,15,19], (9,25), (3,10,14,28)
	Ensure ability to participate to the national level development of healthcare and social welfare services	Cooperation mechanism, resource availability	Rare, property based-intellectual, share information	Asset specificity-task complexity, technology uncertainty-specification	Social capital, structural holes, access the network, motives-common goal	(2,5,16,21), [1,6,19], (9,25), (3,10,14,28)
	Use IT resources and assets efficiently and effectively	IT Unit Costs, ROA	Non-imitable, supplementary to IT, access complementary resources	Transaction frequency, behavior uncertainty-seller competition	Strategic alliance and collaboration, motives-bigger legitimacy, opportunity set	(2,13,16,23), [1,6,19,24], (9,25), (3,14,28)
	Ensure access to specialized capabilities and competencies everywhere in the area	Resource availability	Rare, property&knowledge based, access complementary resources	Asset specificity, transaction frequency, behavior uncertainty-price&quality	Tight relations, actor similarities, functional ties, motives-new resources	(5,16), [1,6,15,19], (9,25), (3,28)
	Ensure availability of equal healthcare and social welfare services everywhere in the area		Valuable, supplementary to non-it, long-term performance	Asset specificity, behavior uncertainty-inter-organizational flexibility	Social capital, access the network, collective sanctions, motives-new resources	(2), [1,6,19,24], (9,25), (3,14,28)
	The development of patient / customer information systems and data storages requires					
	Tighter cooperation on national level	IT governance matrix, resource availability	Non-Substitutable, property based-intellectual, conserve resources	Asset specificity-customer & task complexity, technology uncertainty-specs change	Social ties, tight relations, flow of information, functional ties	(2,5,23), [1,6,15,19], (8,9,25), (3,10,14,28)
	Tighter cooperation on regional level (this area)	IT governacne matrix, cooperation mechanism, resource availability	Non-substitutable, property based-human, conserve resources	Transaction frequency, technology uncertainty-technology improvements	Tight relations, access to social network, flow of information, functional ties, relational trust	(2,4,7), [1,6], (9,25), (10,14,28)
	The creation of jointly agreed data models with sticking to them	Decision making rights, health information governance	Non-imitable, knowledge based-managerial, enhance reliability	Asset specificity-information visibility, behavior uncertainty-"buyer" expectations	Tight relations, collective sanctions, control mechanism, relational trust	(2,11,12,16,23), [1,6,19], (9,22,25), (3,14,28)
	Tighter cooperation between healthcare and social welfare	Past relations, business process costs	Non-substitutable, property based-intellectual, complementary to IT and non-IT	Asset specificity-task & customer complexity, behavior uncertainty-"b." expectations	Social ties, cultural processes, motives-new resources, bigger legitimacity, common goal	(2,5,23), [1,6,19], (9,25), (3,14,28)
	Type of organization (municipal, healthcare district)					
	Geographic area (the area of the research, other parts of the country)					
	Organizational status (H/S manager, H/S expert, IT manager, IT expert), where H/S = healthcare / social welfare					
	Experience in years in H/S managerial positions					
	Experience in years in H/S expert positions					
	Experience in years in IT managerial positions					
	Experience in years in IT expert positions					
	Involvement in the establishment of the inter-organizational IT governance arrangement (=project group member)					
	As a whole the role of IT is generally regarded much too important for healthcare and social welfare services					
	As few as possible funds should be used to healthcare and social welfare IT services					
	In the future IT will be much more important to the development and operation of healthcare and social welfare services					
	My organization is a highly competent deployer of IT for healthcare and social welfare services					
	In my organization IT governance accountabilities are allocated clearly between healthcare / social welfare and IT professionals					
	We have clear measurable objectives for healthcare / social welfare IT services					
	As a whole my organization applies IT so well to healthcare and social welfare services that it would be graded as A or A+ were it considered in terms of educational grading					

References: list of references in (list) are IT governance literature references, in [list] are RBV literature references, in (list) are TCE literature references, and in (list) are social network theory literature references

TAM and E-learning Adoption: A Philosophical Scrutiny of TAM, Its Limitations, and Prescriptions for E-learning Adoption Research

A.K.M. Najmul Islam[1,*], Nasreen Azad[2], Matti Mäntymäki[1], and S.M. Samiul Islam[3]

[1] University of Turku, Finland
{matti.mantymaki,najmul.islam}@utu.fi
[2] Åbo Akademi University, Finland
nasreen.azad@abo.fi
[3] Samsung R&D Institute Bangladesh
sm.samiul@samsung.com

Abstract. TAM and TAM derived theories have been very popular for investigating users' e-learning adoption/post-adoption behavior. However, several philosophical holes as well as a number of limitations of TAM research have been pointed by several leading researchers in the recent years. In this paper, we discuss the philosophical holes and present our reflections and possible prescriptions about these holes while conducting research on e-learning adoption/post-adoption. We also discuss the limitations of TAM research and present prescriptions about how e-learning adoption research can be conducted by addressing these limitations.

Keywords: e-learning, IS continuance model, technology acceptance model, theory of reasoned action.

1   Introduction

The Technology Acceptance Model (TAM) [13] was originally developed from Ajzen & Fisbein's (1980) [2] Theory of Reasoned Actions (TRA). TAM can be viewed as an adaptation of the TRA to the IS discipline. It has been regarded as one of the most influential theories in the IS discipline. It has been widely applied for explaining IT users' intention regarding IT use [32]. TAM researchers have developed a number of extensions to TAM. In addition, they have also contributed in developing the Unified Theory of Acceptance and Use of Technology (UTAUT) [55] and IS continuance model [8, 40] based on the TAM.

TAM has been validated across time, population, and contexts [54]. It is a well-established theory in IS research as well as used in other domain [32]. Venkatesh et al. (2007) [54] argued that TAM has become nearly a law-like model and it often serves as a basis for studies in other areas. TAM and its constructs have been used in areas outside the technology adoption such as information adoption [51], marketing

---

* Corresponding author.

H. Li et al. (Eds.): I3E 2014, IFIP AICT 445, pp. 164–175, 2014.
© IFIP International Federation for Information Processing 2014

[12], and advertising [45]. In addition, TAM has been used as a basis for comparing SEM techniques—PLS vs. LISRELL [10].

TAM and its variants have also been used extensively in e-learning adoption research. For example, Sumak et al. (2011) [50] conducted a meta-analysis with the articles of e-learning adoption and found that 86% studies used TAM, 4% studies used UTAUT, 2% studies used Theory of Planned Behavior (TPB) [3], and 6% studies used other theoretical frameworks to explain users' e-learning systems adoption.

Despite its widespread use, TAM researchers have not carefully scrutinized the philosophical and epistemological foundations of the model [48]. In addition, the researchers have often ignored addressing the limitations of TAM in their adaptation of TAM to their research context. As such it has created two confusions among the scientists in the recent years. The first confusion is related to what extent TAM meets the criteria for scientific theories established for causal, positivistic explanations. The second confusion is related to how the researchers may address the TAM limitations in their research.

Silva (2007) [48] strongly argued that TAM might not be falsifiable. He continued to argue that TAM research is not progressive as the TAM researchers do not question the fundamental foundation of TAM but provide alternative hypotheses when they face anomalies. We believe these arguments are valid due to the fact that very few previous studies have addressed these issues [32, 48]. This research gap motivates this paper. We address these issues in this paper by taking e-learning adoption research as an example. Consequently, we address the following two important TAM related issues in this paper.

- First we present the philosophical holes of TAM research and discuss our reflections with these holes. We discuss possible prescriptions to these philosophical holes for e-learning adoption researchers.
- Second, we present the limitations of TAM research and discuss possible prescriptions that could be applied by e-learning adoption researchers to address those limitations.

2   Addressing Philosophical Holes

There are two major philosophical critiques regarding TAM and its variants in the prior literature [48]. These critiques are identified based on the perspective of two prominent post-positivist philosophers of science: Karl Popper and Irme Lakatos. The first critique asks: Is TAM falsifiable? The second critique asks: Is TAM research progressive? In the next we discuss these questions and present our reflections in answering these questions based on the prior literature.

2.1   Is TAM Falsifiable?

In a particular model, all factors are represented at the same level of aggregation—whereas the actual world is a complex interweaving of different structures at many levels. Such complex world cannot be fully captured with a theory like TAM. Indeed

TAM has been confirmed thousand times in the prior IS literature. Popper (1972) [43] in this regard states that there are no reasons to believe that a theory is scientific only because data—no matter how much of it there is—confirm it. He further states that the reason we find regularities in nature because of a mental habit that makes us jump to conclusions. He advises the scientists to keep their guard up and be suspicious of continuous confirmations. Popper (1972) [43] suggested the scientists to design experiments that aim to falsify the theory to the maximum effects. If they are successful in falsifying all or parts of their theories, they should go back to the drawing board and reformulate a new one. The alternative to this approach is to set up a theory and look for evidences for confirming the theory.

According to Silva (2007) [48], many studies utilizing the TAM targeted confirming the theory instead of falsifying it. It can be hazardous because the world is sufficiently complex and some confirming evidences can be found, no matter how unlikely the theory may be. Following Popper (1972) [43], it can be argued that researchers also need to keep their eyes open and find types of computer adoption behavior that cannot be explained by TAM. However, as discussed in the following it is very difficult to find a computer adoption that cannot be explained by TAM in prior research.

2.1.1   Can a Theory Account for all Type of Human Behavior?

As discussed before TAM was developed from the TRA. According to Ajzen & Fishbein (1980) [2], TRA is very general, and designed for studying virtually any human behavior. If we accept that TRA can account for all types of human behavior, then it is not a scientific theory. Popper (1972) [43] calls such theory as pseudo-science. The difference between a scientific theory and pseudo-science is that a scientific theory is falsifiable—meaning that the theory cannot explain all types of human behavior. Popper (1972) [43] notes the following.

"Every good scientific theory is a prohibition: it forbids certain things to happen. The more a theory forbids, the better it is...."

To evaluate whether TRA is scientific theory, a review of the previous psychological literature is necessary. Ogden (2003) [41] has performed such review among psychological literature. His review revealed that indeed TRA has been found weak in predicting certain behavior. However, he discovered that in such situation instead of rejecting the theory the researchers provided several explanations such as the model should be accepted but the variables were not operationalized properly, the model should be accepted but the sample characteristics may explain the results, etc.

We observed similar things in e-learning adoption research as well when the researchers employed TAM and its variants. For example, Limayem & Cheung (2008) [49] used the IS continuance model (a TAM derived model) and habit to investigate e-learning system users' continued use behavior. They found that their combined model explained only 23% variance of continued use. Instead of rejecting their theory, they described the following.

"We therefore believe that other significant factors (such as socio-cultural and political impacts) may affect students' decisions to continue using the IBLT. In Hong Kong, as in other Chinese communities, social factors have significant impact on usage behavior"

Based the above findings, we argue that the TRA and TAM are falsifiable although the way these have been used in many studies might make an illusion that no data can be collected to falsify the theory. In other word we may argue that this philosophical hole is less related to the TRA or TAM framework and more related to how it has been applied in different studies. The illusion becomes even stronger when we discuss the analytic nature of the relationships among beliefs, attitude, intentions and behavior in the next.

2.1.2   Logical-Connection Argument

TAM and its variants follow the tradition of beliefs-attitude-intention-behavior relation. According to Rosenberg (1995) [46], actions are composed of desires and beliefs and that both provide actions with their meaning. He illustrates this with an example of a person named Smith carrying an umbrella. The action of Smith carrying an umbrella—as a meaningful action—can be explained by Smith's belief that it is going to rain and desire of not getting wet. Thus, the intention of carrying an umbrella can be stated in terms of Smith's beliefs and desires. It implies that there cannot be actions without intentions.

Silva (2007) [48] referring to Rosenberg's example stated that Smith might carry an umbrella for different reasons than his belief in imminent rain and his desire to not get wet. He can carry an umbrella because it is part of his attire or because he wants to use it as a weapon. Following this, it is clear that carrying an umbrella will always linked to intentions, and identifying them will not predict the action—instead, they render an action its meaning. Hence, beliefs, intentions, and self-reported behavior measured in a cross-sectional survey are linked by definition. This is called the logical connection argument [4, 39].

According to the logical connection argument, intentions cannot predict behavior as these are not linked contingently but analytically. The difference between these two types of connections is that a contingent entity depends on natural process to occur, while an analytic one does not depend on natural process [19]. An analytic truth is true by definition [41]. A chemical reaction can be regarded as contingent entity as it requires the conjunction of different natural factors. On the other hand stating that a rectangle has four sides is an analytical truth. Empirical science based on experiments and observation can deal only with contingent entities.

Ogden (2003) [41] observed that the relationships in TRA are often analytical in nature. His review revealed that researchers often measure different constructs with similar statements. For example, he observes that researchers attempted to correlate perceived behavioral control with behavioral intentions, while both were measured using similar questions. Hence, finding a high correlation between these two constructs is not surprising. This makes the theory analytical. A good theory should avoid analytical truth, otherwise it will be tautological [41]. In such a case belief, attitude, and intention cannot be linked causally. Silva (2007) [48] states the problem of this in the following way.

"The problem is that intentions, stated in terms of desires and beliefs, constitute only a re-description of the action they are thought to be predicting."

It implies that when beliefs, attitude, and intentions are linked analytically, the causal relationships between these cannot be tested empirically and cannot be subjected to falsification. Hence, in such situation the causal links of TAM cannot be regarded as scientific according to Popper's [43] classification.

According to our critical evaluation, TAM's relationships have not been analytical in prior literature. For example, Lee et al. (2003) [32] reported examples of many studies where TAM relationships were inconsistent. We argue that if the relationships are truly analytical, then such inconsistency should not be visible. But we accept that adoption researchers are required to be more careful in their research design, especially while operationalizing the constructs of their models to avoid analytical nature. For example, the original TAM was developed to predict users' organizational IS use and cautious should be taken when operationalizing the constructs to the e-learning context. We present two suggestions for the e-learning adoption researchers in relation to construct operationalization.

First, researchers should avoid similar statements in measuring different constructs. In addition, they should present the questions in a random fashion. Second, researchers should consider using objective measures of both beliefs and behavior to test the theories like TAM. Neuroscience approaches [15] can be used to collect objective data. Both suggestions can avoid analytical nature of the constructs to some extent.

2.2   Is TAM Research Progressive?

The original TAM had only two beliefs: perceived usefulness and perceived ease of use to predict IS use. When TAM is applied to explain adoption of a mobile service, it can be thought that the users adopt mobile services due to its usefulness and ease of use. But let's think about the adoption of an Enterprise Resource Planning (ERP) system in an organization. Although the system might be difficult to use and disruptive to employees' work, they use it. The TAM researchers argued that such adoption could be explained by subjective norm—that is the influence of authorities. The original TAM model did not include subjective norm, but latter whenever researchers found that TAM may not be able to explain a particular adoption, they add more variables to it.

This approach can be explained by the concept of research programme proposed by Lakatos (1970) [29]. A research programme is an organic unity, which contains both rigid and flexible components—essential, structural components as well as non-essential components. The essential structural components are the hard-core and positive heuristic of the research programme. The hard core consists of a set of theoretical assumptions to which a community of scientists is committed. The committed scientists will defend the credibility of the hard core against any threats posed by others. Yet the research programme does contain or generate components, which could be given up or replaced without abandoning the hard core. Non-essential, replaceable components of the research programme are called protective belt of the research programme. The protective belt can be viewed as the auxiliary hypotheses in defense of the hard core.

For TAM derived research programme the hard core is the basic TAM model with beliefs-attitude-intention relation. The protective belt is the additions of different researchers committed to the TAM derived research. New constructs have been added and auxiliary hypotheses have been offered to explain unexpected results without questioning the hard core. For example, several TAM studies [17, 37] have noticed that perceived ease of use has not been consistently linked to adoption. These studies explained the anomalies by suggesting that the role of perceived ease of use depends on the task, which is an auxiliary hypothesis according to Silva (2007) [48]. Lee et al. (2003) [32] in their review found that about 24% prior studies did not find a significant relationship between perceived ease of use and behavioral intention. However, many of these studies did not challenge the TAM, instead provided auxiliary explanations. Silva (2003) [48] noted the following in relation to this.

"In the light of Lakatos methodology of scientific research programmes, I argue that the complementary constructs and additional theoretical explanations were added by TAM researchers to protect the hard core. The additions can be considered auxiliary hypotheses that have been incorporated in the protective belt. In this sense, it is also worth mentioning that in my reading of TAM literature, I could not find papers that challenged the hard core."

Following Popper (1972) [43], such auxiliary hypotheses can be regarded as ad hoc and eventually a bad thing while following Lakatos (1970) [29], it can be argued that adjusting and developing protective belt is not necessarily a bad thing for a research programme. We argue that instead of asking whether a theory is falsifiable, the more important question is to ask whether the research programme is progressive or degenerative. A research programme is progressive if it is able to discover novel facts, develops new experimental techniques, and performs precise predictions. A research programme is degenerative if it is not able to produce novel facts by changing the protective belt. In such cases, the added hypotheses of the protective belt can be considered as ad hoc and not acceptable according to Lakatos (1970) [30].

According to our critical evaluation, the TAM derived research program has been progressive. For example, TAM researchers have added several new constructs to the original TAM such as age, gender, prior experience, management support, and voluntariness to explore the boundary conditions for TAM [32, 48]. Researchers have discovered that the psychological motivation behind initial use and subsequent use is different and thus have put more importance on post-adoption behavior than adoption behavior [8, 26, 27]. In this regard, the IS continuance model has been developed from the Expectation-Confirmation Theory (ECT) [42] to study post-adoption behavior. In addition, leading IS researchers also pointed that there are still many unexplored research areas regarding IS adoption and use, such as testing the effect of IT artifact's design characteristics on perceived ease of use, perceived usefulness and use [6, 7], investigating actual usage and its relation with objective performance measures [6, 32] and exploring organizational and societal adoption and use of IS [32] to name only a few.

3   TAM Limitations and Possible Prescriptions

There are a number of other limitations of TAM derived studies as described by leading IS researchers [5, 6, 18]. TAM researchers must need to address these limitations for progressing TAM research. In the following we discuss these limitations and provide possible prescriptions in order to overcome these limitations in an e-learning adoption study.

3.1   TAM Lacks Design and Implementation Constructs

Benbasat & Barki (2007) [6] argued that TAM lacks investigating and understanding both design- and implementation-based antecedents of IT adoption and acceptance. The beliefs: perceived usefulness and perceived ease of use are regarded as black boxes with little research effort into investigating what makes a system useful or easy to use. As such TAM studies often lack actionable guidance for practitioners.

Indeed, TAM 2 and TAM 3 investigated the possible determinants of perceived usefulness and perceived ease of use. However, it was assumed that these two beliefs would always mediate the effects of other beliefs on behavioral intention or actual behavior. Following these assumption researchers rarely tested the direct effect of other variables on behavioral intention or actual behavior. Benbasat (2010) [7] argued that to provide more practical implications for designers and managers, researchers must need to investigate the potential effects of design and implementation characteristics related variables of an IT artifact on the construct of interest such as attitude, behavioral intention and actual behavior.

In fact prior research has found that there are variables other than perceived ease of use and perceived usefulness, which may have significant impact on behavioral intention [21]. Utilizing only two beliefs have made TAM parsimonious. However, parsimony is an Achilles' heel for TAM. In practice, although TAM predicts behavioral intention using only two beliefs, but there could be more beliefs in a particular context.

The problem related to lack of variables is even more severe for the context of e-learning. TAM does not contain any specific variable related to e-learning. When researchers use TAM model to investigate e-learning system users' adoption behavior they definitely need to use variables related to e-learning. However, very few prior studies have done so [22, 23, 24]. Most of the studies used TAM in such a format that does not contain any e-learning context specific variables.

We believe that e-learning adoption researchers should consider potential design and implementation variables related to e-learning for employing TAM and its variants in their research. The researchers should keep in mind that e-learning systems are distinct from general IS at least to some extent. For example, Shee & Wang (2008) [47] argued that an e-learning system is a highly user-oriented system that focuses on the content and how it is presented. An e-learning system offers educators and learners "possibilities", rather than "ready to use" resources. In this regard, while general IS elicits performance from individual users, e-learning is based on the cooperation between educators and students. There are many factors that may cause users' dissatisfaction and rejection of e-learning system use such as lack of cues, lack of face-to-face contact, non-verbal communication, isolation, problems with hardware/software, and network connectivity [9]. Additionally, educators' roles and teaching models also affect students' learning outcome [11].

Following these, we believe that e-learning adoption researchers using TAM should investigate the effect of such factors on perceived usefulness, perceived ease of use, attitude, behavioral intention, and actual behavior. Such research will bring valuable implications for both practitioners and researchers.

3.2   Limited Understanding of Behavior

TAM and its variant theories uncritically accept the association between intention and actual behavior. Based on this, majority of prior research has investigated behavioral intention as the final dependent variable. The idea is that if behavioral intention is high, it will automatically improve use behavior. It is assumed that use behavior is driven by conscious intentions that result from a rational decision-making process involving beliefs, expectations, reflection on past experience, etc. and emotion such as satisfaction, frustrations, etc.

However, behavioral intention may not predict behavior. For example, De Guinea & Markus (2009) [14] argued that emotion may also drive IS use directly. It is because: a) that the connection between emotion and behavior can occur without a person being consciously aware of the connection, and b) that the effect of emotion may not create a particular behavioral intention, but rather to derail a previously formed behavioral intention about IS use. It suggests that sudden intense emotions, such as frustration associated with a system crash or the pleasure aroused while using an IS, may be more important in its influence on behavior than intention which is driven by stable attitudes and expectations. In addition, it is often argued that frequently performed behavior becomes automatic or habitual, and it ultimately reduces the impact of intention on use [28, 49].

The above argument raises one important issue – that is intention may not predict behavior as emotions and beliefs might have stronger influence on behavior. Hence, it becomes perhaps more important to investigate usage behavior rather than intention. However, very few prior studies have investigated usage behavior and its relation with intention [32, 16, 24].

Indeed some studies measured behavior. However, most of these studies measured self-reported usage behavior instead of actual usage behavior. Self-reported usage is assumed to be a reasonable predictor of actual system usage in adoption studies [1, 25]. However, several studies have cautioned in the use of self-reported usage instead of actual usage [31, 32, 44, 49, 52]. For example, Straub et al. (1995) [49] found that research based on self-reported usage shows distinctly different results from that of actual usage. In addition, self-reported usage was also found to be the major reason for common method bias [20].

In addition, the self-reported usage itself has largely been viewed as a 'black box', and hence understanding the situation specific usage behaviors is limited [22, 53]. For example, e-learning services can be used in many situations such as at school, home, and even while moving. Thus, we suggest researchers open the usage 'black box' into situation specific actual behaviors in their future research.

3.3   Missing Adoption/Usage Outcomes

TAM assumes that more use is better. In other word more utilization of a technology increases performance. Following this prior studies have put highest importance in explaining users' behavior with the target system. These studies investigated possible antecedents and determinants of system use behavior. As such these studies often ignored the outcome of system use. Many studies argued that more use might not necessarily improve individuals' or organizational performance. For example Islam (2013) [24] found that heavy usage of an e-learning system might not necessarily help students in their study to achieve better academic performance. However, only a few prior e-learning adoption studies verified the relation between use and learning outcomes.

Indeed, few studies have gone beyond use to explore the factors associated with learning. McGill & Klobas (2009) [38] found that e-learning system utilization influences perceived impact on learning. Lee & Lee (2008) [33] revealed that a number of e-learning environment quality related variables affect satisfaction with e-learning. In turn, satisfaction was found to influence academic achievement. Liaw (2008) [34] found high correlation between intention to use e-learning and e-learning effectiveness. These studies provide some empirical support about the possible relationships between e-learning system use and e-learning outcomes. However, these studies have been conducted with a variety of outcome variables that use different explanatory variables and this has led to models that offer only weak theoretical support. Thus, these studies fall short in explaining the relationship between the antecedents of adoption and use of e-learning systems and their use outcomes, and the relationship between e-learning system use and use outcomes.

We suggest researchers develop complete nomological network taking into account e-learning system usage antecedents, usage, and performance outcomes in the future. Performance outcome variables could be related to both teaching and learning. Examples of teaching performance related variables are planning, managing, instructing, assessing, and collaborating. Example of learning performance related variables are perceived learning, and grade.

4   Conclusion

This paper conducted a scrutiny of TAM and its variants in relation to e-learning adoption research. We presented two types of critique of TAM: philosophical holes and limitations regarding missing variables. Based on our critical evaluation, TAM has served the IS researchers as a theoretical model that speaks to the unique nature of information systems. From philosophical perspective, first we asked the question: Is TAM falsifiable? We observed from prior literature that TAM is falsifiable, although the way the results have been reported in the prior literature might make an illusion that TAM is not falsifiable. Second, we asked the question: Is TAM research progressive? Again, from the prior literature, we observed that TAM research is progressive in a variety of contexts including e-learning. Overall, we found that philosophically there is no problem in applying TAM and its variants in studying e-learning adoption. However, we think that TAM contains deceptively straight-forward constructs and measures. Thus, we believe that TAM should be revisited to ensure design, usage, and outcome constructs have been measured in the best possible way by e-learning adoption researchers. The researchers should develop sophisticated conceptualizations of e-learning outcomes as well as what system usage means in a specific research contexts.

References

1. Agarwal, R., Prasad, J.: Are individual differences germane to the acceptance of new information technologies? Decision Sciences 30(2), 361–391 (1999)
2. Ajzen, J., Fishbein, M.: Understanding attitudes and predicting social behavior. Prentice-Hall, Englewood Cliffs (1980)
3. Ajzen, I.: The theory of planned behavior. Organizational Behavior and Human Decision Processes 50(2), 179–211 (1991)
4. Anscombe, G.E.M.: Intention, 2nd edn. Basil Blackwell, Oxford (1957)
5. Bagozzi, R.R.: The legacy of the technology acceptance model and a proposal for a paradigm shift. Journal of the Association of Information Systems 8(4), 244–254 (2007)
6. Benbasat, I., Barki, H.: Quo vadis, TAM? Journal of the Association of Information Systems 8(4), 211–218 (2007)
7. Benbasat, I.: HCI research: Future challenges and directions. AIS Transaction on Human-Computer Interaction 2(2), 16–21 (2010)
8. Bhattacherjee, A.: Understanding information systems continuance: An expectation-confirmation model. MIS Quarterly 25(3), 251–370 (2001)
9. Buckley, K.M.: Evaluation of classroom-based, web-enhanced, and web-based distance learning nutrition courses for undergraduate nursing. Journal of Nursing Education 42(8), 367–370 (2003)
10. Chin, W.W., Todd, P.A.: On the use, usefulness and ease of use of structural equation modeling in MIS research. MIS Quarterly 19(2), 237–246 (1995)
11. Coppola, N.W., Hiltz, S.R., Rotter, N.G.: Becoming a virtual professor: Pedagogical roles and asynchronous learning networks. Journal of Management Information Systems 18(4), 169–189 (2002)
12. Dabholkar, P.A., Bagozzi, R.P.: An attitudinal model of technology-based self-service: Moderating effects of consumer traits and situational factors. Journal of Academy of Marketing Science 30(3), 184–201 (2002)
13. Davis, F.D.: Perceived usefulness, perceived ease of use, and user acceptance of information technology. MIS Quarterly 13(3), 319–340 (1989)
14. De Guinea, A.O., Markus, M.L.: Why break the habit of a lifetime? Rethinking the role of intention, habit, and emotion in continuing information technology use. MIS Quarterly 33(3), 433–444 (2009)
15. Dimoka, A., Bagozzi, R., Banker, R., Brynjolfsson, E., Davis, F., Gupta, A., Riedl, R.: NeuroIS: Hype or Hope? In: Proceedings of the 30th International Conference on Information Systems, pp. 1–11 (2009)
16. Freeze, R.D., Alshare, K.A., Lane, P.L., Wen, H.J.: IS success model in e-learning context based on students' perceptions. Journal of Information Systems Education 21(2), 173–184 (2010)
17. Gefen, D., Straub, D.W.: The relative importance of perceived ease of use in IS adoption: A study of e-commerce adoption. Journal of the Association of Information Systems 1(8), 1–28 (2000)
18. Goodhue, D.L.: Comment on Benbasat and Barki's "Quo vadis TAM" article. Journal of the Association of Information Systems 8(4), 219–222 (2007)
19. Honderich, T. (ed.): The Oxford companion to philosophy. Oxford University Press, Oxford (1995)
20. Igbaria, M., Zinatelli, N., Cragg, P., Cavaye, A.L.M.: Personal computing acceptance factors in small firms: A structural equation model. MIS Quarterly 21(3), 279–305 (1997)

21. Islam, A.K.M.N.: The determinants of the post-adoption satisfaction of educators with an e-learning system. Journal of Information Systems Education 22(4), 319–332 (2011)
22. Islam, A.K.M.N.: The role of perceived system quality as the educators' motivation to continue e-learning system use. AIS Transaction on Human-Computer Interaction 4(1), 25–43 (2012a)
23. Islam, A.K.M.N., Mäntymäki, M.: Continuance of professional social networking sites: A decomposed expectation-confirmation approach. In: Proceedings of the International Conference on Information System (ICIS 2012), Orlando, Florida, USA (2012)
24. Islam, A.K.M.N.: Investigating e-learning system usage outcomes in the university context. Computers & Education 69, 387–399 (2013)
25. Jackson, C.M., Chow, S., Leitch, R.A.: Toward an understanding of the behavioral intention to use an information system. Decision Sciences 8(2), 357–389 (1997)
26. Jasperson, J.S., Carter, P.E., Zmud, R.W.: A comprehensive conceptualization of post-adoptive behaviors associated with information technology enabled work systems. MIS Quarterly 29(3), 525–557 (2005)
27. Karahanna, E., Straub, D.W., Chervany, N.L.: Information technology adoption across time. MIS Quarterly 23(2), 183–213 (1999)
28. Kim, S.S., Malhotra, N.K., Narasimhan, S.: Two competing perspectives on automatic use: A theoretical and empirical comparison. Information Systems Research 16(4), 418–432 (2005)
29. Lakatos, I.: Falsification and the methodology of scientific research programmes. In: Lakatos, I., Musgrave, A. (eds.) Criticism and the Growth of Knowledge, pp. 91–196. Cambridge University Press (1970)
30. Lakatos, I.: The methodology of scientific research programmes: Philosophical papers, vol. 1. Cambridge University Press, Cambridge (1978)
31. Lederer, A.L., Maupin, D.J., Sena, M.P., Zhuang, Y.: The technology acceptance model and the world wide web. Decision Support Systems 29(3), 269–282 (2000)
32. Lee, Y., Kozar, K.A., Larsen, K.: The technology acceptance model: Past, present, and future. Communications of the Association for Information Systems 12, 752–780 (2003)
33. Lee, J.-K., Lee, W.-K.: The relationship of e-Learner's self-regulatory efficacy and perception of e-Learning environment quality. Computers in Human Behavior 24(1), 32–47 (2008)
34. Liaw, S.-S.: Investigating students' perceived satisfaction, behavioral intention, and effectiveness of e-learning: A case study of the Blackboard system. Computers & Education 51(2), 864–873 (2008)
35. Limayem, M., Cheung, C.M.K.: Understanding information systems continuance: The case of Internet-based learning technologies. Information & Management 45(4), 227–232 (2008)
36. Limayem, M., Hirt, S.G., Cheung, C.M.K.: How habit limits the predictive power of intention: The case of information systems continuance. MIS Quarterly 31(4), 705–737 (2007)
37. Ma, Q., Liu, L.: The technology acceptance model: A meta-analysis of empirical findings. Journal of Organization and End User Computing 16(1), 59–72 (2004)
38. McGill, T.J., Klobas, J.E.: A task-technology fit view of learning management system impact. Computers & Education 52(2), 496–508 (2009)
39. Melden, A.I.: Free Action. Routledge and Kegan Paul, London (1961)
40. Mäntymäki, M., Islam, A.K.M.N.: Social virtual world continuance among teens: Uncovering the moderating role of perceived aggregate network exposure. Behavior & IT 33(5), 536–547 (2014)

41. Ogden, J.: Some problems with social cognition models: A pragmatic and conceptual analysis. Health Psychology 22(4), 424–428 (2003)
42. Oliver, R.L.: A cognitive model of the antecedents and consequences of satisfaction decisions. Journal of Marketing Research 17(11), 460–469 (1980)
43. Popper, K.: Conjectures and refutations: The growth of scientific knowledge. Routledge & Kegan Paul, London (1972)
44. Rawstorne, P., Jayasuriya, R., Caputi, P.: Issues in predicting and explaining usage behaviors with the technology acceptance model and the theory of planned behavior: When Usage is Mandatory. In: Proceedings of the 21st International Conference on Information Systems (ICIS 2000), Atlanta, GA, USA (2000)
45. Rodgers, S., Chen, Q.M.: Post-adoption attitudes to advertising on the internet. Journal of Advertising Research 42(5), 95–104 (2002)
46. Rosenberg, A.: Philosophy of social science. Westview Press, Oxford (1995)
47. Shee, D.Y., Wang, W.S.: Multi-criteria evaluation of the web-based e-learning system: A methodology based on learner satisfaction and its applications. Computers & Education 50(3), 894–905 (2008)
48. Silva, L.: Post-positivist review of technology acceptance model. Journal of the Association for Information Systems 8(4), 255–266 (2007)
49. Straub, D., Limayem, M., Karahanna, E.: Measuring system usage implications for IS theory testing. Management Science 41(8), 1328–1342 (1995)
50. Sumak, B., Hericko, M., Pusnik, M.: A meta-analysis of e-learning technology acceptance: The role of user types and e-learning technology types. Computers in Human Behavior 27(6), 2067–2077 (2011)
51. Sussman, S.W., Siegal, W.S.: Information influence in organizations: An integrated approach to knowledge adoption. Information Systems Research 14(1), 47–65 (2003)
52. Szajna, B.: Empirical evaluation of the revised technology acceptance model. Management Science 42(1), 85–92 (1996)
53. Teo, T., Schaik, P.: Understanding the intention to use technology by pre-service teachers: An empirical test of competing theoretical models. International Journal of Human Computer Studies 28(3), 178–188 (2012)
54. Venkatesh, V., Davis, F.D., Morris, M.G.: Dead or alive? The development, trajectory and future of technology adoption research. Journal of the Association of Information Systems 8(4), 267–286 (2007)
55. Venkatesh, V., Morris, M.G., Davis, G.B., Davis, F.D.: User acceptance of information technology: toward a unified view. MIS Quarterly 27(3), 425–478 (2003)

The Overall Design of Digital Medical System under the Network Environment

Yuping Zhou[1], Yicheng Zhou[2], and Min Yu[3,*]

[1] The Collage of Information Science and Technology, Hainan Normal University,
Haikou, China
zypnew@qq.com
[2] First Affiliated Hospital of Harbin Medical University, Harbin, China
yichengzhou2011@163.com
[3] Academic Affairs Office, Hainan Normal University, Haikou, China
704222535@qq.com

Abstract. The wireless sensing network technology, communication technology, password technology, etc. were adopted to carry out the overall design to the digital medical system under the network environment, in order to realize the functions of the remote medical diagnosis, health care, locating and tracking, emergency calls and first aid. In this paper, the design method of the community and personal digital medical system was expounded and the overall design scheme of the digital medical system was provided, including the function module design such as the user terminal system, safety call system, electronic medical records system, hospital information management system, diagnosis expert system and emergency center control system.

Keywords: digital medical treatment, network environment, system design.

1 Introduction

With the rapid development of network technology and the continuous improvement of people's living standard, people have a higher request to the life quality; therefore, people have a higher request to the daily health care service. The traditional medical model of the diagnosis and treatment such as face to face between doctors and patients cannot meet the requirements of people. People's demand for health care is growing rapidly and the emergency medical treatment is required higher and higher.

Mobile digital hospital mainly involved in a number of sub-systems such as the hospital wireless clinical information systems, intelligent expert clinics, health tracking service system, remote medical monitoring system, community health telematics system, etc. It can create an all-round, new medical services platform for patients, in order to promote the establishment of new hospital service model. With the development of digital communications technology, the digital medical in the network environment will give people more and more help and services. Mobile digital hospital will become more and more common.

---

* Corresponding author.

H. Li et al. (Eds.): I3E 2014, IFIP AICT 445, pp. 176–184, 2014.

Wireless sensor network is the one that is organized by the sensor nodes in the way of wireless communication, in particular application environment. The sensor nodes were mainly used to collect data, and then, the collected data were sent to the network via a wireless sensor network, and finally received by the particular application system. Wireless sensor network is a new mode of information access and information processing.

It is very important to apply wireless sensor networks in the community and personal digital medical and to study out the digital medical system for the remote medical treatment, remote diagnosis, remote medical care, real-time nursing and emergency medical treatment. The system has advantages of the high reliability, high safety, high efficiency and humanization, etc. In this paper, the design of digital medical system under the network environment was introduced, including the system structure, system hardware platform, system software platform and its main function modules.

2   Study Background

A complete digital medical system should provide the personalized, digital medical service and health management platform for community residents. The platform system should include services such as the prevention, diagnosis, health care, medical care, first aid and interactive, etc. Therefore, it is necessary to comprehensively utilize the network technology, communication technology, control technology, and some medical equipment terminals to provide the medical service, health care service, medical care services, emergency services and health services for the residents.

Through the digital medical service system, community residents will get fast and convenient medical services. Residents can get SMS health information remind, remote medical diagnosis, remote health care, emergency treatment, expert interaction, online consultation and other services through fixed network or wireless network technology. In 1906, Wilhelm Einthoven, the inventor of the electrocardiogram, successfully carried out the remote consultation experiment through telephone lines, and become the first of the telemedicine development. In 1967, a radiologist in the United States established the interactive telemedicine system, which was the first to realize the doctor-patient interaction of remote medical system [1]. In 2006, 306th hospital of the people's liberation army carried out the wireless clinical medical to patients through the hospital local area network, and the wireless PDA was applied to query and monitor for patients sick signs, have a prescription, issue orders, make the mobile medical successfully by using the local area network, which greatly improved the efficiency of the doctor's work [2]. In 2012, Huaxiang and others developed "hospital information management system", whose main function was to realize the computerized information management of the people-money-goods such as health information, financial information, and decision-making information. The C/S mode, prototype method, structured system analysis method and object oriented method, etc were used in the system. The SQL Server 2000 database development technology was used as the database [3]. Digital medical treatment is a new type of modern medical mode, in which the modern information technology and computer technology are applied to the whole medical process. It is the management goal and development direction of the modern public health. A typical digital medical model is mainly composed of 3 big parts such as the digital medical equipment, hospital network & management information system, and remote medical service, of which the digital medical and surveillance equipments are the important foundation of digital medical [4]. According to the hospital digitization development experience at home and abroad, the development of the digital hospital can be divided into three stages: the management digital stage, medical digital stage and digital hospital stage featured as the regional medical [5, 6]. At present, the hospital digital development and construction has entered the third stage in the developed countries such as the United States, Japan and Germany. The construction of digital medical treatment in our country started late. At present, it is in the initial phase of the second stage.

In conclusion, medical digital system developed by the network technology has been the trend of The Times in the rapid development today of the digital technology and network technology.

3   System Research and Design

3.1   Main Function of the System

The main function of this system is: medical diagnosis, health care, diagnostic tracing and emergency medical treatment, etc. The security, real-time performance, flexibility and intelligent of the system in the development of the system should be considered. The system security includes the security of data transmission, information storage security and the security of the information access. The real-time performance refers to collecting a number of physiological data to users in time by using high precision medical sensors. The data collected will be transmitted to the medical diagnosis and emergency centers in time through a wireless network or mobile network. Medical system will analyze and process the data in time to get a result, which will be feed backed to the users in time. The flexibility can be obtained by using medical sensor with a wireless connection, which is easy to carry and remove. The positioning method can be selected flexibly according to customer's actual environment. The intelligent means that the system can analysis and process the physiological data collected by sensors to judge the optimization treatment scheme, and to provide efficient health services through scheduling the relevant units and personnel in the fastest time for the remote medical care and emergency medical first aid.

3.2   System Frame Structure

The frame structure of the system was divided in three levels: the business platform, support platform and foundation platform. (1) The business platform included the user terminal system, security call system, electronic medical records system, hospital information management system, diagnosis expert system and emergency dispatch control system. This layer was the application level to realize the information resource digitalization, the business processing networking, management and decision scientization. (2) The support platform provided the system WEB server, data access, transaction processing, security management and other functions. This part provided the supporting environment of the application software system. It provided a variety of services and operational environment for the realization of the function of the application system, and guaranteed the information sharing among systems and among modules. (3) The foundation platform was composed of the mobile and internet platform, database, WEB services, and application services. It provided a support to the system hardware and software. The framework diagram of digital medical system was shown in Figure 1.

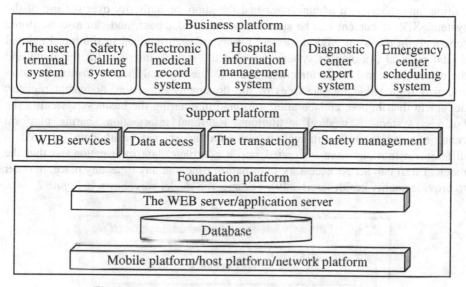

Fig. 1. The framework diagram of digital medical system

3.3 The Environment of the System Software and Hardware

The hardware involved in this system was divided into two categories, of which the first was the one about the health care, such as: the pulse sensor, blood pressure sensor, oxygen sensor, temperature sensor, acceleration sensor, electrocardiogram equipment and electroencephalograph, etc. The second type was the one about the computer and network, such as: personal computers, laptops, IPTV, RFID equipment, intelligent mobile phones with the positioning function, PDA, and other equipment.

The input, processing and output flows of an application were separated into the corresponding model layer, view layer and control layer in MVC (model-view-control) design patterns according to the model M, view V and control C mode [7]. The application program in the design pattern was divided into the model, view and controller of three different parts, of which each part had its corresponding different functions.

The application System in community hospitals, medical diagnostic center, and the emergency center were all based on the J2EE platform, but the business logic encapsulated in the EJB containers in the business layer was different. There were J2ME, Windows, Symbian, Mac OS, Android OS, Web OS and OMS platform in the software environment

3.4   The System Model

Operations related to a system administrator in the background management terminal of the digital medical system were provided, including the system data update and editing, and the system administrator had the supreme authority over the use of the system. XDS document can be stored in a central database, and can also be stored separately in the local XDS document library of each medical institution. Regional digital medical information system provided the service interface for the client and the regional medical institutions front-end [8]. The regional medical institution front-end was responsible for interfaces of the document extraction, document register, document distribution, collaboration information entering the business system in the business system of medical institutions. Regional information sharing platform centering on resident health records was established, and health data scattered in different institutions were integrated into a complete data information, so that the medical staff can access necessary information timely at any time, any place, in order to provide quality health services [9]. The specific model was shown in Figure 2.

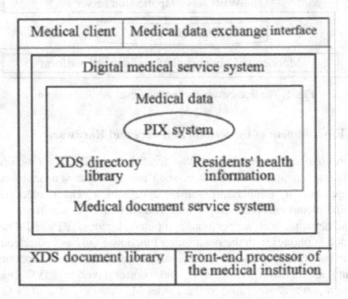

Fig. 2. The specific model of the system

3.5   Application System Framework

An application system framework included the following five layers: client layer, presentation layer, mobile layer, business layer and EIS (enterprise information system) layer.

Client layer: the Web was the main way of J2EE client layer in the system. In addition, there were ways such as the mobile Internet (WAP/HTTP), SMS (CMPP/SGIP), etc.

Presentation layer: the presentation layer mainly included the JSP (Java Server Pages) pages and Java Servlet components of two parts, of which the JSP page programming design realized the view function in the MVC pattern, and the servlet components used in the background service program realized the control function in the MVC pattern.

Mobile layer: this system realized the function of sending and receiving short message via SMS access modules in the mobile layer. Related physiological data, diagnostic results and suggestions, health care information, etc can be sent and received through SMS access modules.

Business layer: EJB was used in the business layer of the system to encapsulate specific business and achieve the model function of the MVC pattern.

EIS (enterprise information system) layer: in this system, ERP and CRM in EIS layer were the existing system of the community hospital, medical diagnostic center, and emergency center.

In order to further improve the effectiveness of medical diagnosis, monitoring control window can be enabled and the current patient physiological information were examined in the current window [10]

4   System Function Module

The main function modules of this system included: remote medical diagnosis module, health care module, positioning and tracking module, emergency calls and emergency module.

4.1   Remote Medical Diagnosis Module Function

The users needing help can be diagnosed and cured in face to face by doctors and experts in the community hospitals and diagnostic center through IPTV, home computers and mobile terminal. The process was: the user used the dynamic identity to be authentication login and enter the system in the terminal (computer or mobile phone), submitted the consultation application to the hospital diagnosis center, then the application was checked, and the doctor and consultation specific time were arranged after the consultation application was received by the hospital diagnostic center. The hospital diagnosis center will send the consultation information to the applicant or the user by calling system. The confirmation message will be replied by the applicant to the hospital diagnosis center after the consultation arrangement information received. Users were diagnosed in the video by the doctor in the diagnostic center by using the video system of a mobile phone or family IPTV. The doctor in the diagnostic center made diagnosis based on user's various physical parameters, electronic medical records and the information obtained through the video. After diagnosis, the doctor filled out the diagnosis report and submitted it to the server. Then, the diagnostic center server sent the diagnosis result to the user. Medical common human physiological parameters were collected through different sensors selected based on different human physiological parameters characteristics [11].

Telemedicine systems can be divided into the acquisition of physiological parameters (completed by physiological parameters collection terminal), data transmission (completed by wireless communication network) and remote monitoring (completed by remote monitoring center server terminal) and other modules, according to their functions and processes [12]. The service system of the regional health information-sharing platform provided service interface for clients of regional health information sharing platform and front-end processor of regional medical institutions [13].

4.2   Health Care Function Module

Patients collected the physiological data information with carried medical sensors, and then sent it to the medical diagnosis center. After the medical diagnostic center received the patient's physiological data, the patient's electronic medical record was accessed. Furthermore, both the physiological data information and electronic medical records were sent to the expert system. Based on the patient's electronic medical record information and patient's physiological data, expert system made a scientific judgment, and gave a guideline. If the treatment were needed, it would be sent to the information management module together with the doctor diagnosis information. After the doctor information management module received the diagnosis results from the expert system, the diagnosis and treatment doctor was recommended for the patient, and the recommendation information was sent to the transceiver module. Information transceiver module put forward the advice information to patients. After the treatment information was confirmed by patients, the treatment time and location were confirmed by the doctor information management module, and the detailed treatment information was sent to the medical center. After the medical center received the information, the detailed treatment information was sent to patients. At the same time, the electronic medical records of patients were updated.

4.3   Positioning and Tracking Module Function

When users or patients were within the community, the user's accurate location information can be obtained through their carried sensors. This localization method is mainly aimed at children, the elderly and disabled. When patients or monitored person did activities in community, the physiological data information was collected in real time through the carried sensors, and sent it to the diagnosis center. The diagnostic center will send its identity ID to the emergency medical center, based on the monitored person's phone number and the RFID electronic tag. Based on the identity ID, the emergency center accessed its electronic medical records and related data in the electronic medical record system, and searched rapidly the right doctors and medical staff. At the same time, the information will be sent to the person's family through the security call system. If necessary, the nearest ambulance and community hospitals will be notified to get the treatment for the patient in the fastest time. The above process requires that the same patient ID be identified with different code in different medical institutions. When the patient referral was happened between different medical institutions, referral or collaborative information need be exchanged to share medical documents. When sharing medical documents, the first is to accurately identify the patient ID, which needs a cross-reference system to link the patient identification code in different medical institutions through the index. When a system needs to be accessed, the patient identification code can be provided in the system [14].

4.4   Emergency Calls and Emergency Module Function

When a user or patient had an emergency symptom or sudden emergency, the position information of the person can be obtained in the fastest time through positioning and tracking function. The emergency medical center distributed the ambulance or notified the nearest ambulance to rush to the scene. The medical first aid center accessed the person's electronic medical record information from the electronic medical record system, and sent it to the hospital. At the same time, the injured family information was found, and the person's accident and location information were sent to their relatives through the security call system. The emergency medical center forwarded the person's real-time physiological data and electronic medical records information to the hospital, in order to facilitate making the most effective treatment plan. The emergency medical center determined whether need to inform the police according to the accident type information. If necessary, the nearest police will be informed. The emergency medical center found the injured insurance information, notified the insurance company to is prepared to compensate. Hospitals recorded the relevant medical procedures and information to the person's electronic medical records in the whole course of the treatment. Then, the data in the electronic medical record system was updated.

5   Discussion and Implication

The wireless sensing network technology, communication technology, password technology and network technology were integrated in this study system to realize the remote medical treatment, health care, positioning and tracking, emergency calls and first aid, etc. In the paper, details about the system hardware and software platform, system structure and the framework of system function, etc. were illustrated separately in the paper. How to apply the computer technology, network technology and communication technology into the specific examples in the real life of people was elaborated through the general design of digital medical system under the network environment. It fully embodies what roles the modern technology can play in today's life. The design of the system is feasible in theory, but in the actual operation and test process, there may be a few problems needed to be solved. It will be continuously developed and improved in the future.

References

1. Egan, G.F., Liu, I.Q.: Computers and Networks in Medical and Healthcare Systems. Computers in Biology and Medicine 25, 355–365 (1995)
2. Li, G.: Design and implementation of mobile medical emergency self-help system based on the 3 G. Shanghai (2009)
3. Hua, X.: Design and development of hospital information management system. Chengdu (2012)
4. Cheng, N.: Digital medical. Digital life 3, 33 (2001)
5. Li, H.: Digital hospital——hospital mode in the future. Medical and Health Care Equipment 12, 126–127 (2003)
6. Li, L.: Digital hospital——The necessity of the hospital modernization developing. China Journal of Modern Medicine 21, 153–155 (2004)
7. John, D.: Model-View-Cont roller (MVC) Architeture (2000),
   http://www.jdl.co.uk/bridfings/MVC.pdf
8. Bai, J.: Intelligent Community Health China medical device information 2, 14–16 (2000)
9. Liu, G.: The Design and Development of the Transmission Platform of Community Telemedicine System based on Web. Tianjin Medical University (May 2009)
10. Wu, Q.: Remote Medical Monitoring System based on Wireless Sensor Networks. University of Electronic Science and Technology (February 2006)
11. Yang, Y., Wang, L.: Architecture for body sensor networks. In: IEEE Proceedings of the Perspective in Pervasive Computing, vol. 03, pp. 23–28 (2005)
12. Zhao, Z., Cui, L.: A Remote Health Care System Based on Wireless Sensor Networks. Information and Control (2), 265–269 (2006)
13. Bai, J.: Intelligent Community Health Care System. China Medical Devices Information (2), 14–16 (2000)
14. Li, Z., Gong, X., Yuan, C.: The Development Status and Problems of Community Health Service in China. Chinese Primary Health Care (11), 31–32 (2007)

Improving Physical Activity and Health with Information Technology

Eija Koskivaara

Turku School of Economics, University of Turku, Finland
eija.koskivaara@utu.fi

Abstract. Physical inactivity and overweight/obesity kill 6 million people yearly [1]. Regular physical activity (PA) such as walking, cycling, or participating in sports has significant benefits for health and weight-loss maintenance. It reduces of the risk of diseases, e.g. diabetes, depression, or helps weight controlling. This one year case study explores how daily monitoring of objective PA and weight effects on body mass index (BMI) -value when the target is to achieve globally accepted normal BMI-value. The study aims to learn by cost-effective modelling how improvements in wellbeing and health on an individual level occur with the help of information technology gadgets.

Keywords: physical inactivity, activity monitoring, improvements in health and wellbeing.

1 Introduction

World Health Organization (WHO) has reported physical inactivity as the fourth leading risk factor for global mortality causing an estimated 3.2 million deaths globally. Indeed the number of deaths per year increases to six million when also overweight and obesity are counted together with physical inactivity [1]. This is a tremendous figure - especially - as it is mostly related to our living habits.

Regular physical activity (PA) such as walking, cycling, or participating in sports has significant benefits for health and weight-loss maintenance [2]. Although, the nature between objectively measured of physical activity and abdominal fat distribution has not been well characterized [3], previous studies have firmly shown that PA reduces of the risk of diseases and reduce mortality and extend life expectancy [4, 5]. But the challenge is: How do we change our way of living? How can we be more physical active in modern information society? Indeed, we need to find practical solutions how this is possible to do with the help of ICT gadgets.

To be more active is challenging as at the same time there are several new attractive leisure time thefts such as Angry Birds, Facebook, You Tube, Play Stations, Wiis, and virtual games on internet. As technology devices and services are penetrating the society into deeper levels, the need for studying their usefulness for physical activity and wellness becomes imperative. Modern technology and popularization of internet has brought a variety of applications aiming at promoting personal health and

H. Li et al. (Eds.): I3E 2014, IFIP AICT 445, pp. 185–194, 2014.

wellness available for layman, such as pedometers, heart rate monitors, and multidimensional accelometers.

The PA is a well-studied field within healthcare research internationally. We have reviewed literature and found around two hundred research journals where technology have been used for improving the individual ability to get encouraged for achieving recommended levels of exercise and physical activity [6]. However, in most of these research articles the research design and sample is based on treatment of some disease. In order to increase the physical activity of population we need to understand how normal layman use information and communication technology (ICT) devices to support their physical activity and to improve their health and wellbeing.

In this one year case study our focus is on exploring how daily monitoring of PA, physical exercise (PE) and weight (WE) effects on Body Mass Index (BMI) -value. The BMI is commonly used to classify underweight, overweight and obesity in adults. It is a simple index of weight-for-height. The BMI is defined as the weight in kilograms divided by the square of the height in meters (kg/m2). The international classification of adult BMI for normal range is from 18.50 to 24.99. The wellbeing target of the current study is to achieved globally accepted normal BMI (body mass index) -value. The assumption is that with the normal range BMI value the life time expectancy is superior and quality of live is better than with overweight situations: i.e. health and wellbeing of an individual improves when BMI value of overweight/obese decreases.

The primary contribution of the study presents how information and communications devices and the data they provide can be used to improve our everyday health. In our study model we use this data to support achieving normal BMI value. The success of the model is judged based on achieved BMI target.

The structure of the paper is organized as follows. Section two focuses on research background: 1) the use of information and communication technology (ICT) for health and well-being, and 2) global trends behind the study. In section three the research design is described. The results are presented in section four. Conclusions and research limitations as well as the future research directions end the papers.

2   Research Background

2.1   ICT Used in or Proposed to Use for Improve Health

ICT has been proposed to improve health in many ways and in different levels but there are communication and integration challenges. Analysis of big data on population level is one approach. Electronic health records could improve population health by including better understanding of the level and distribution of disease, function, and well-being within populations [7]. When an individual get access to his or her own data electronic health records it can be called personal health record which enables patients to access their health information and improves care quality by supporting self-care [8]. However, then personal health records need to be integrated with physicians' electronic health records systems and provide shared access both ways in addition to secure e-mail communication and educational modules [9].

Delivering health related data via internet and establish kiosk and centers have worked very well in developing countries [10, 11]. However, delivering healthcare information totally freely, for example via You Tube, requires to design some kind of interventions to enable consumers to critically assimilate the information with more authoritative information sources to make effective healthcare decisions [12].

On the other hand, implementations of new information systems have faced difficulties, especially when it changes dramatically the well-established business models in the field. For example, in Europe implementations of electronic prescriptions have taken more time than expected [13, 14].

On individual level there has been several approached on using technology to health conditions. A systematic literature review of mobile health technologies reveal that they have potential to be used as tools for the prevention and treatment of overweight and obesity, particularly with mobile phones and texting, which are already used daily by most of the population [15]. Based on another systematic review, there is an argument that despite the bold promise of mHealth to improve health care, much remains unknown about whether and how this will be fulfilled [16]. Electronic lifestyle activity monitors are commercially popular and show promise for use in public health interventions and provide feedback via an app in computer or mobile [17].

Using the PA devices provides more precise data than a subjective self-assessment. The use and feasibility of physical activity promoting websites and applications have been studied with encouraging results [18,19]. However, little is known about how objective physical activity assessment on 24/7 basis effects health. In this study, we observe objective PA assessment together with self-reporting PE, WE, and measure the success, i.e. health effects, of the project with the change of the BMI value. The feasibility of high intensity PA value was confirmed with self assessment dairy of physical exercise. The study aims to model a cost-effective way for improving well-being and health on individual level without any communications or integrations to health professionals.

2.2 Global Trends in World Health

There are convincing evidence that a sedentary and unfit way of living increase the risk of numerous chronic diseases and conditions and even decreases longevity [21]. A physical inactivity has been one of the highest leading global risks for mortality in the world already for some time [21]. Physical activity is defined as any bodily movement produced by skeletal muscles that require energy expenditure. Physical inactivity causes an estimated 3.2 million deaths globally [1].

Overweight and obesity are defined as abnormal or excessive fat accumulation that presents a risk to health. A crude population measure of obesity is the body mass index (BMI), a person's weight (in kilograms) divided by the square of his or her height (in meters). A person with a BMI of 30 or more is generally considered obese. A person with a BMI equal to or more than 25 is considered overweight. Obesity has reached epidemic proportions globally, with at least 2.8 million people dying each year as a result of being overweight or obese [1].

The physical inactivity and obesity are modern rising risk factors and they can be found in everywhere, i.e. in high, middle and low income countries. Together they kill about 6 million persons per year. In order to avoid this, WHO has been launching "Global Strategy on Diet, Physical Activity and Health".

Being physically active is a major contributor for both physical and mental wellbeing [22]. PA has many scientifically proven health enhancing effects and the PA is extremely effective in preventing and treating just lifestyle connected diseases. Studies also confirm that long term physical activity e.g. walking is associated with significant better cognitive function and reduced risk of dementia [23, 24]. A major goal for public health is to identify evidence-based interventions to promote PA in populations [25]. This includes research on how ICT could be used to promote PA in our everyday life.

Current global PA guidelines given by WHO for adults accumulate at least 150 minutes of moderate-intensity aerobic physical activity throughout the week or at least 75 minutes of vigorous-intensity aerobic PA throughout the week or an equivalent combination of moderate- and vigorous-intensity activity [26]. The duration of PA should be at least 10 minutes per time. And one should involve at least two or more muscle-strengthening activities per week. Globally, around 31% of adults were insufficiently PA in 2008 according to Global Health Observatory of WHO. The lack of PA has effect on public health as there is evidence that inactive individuals have higher risks for many lifestyle diseases such as coronary heart disease, high blood pressure, stroke, type 2 diabetes, metabolic syndrome, colon and breast cancer, and depression. These risks could be significantly reduced with preventive behaviors, such as improving nutrition and enhancing in regular PA [27].

In many countries the PA has been promoted by recommendations. The PA recommendations differ from country to country, although the recommendations for adults of WHO are probably most well-known. Many persons want to be more physically active, but achieving sustainable changes in lifestyles can be challenging and behavior determinants differ based on individuals and environments [28]. Despite many benefits of the PA, initiation and maintenance rates in the general population have been rather disappointing [29]. And based on the resent figures we still need methods and implementation of successful PA interventions.

ICT-embedded health and wellness services have suggest empowering people to manage their health [30]. Indeed, research evidence suggests that individuals who exercise are more likely to maintain weight losses [31].

3   Monitoring of Physical Activity and Weight during the Study

The ICT has penetrated into our lives to a level where it has started to show as an integrated part of our bodies and ways of living. It has reshaped our habits. However, information and communication technology along with effective decision making combining motivational and environmental factors, can definitely improve our health level. One of the major contributing parts of physical activity is technology tools and services, such as: pedometer, accelerometer, heart rate monitor, social networking, sport gaming and devices, computer based counseling system, global positioning technology, mobile entertainment electronics. In this study, we are more interested in use and awareness of activity monitoring tool.

Activity monitors provide a means to examine the intensity, frequency, and duration of PA. The knowledge of daily activity may motivate some of us to be more active. The purpose of this study is to analyze and compare the change of PA per day by using Polar Active [32] activity monitor for one year study period. Polar Active was chosen for this study because of its features. For example, it measures activity 24/7, it divides activity to different zones, and it contains daily activity target feature. The technology for the target calculations was the most important feature for selecting Polar Active. Primary, Polar Active has been developed for Physical Education purposes of students: The tacit educational aspect is, indeed, one of the issues in this study context: Are we able to change our living habits with the help of technology? Are we able to improve our wellbeing or health by using technology daily to support our change of living habits?

Polar activity technology detects and filters activity intensity, and calculates it to MET (Metabolic Equivalent of Task, or simply metabolic equivalent, a physiological measure expressing the energy cost of physical activities) values. In Polar activity technology METs are used to accumulate time in the five different activity zones: very easy (1-2 MET), easy (2-3.5 MET), moderate (3.5-5 MET), vigorous (5-8 MET), and vigorous+ (>8 MET) (Table 1).

Table 1. Activity zones, MET values, and zones that add up active time in Polar Active

ZONE	MET	ACTIVE TIME
Vigorous+	>8	x
Vigorous	5-8	x
Moderate	3.5-8	x
Easy	2-3.5	
Very easy	1-2	

The calculation of active time (≥3.5 MET) has been patented by Polar. Active time is the sum of the times spent in the 3 upper zones. In this case study, we are interested in this MET over 3.5 values per day. Indeed, the data can be downloaded to appropriate Polar web service where the activity zones and also sleeping time will be visualized. In the web service, also sleeping time or more accurately time in bed can be calculated.

In Polar activity monitors (Polar Active, FA20, AW200), one dimensional (1D) acceleration is measured. When comparing 1D to 3D measurement in accelerometers in general, it has been shown that 3D does not significantly improve the prediction of energy expenditure compared to 1D [32]. In Polar devices, 1D also provides longer battery lifetime.

The data is analyzed in 30 s epochs, and all epochs above 3.5 MET accumulate active time. Typical activity for moderate 3.5-5 MET zone is brisk walking. For vigorous 5-8 MET zone typical is playing and games e.g. playground games and rope jumping. Basketball, football and soccer usually are 7-8 METs. Typical vigorous+ activity is running fast. Sedentary activities (e.g. screen time) accumulate very easy zone. All the features and calculations of the monitor apply to all age groups from children to adolescents to adults.

Activity is counted in METs that express energy expenditure and are multiples of resting metabolic rate (1 MET=BMR). Calories are expressed as total kilocalories summing the daily activity calories and the user's basic metabolic rate through day and night. Steps are accumulated when activity is detected when cadence exceeds 70 steps per minute.

The acceleration based measurement does not measure accurately all activity modes. For example cycling, indoor cycling, weight training (gym) and 'light' aerobics do not give accurate METs, calories or steps.

Plus the activity values given by Polar activity monitor, the daily physical activity or actually physical exercise (PE) was also traced by minutes with watch. They daily weight was measured with Omron Body Composition Monitor BF500 at the same time of day with the same clothing, i.e. every morning before breakfast.

All the daily values were collected to the metadata file for further analyses from one year trim down case study of overweight middle-aged blue-collar woman. Data is gathered and analyzed based on daily PA (>3.5 MET, Polar Active) and physical exercise (PE) in minutes, and morning weight (kg). The PA and PE counselling was similar: avoid two successive days of non-PA/PE. The day was non-PA if the value was below 60 minutes. The trim down target was set to -1 kg/month.

4   Results

In 87 days PA was below one hour in two or more successive days (24 %). In 96 days there was no PE in two or more successive days (26 %). In 60 days both these values were below the target (16 %). In 115 days the morning weight was not measured because of work or holiday (32 %). The correlation between the change of weight and PA (-0.0343857) was higher than the correlation between the change of weight and PE (-0.01237). BMI-value decreased (29.0 -> 26.7), but is still 1.8 above normal value. Trim down project was 50 % successful (-6kg/12kg). Figure 1 shows daily PE and PA. Figure 2 shows change of weight during the study time.

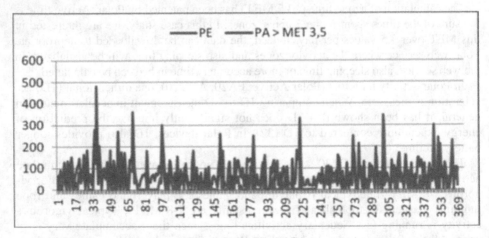

Fig. 1. Daily PE and PA

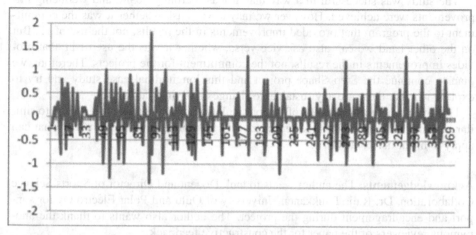

Fig. 2. Change in weight (kg/day)

Monitoring daily PA, PE and morning weight do have positive effects on BMI trim down project. Daily feed-back such as PA value keeps trim down project going on. A moderate target as trim down one kilogram per month is challenging when it is based on only changes in PA. But we need simply models to increase PA and decrease BMI worldwide.

5 Conclusions, Research Limitations and Future Research Directions

Physical inactivity and obesity are leading global risks for mortality in the world. One of the major contributors for increasing the physical activity and decreasing obesity could be information and communication technology tools and services. This case study tried to develop an easy and implementable but effective ICT supported counselling interaction program for PA supported weight controlling program. This is important both for society and individuals. We all need urgently new approaches and tools to reshape our habits or create new ones in a rapidly changing world. Actually, we need different kind of solutions for keeping our daily PA on recommended level. In some cases we need education and training to understand what the PA recommendations are and how they are reached or how to use ICT tools to improve our health and fitness.

For investigating and understanding this phenomenon we have created a Step-Shape –project. This case study is one part of the project where we try to understand and learn the effects of IT use on health and fitness goals. This is a concreate, practical context and individual depending knowledge and therefore case study approach suits for it by giving us a possibility to learn something new.

The study was successful in a way that at least partially health and wellbeing improvements were achieved. However we may always ask whether it was the commitment to the program that provided improvements in the results, not the use of IT. But on the other hand we can also ask vice versa, whether it was the use of IT that provides improvements in the results, not the commitment for the projects. Therefore, we plan to continue the Step Shape project and this longitudinal case study and try to explore whether there are any sustainable changes in living habits.

However, whether the results of this particular case study can be transferred to similar situations are open. It is the reader, not the researcher, who determines what can apply to his or her context.

Acknowledgements. The author wants to tank Docent and Director of Sports Science Collaboration, Dr. Raija Laukkanen University of Oulu and Polar Electro Oy for support and encouragement during the project. The author also wants to thank the anonymous reviewers of the paper for the constructive feedback.

References

1. WHO 2015, "Global Strategy on Diet, Physical Activity and Health", http://www.who.int/dietphysicalactivity/factsheet_inactivity/en/ (retrieved July 14, 2014)
2. Ekelund, U., Besson, H., Luan, J., et al.: Physical activity and gain in abdominal adiposity and body weight: prospective cohort study in 288,498 men and women. American Journal of Clinical Nutrition 93(4), 826–835 (2011)
3. Philipsen, A.: Associations of Objectively Measured Physical Activity and Abdominal Fat Distribution Medical Science and Sports Exercise (September 9, 2014)
4. Vanhees, L., Geladas, N., Hansen, D., et al.: Importance of characteristics and modalities of physical activity and exercise in the management of cardiovascular health in individuals with cardiovascular risk factors: recommendations from the EACPR. Part II. European Journal Preventive Cardiology 19(5), 1005–1033 (2012)
5. Wen, C.P., Wai, J.P.M., Tsai, M.K., et al.: Minimum amount of physical activity for reduced mortality and extended life expectancy: a prospective cohort study. The Lancet 378(9798), 1244–1253 (2001)
6. Tudor-Locke, C., Hart, T.L., Washington, T.L.: Expected values for pedometer-determined physical activity in older populations. International Journal of Behavioral Nutrition and Physical Activity 6, 59 (2009)
7. Friedman, D.J., Parrish, R.G., Ross, D.A.: Electronic health records and US public health: current realities and future promise. Am. J. Public Health 103(9), 1560–1567 (2013)
8. Ahmadi, M., Jeddi, F.R., Gohari, M.R., Sadoughi, F.: A review of the personal health records in selected countries and Iran. Med. Syst. 36(2), 371–382 (2012)
9. Johansen, M.A., Henriksen, E.: The Evolution of Personal Health Records and their Role for Self-Management: A Literature Review. Stud. Health Technol. Inform. 205, 458–462 (2014)
10. Joshi, A., Puricelli Perin, D.M., Arora, M.: Using Portable Health Information Kiosk to assess chronic disease burden in remote settings. Rural Remote Health 13, 2 (2013)

11. Zakar, R., Zakar, M.Z., Qureshi, S., Fischer, F.: "Harnessing information technology to improve women's health information: evidence from Pakistan". BMC Womens Health 14 (September 4, 2014)
12. Madathil, K.C., Rivera-Rodriguez, A.J., Greenstein, J.S., Gramopadhye, A.K.: Healthcare information on YouTube: A systematic review. Health Informatics J. (March 25, 2014)
13. Hammar, T., Ohlson, M., Hanson, E., Petersson, G.: Implementation of information systems at pharmacies - A case study from the re-regulated pharmacy market in Sweden. Res. Social Adm.Pharm. (August 2014)
14. Mäkinen, M., Rautava, P., Forsström, J., Aärimaa, M.: Electronic prescriptions are slowly spreading in the European Union. Telemed. J. E Health 17(3), 217–222 (2011)
15. Sarno, F., Canella, D.S., Bandoni, D.H.: Mobile health and excess weight: a systematic review. Rev. Panam. Salud Publica 35(5-6), 424–431 (2014)
16. Peiris, D., Praveen, D., Johnson, C., Mogulluru, K.: Use of mHealth Systems and Tools for Non-Communicable Diseases in Low- and Middle-Income Countries: a Systematic Review. J. Cardiovasc. Transl. Res (September 11, 2014)
17. Lyons, E.J., Lewis, Z.H., Mayrsohn, B.G., Rowland, J.L.: Behavior change techniques implemented in electronic lifestyle activity monitors: a systematic content analysis. J. Med. Internet Res. 16(8) (August 15, 2014)
18. Vandelanotte, C., Kirwan, M., Rebar, A., Alley, S., Short, C., Fallon, L., Buzza, G., Schoeppe, S., Maher, C., Duncan, M.J.: Examining the use of evidence-based and social media supported tools in freely accessible physical activity intervention websites. Int. J. Behav. Nutr. Phys. Act. 11, 105 (2014)
19. Al Ayubi, S.U., Parmanto, B., Branch, R., Ding, D.: A Persuasive and Social mHealth Application for Physical Activity: A Usability and Feasibility Study. JMIR Mhealth Uhealth 2(2), e25 (2014)
20. Blair, S., Haskell, W.: Objectively measured physical activity and mortality in older adults. The Journal of American Medical Association 296, 216–218 (2006)
21. WHO 2009, "GLOBAL HEALTH RISKS Mortality and burden of disease attributable to selected major risks". WHO Library Cataloguing-in Publication Data. WHO Press
22. Das, P., Horton, R.: Rethinking our approach to physical activity. The Lancet 380(9838), 189–190 (2012)
23. Abbot, R., et al.: Walking and dementia in physically capable elderly men. The Journal of American Medical Association 292, 14447–1453 (2004)
24. Weuve, J., et al.: Physical activity, including walking, and cognitive function in older women. The Journal of American Medical Association 292, 1447–1453 (2004)
25. Pratt, M., Sarmiento, O.L., Montes, F., Ogilvie, D., Marcus, B.H., Perez, L., Brownson, R.: The implications of megatrends in information and communication technology and transportation for changes in global physical activity. The Lancet 380 (9838), 828–293 (2012)
26. WHO 2014 "Physical Activity and Adults. Recommended levels of physical activity for adults aged 18 - 64 years", http://www.who.int/dietphysicalactivity/factsheet_adults/en/index.html
27. Warburton, D.E.R., Nicol, C.W., Bredin, S.S.D.: Health benefits of physical activity: the evidence. Review2 CMAJ 174(6) (March 14, 2006), doi:10.1503/cmaj.051351
28. Sherwood, N.E., Jeffery, R.W.: The behavioral determinants of exercise: implications for physical activity interventions. Annu. Rev. Nutr. 20, 21–44 (2000)
29. Estabrooks, P., Glasgow, R., Dzewaltowaki, D.: Physical activity promotion through primary care. The Journal of American Medical Association 289, 2913–2916 (2003)

30. Anita, H., Kirsikka, K., Henri, H., Niilo, S.: Rethinking health: ICT-enabled services to empower people to manage their health. IEEE Reviews in Biomedical Engineering 4, 119–139 (2011)
31. Baker, C.W., Browell, K.D.: Physical activity and maintenance of weight loss: physiological and psychological mechanisms. Physical Activity and Obesity, 311–328 (2000) ISBN 0-88011-909-8
32. Brugniaux, J.V., Niva, A., Pulkkinen, I., et al.: Polar Activity Watch 200: a new device to accurately assess energy expenditure. Br. J. Sports Med. 44, 245–249 (2010)

A Novel Regional Cloud Digital Library Network Based on Mobile Ad Hoc Networks

Zhiming Zhang[1] and Wei Zhang[2,*]

[1] Department of Information Engineering, Engineering University of CAPF, China
4406875@qq.com
[2] Refree of The 210th Institute of the Sixth Academy of CASIC, China
zwzy0717@163.com

Abstract. At present the digital library has entered into the period of cloud computing. The cloud digital library is a kind of virtual library, which is built upon Internet, and uses the cloud computing technology to provide services for readers. For the status and developing trend of cloud digital libraries, a regional cloud digital library network based on the mobile ad hoc network is proposed, the network architecture is designed, and its performance is evaluated by network simulation software NS-2. The result shows that the performance of the network is good, and it is feasible. The research result in the paper is valuable for the study and application of cloud digital libraries.

Keywords: cloud digital library, cloud computing, mobile Ad hoc network, region network architecture.

1 Introduction

The cloud computing technology is one of the most important technologies in information technology domain these years. In 2009, cloud computing was defined by National Institute of Standards and Technology (NIST) as a model in which networks were used to provide rapid and convenient services for a series of shared computer resources, nonetheless the administration cost in demand and suppliers' interaction cost are minimum 1. It leads the development of industry and society informatization, along with the popularity of Internet of things and mobile Internet; the permeation of cloud computing in various industries was becoming increasingly apparent 2. After experiencing the period of Internet, grid mesh and Web2.0 by sequence, digital library is entering into the period of "cloud computing". When using cloud computing, library services can reduce costs and improve efficiency greatly, according with the library's development needs. The largest organization for library cooperation in the world, Online Computer Library Center (OCLC) has already used cloud computer technology to establish a cloud digital library named OCLC Worldcat to provide services for readers in different countries 3. China Academic Library & Information System (CALIS) is being constructed in the third stage, planning for building multilevel

---

* Corresponding author.

H. Li et al. (Eds.): I3E 2014, IFIP AICT 445, pp. 195–203, 2014.

sharing center using cloud computing technology and realizing the localized and low cost college digital libraries 4.

During the past few years, some researchers have already investigated the theory and application of cloud digital library. Ref 5 surveys the advance on basic theory of cloud libraries, and points out the study on theory and application of cloud library includes five phases, i.e., consumers, resources, cloud services, cloud platform and cloud library administrators. Taking the cloud digital library in Shanxi University of Finance and Economics for example, ref 6 discusses the basic architecture and function of the cloud digital library, and explores the approaches and methods by which traditional libraries enter into the cloud digital libraries. Ref 7 proposes that the application of the cloud computing technology in libraries includes three developing phases, i.e., digital library, regional cloud library and total cloud library, and analyzes the transforming direction of traditional library and the challenges of the development of cloud libraries. Ref 8 proposes a novel service-oriented and layered regional cloud library, designs the architecture, which includes consumer layer, access layer, application portal layer, application layer, supporting tool layer, basic technology layer and cloud resource layer. However, in the above references, the authors only discuss the theory and application of the cloud digital library, or design the system model, without further simulation of the performance of the system model.

Based on the former work, the theory and application of the cloud digital library are further studied in the paper. Firstly, the concept and architecture of the cloud digital library is overviewed. Then, for the status and developing trend of cloud digital libraries, a regional cloud digital library network based on MANET is proposed, the network architecture is designed, and its performance is evaluated by network simulation software NS-2. The result shows that the performance of the network is good, and it is feasible. The research result in the paper is valuable for the study and application of cloud digital libraries.

2   Overview of the Cloud Digital Library

2.1   Concept of the Cloud Digital Library

Cloud digital library can be defined as a virtualization library based on Internet and providing all kinds of services for readers using the cloud computing technology. In other words, cloud digital library is library facility and service constructed by the cloud computing technology. The cloud digital library integrates the digital resources in several libraries by a cluster of parallel computers in a large-scale library. Thus, the cloud digital library can search resources and process data rapidly and conveniently, and can be accessed by users on demand. In the traditional library, different libraries cannot share each resource due to the disparity of every system. Whereas the cloud digital library can expand information services, change the library service mode, meet the personalized need of users, and thus bring vast opportunity to its development 6.

2.2   Advantages of the Cloud Digital Library

The cloud digital library can not only avoid repeated construction to achieve full sharing of the resources, but also improve network performance and service efficiency greatly. Specifically, the advantages are as followed 9:

(1) Reduce construction cost and improve operating efficiency. On the basis of cloud computing, the general medium or small scale libraries will not need to spend a lot of money on expensive hardware. They only need to construct their own cloud computing or ask the providers to do the construction, and the software could be upgraded and maintained online by providers. At the same time, the backward devices in library could still be fully used, such as to do the simple I/0 interactive computing. Also, we are no longer to worry about problems like data loss, computer virus or server being down, for there are millions of severs in "cloud". Therefore, if one sever goes wrong, the others could continue to work instantly, thus providing the most reliable and safest data storage center to the libraries.

(2) Bring down investment on repeated construction and realize resource sharing. The library which apply cloud computing could co-construct the information commons to share the information and resources with each other. All the digital library resources in the world could be gathered into the storage server of "cloud". The library administrator only needs to administrate, classify the resources and set out the matching visiting rules, and as long as following the rules, the users can obtain the digital resources from every corner of the world simply by entering the key word. The information resources could be fully shared in this case.

(3) Offer customized and personalized service. The current digital library for users could neither allocate resources according to the users' demands, nor provide individualized services. To provide resources according to the needs, and to charge per amount of usage are the outstanding merits of cloud computing. On the basis of cloud computing, the digital libraries will develop in a more personalized, liberalized and diversified direction. Every user could use the applications and digital resources gathered in the "cloud" to construct his personal digital library.

2.3   Architecture of the Cloud Digital Library

At present the architecture of cloud computing can be classified into three categories 10: SaaS (Software as a Service), PaaS (Platform as a Service) and IaaS (Infrastructure as a Service). SaaS is a method of supplying complete applications for the Internet as a service, such as Google Docs, Gmail and Salesforce.com 11. To develop and deploy custom applications, PaaS provides a platform, such as Google App Engine 12 and Microsoft Windows Azure 13. IaaS is a way of providing storage and elastic computing resources on demand, such as Amazon's Simple Storage Service (S3) 14, Elastic Compute Cloud (EC2) 15, and several open source implementations, for example, Eucalyptus 16 and OpenStack 17.

The architecture of the cloud digital library is shown in figure 1. According to the status of libraries and the service provided by cloud computing, its architecture can be generally classified into five layers, i.e., application layer, platform layer, data layer, hardware virtualization layer and infrastructure layer. Every layer is consisted of the corresponding cloud computing service, and can provide the service that the digital library can provide. For example, SaaS is used to constitute the application software, such as Platform for Library Content Selection, Automated Management System; PaaS

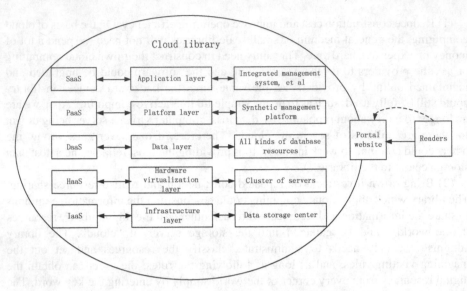

Fig. 1. Architecture of regional cloud digital library

is used to develop application service platform, and provides database services of the cloud digital library; Data as a Service (DaaS) is used to integrate the literatures in every library; Hardware as a Service (HaaS) is used to form the cluster of servers, owning the effective ability of flexible computing; IaaS is used to construct the storage and data center segment of the cloud digital library. Finally, the cloud digital library based on Internet is established. On the other hand, the portal website is created, and readers can enjoy the services provided by the cloud digital library 18.

3   Regional Cloud Digital Library Network Based on MANET

3.1   Concept of Regional Cloud Digital Library

The construction and development of cloud digital libraries is a progressive process. The cloud digital library will develop gradually with the progress of the cloud computing-related technology and philosophy, and transit from the regional cloud digital library to the cloud digital library totally based on Internet gradually. At present, cloud computing technology is in the stage of preliminary research and application. Therefore, regional readers need-oriented networking services based on Internet are suitable to be carried on, and the regional cloud digital library can be constructed.

The regional cloud digital library is based on a digital library whose construction and service are senior in a certain region. Through the synthetic integration of technology, resource and service, the efficient software, hardware and administration platform of digital resource in digital libraries are established to provide every service of cloud digital libraries for the users in the region. The main function of the regional cloud digital library includes: (1) to provide literature-related service for the users in the region; (2) to provide software and hardware platform of digital libraries for other institutes in the region; (3) to provide literature, software and hardware platform, and administration integrated service for the users in the region 8.

3.2   Architecture of Regional Cloud Digital Library Network Based on MANET

The user network of the regional cloud digital library in the paper adopts Mobile Ad hoc Network (MANET). MANET is a kind of mobile wireless network, which is consisted of mobile nodes and does not rely on the network infrastructure. In MANET, nodes exchange data by their wireless sending and receiving equipments. When nodes are beyond their communication range, the multi-hop communication will be accomplished by the relay of other nodes. In the paper, MANET is used in the cloud digital library to making full use of advantages of flexibility and high efficiency.

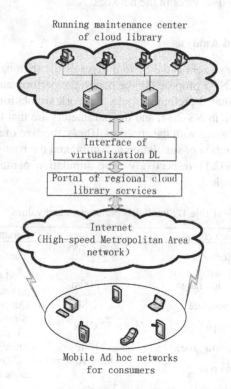

Running maintenance center
of cloud library

Interface of
virtualization DL

Portal of regional cloud
library services

Internet
(High-speed Metropolitan Area
network)

Mobile Ad hoc networks
for consumers

Fig. 2. Architecture of regional cloud digital library network based on MANET

The architecture of the regional cloud digital library network based on MANET is shown in figure 2. It can be divided into five layers, i.e., consumer MANET layer, Internet or high-speed MAN layer, portal website access layer, virtualization DL interface layer, and running maintenance center layer. The basic workflow of the network is: (1) the service portal is formed through integration of relative resources by the regional cloud digital library, and the user in the region can access the portal website through MANET using all kinds of network terminal; (2) Interface of virtualization DL adopts Service-Oriented-Architecture (SOA) and provides network service interface for portal website access layer; (3) the virtualization services of hardware, software and resources are realized through the virtualization technology of cloud computing by running maintenance center of the region cloud digital library, and the normal working of the network can be ensured.

Because of the security threat in data storage, reliability of cloud platforms and sustainability of services, user authority management, and virtualization, et al, faced by cloud digital libraries 19, the network we proposed in the paper adopts the scheme of static data encryption to manage the core data in cloud storage areas, the encryption and digital signature technologies in data's storage and transmission process, and Privilege Management Infrastructure (PMI) to control the users' authority for resources to ensure the security the data and services in the network.

3.3   Simulation and Analysis

Since MANET for consumers is the bottom layer of the regional cloud digital library network based on MANET proposed in the paper, the performance of the network will be simulated and evaluated preliminarily by network simulation platform NS-2. The network model is setup in NS-2.34, and the parameters are that the number of users is 10 to 50, every node moves with the speed of 10m/s, the size of simulation scenario is 5000×5000m2, the duration of simulation is 10min, and the routing and MAC protocol are DSR and IEEE 802.11 respectively. The simulation parameters and values are shown concretely in table 1.

Table 1. Simulation parameters and values

Parameters	Values	Parameters	Values
Simulation area	5000×5000m2	Channel type	Wireless Channel
Communication range	300m	MAC protocol	IEEE 802.11
Routing protocol	DSR	Queue type	PriQueue
Flow type	CBR	Simulation time	600s
Maximum moving speed	10m/s	Queue length	50
Antenna type	Omni-Antenna	Channel capacity	100Mps
Propagation	TwoRay-Ground	Data rate	50Mbps
No. of users & Maximal No. of connection	10 & 3, 20 & 6, 30 &9, 40 & 12, 50 & 15		

After simulation, the packets delivery rate, average end-to-end delay, and route costs of the network with the number of the users in MANET are calculated, which are shown in figure 3 to 5. From figure 3 to 5, it can be found that every performance indicator of the network is good enough to meet the need of users to the cloud digital library. On the other hand, simulations show that with the increase of the number of users in MANET, the packets delivery rate decreases, the average end-to-end delay and route costs rise, which indicate the worsening of performance of the network. The reason is that when the number of users increases, the total traffic volume of the network increases, while the bandwidth of the network does not change. Therefore, the packet loss rate, end-to-end delay and route costs increase at the same time.

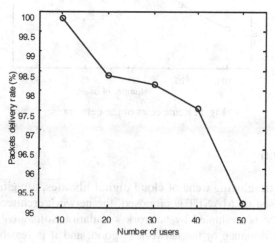

Fig. 3. Packets delivery rate of the network

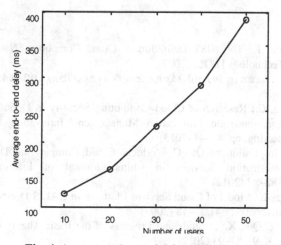

Fig. 4. Average end-to-end delay of the network

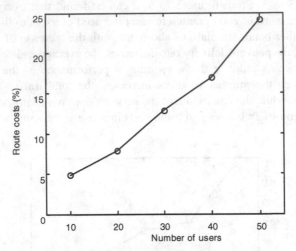

Fig. 5. Route costs of the network

4   Conclusion

For the status and developing trend of cloud digital libraries, a regional cloud digital library network based on MANET is proposed, the network architecture is designed, and its performance is evaluated by network simulation software NS-2. The result shows that the performance of the network is good, and it is feasible. The research result in the paper is valuable for the study and application of cloud digital libraries.

References

1. Mell, P., Grance, T.: The NIST Definition of Cloud Computing. National Institute of Standards and Technology [J/OL] (2009),
   http://csrc.nist.gov/publications/nistpubs/800-145/
   SP800-145.pdf
2. Meng, Q., Gong, C.: Research of Cloud Computing Security in Digital Library. In: 6th International Conference on Information Management, Innovation Management and Industrial Engineering, pp. 41–44 (2013)
3. Wen, Z.: The Inspiration of OCLC Worldcat Cloud Computering Digital Library on Government Information Services in China. Journal of Information Resources Management 4, 82–87 (2012)
4. Wang, W., Chen, L.: Model of Cloud Services Platform in CALIS Digital Library. Journal of Library in University 27(4), 13–18 (2009)
5. Zhang, X., Li, C., Qin, X.: A Study and Progress of the Basic Theory of Cloud Library. Library Tribune 32(9), 87–93 (2012)
6. Wang, H.: Architecture and Implementation of the "Cloud Library" Platform. Theory and Practice of Intelligence 33(10), 108–112 (2010)

7. Hu, X.: Conception of Cloud Library. Theory and Practice of Intelligence 33(6), 29–32 (2010)
8. Hu, X., Shen, H., Zhang, Z.: Research on Construction of Regional Cloud Digital Library. Theory and Practice of Intelligence 34(2), 77–84 (2011)
9. Wei, M., Wang, F., Xu, X.: Development of Digital Libraries on the Basis of Cloud Computing. In: IEEE International Conference on Computer Science and Service System, p. 304 (2012)
10. Lu, W., Zheng, L., Shao, J., et al.: Digital Library Engine: Adapting Digital Library for Cloud Computing. In: IEEE Sixth International Conference on Cloud Computing, pp. 934–941 (2013)
11. Salesforce Customer Relationships Management (CRM) System, http://www.salesforce.com/
12. Google AppEngine, http://code.google.com/appengine
13. Windows Azure Platform, http://www.microsoft.com/windowsazure/
14. Amazon Simple Storage Service, http://aws.amazon.com/s3/
15. Amazon Elastic Compute Cloud, http://aws.amazon.com/ec2/
16. Nurmi, D., Wolski, R., Grzegorczyk, C., et al.: The Eucalyptus Open-Source Cloud-Computing System. In: CCGRID 2009 (2009)
17. OpenStack: The open source, open standards cloud, http://openstack.org
18. Vaquero, L.M., Rodero-Merino, L., Caceres, J., et al.: A Break in the Clouds: Towards a Cloud Definition. ACM SIGCOMM Computer Communication Review 39(1), 50–54 (2009)
19. Ma, X.-T., Chen, C.: Study on Cloud Computing Security Analysis and Management Strategy for Digital Library. Intelligence Science 29(8), 1186–1191

Process Design of Digital Platform for China's Industrial Investment Fund

Xuan Yang[1,*], Jin Chen[1], and Yanbo J. Wang[2,3]

[1] School of Information Technology & Management,
University of International Business and Economics, Beijing, China
hlyangxuan@126.com
chenjin@uibe.edu.cn
[2] Department of Development Planning, China Minsheng Banking Corp., Ltd.
[3] Institute of Finance and Banking, Chinese Academy of Social Sciences, Beijing, China
wangyanbo@cmbc.com.cn

Abstract. Nowadays, the Industrial Investment Fund is booming in China. Though it has achieved big success till now, there are still some key problems troubling the further development of Industrial Investment Fund. Two major challenges are the recognition of appropriate industries and also related companies. In order to solve such problem, we propose a design for the digital platform of Industrial Investment Fund. In the process of this platform, Web Text Extraction and Data Mining techniques are employed in order to help investors to make decisions in a Big Data Analysis manner.

Keywords: Industrial Investment Fund, Digital Platform, Process Design, Data Mining, Web Text Extraction.

1   Introduction

In recent years, Industrial Investment Fund is booming in China. According to *Interim Management Measures of China's Industrial Investment Fund* published by Chinese government, Industrial Investment Fund is defined as a collective investment system, which makes equity investment on the unlisted enterprises. The main investment approaches of China's Industrial Investment Fund include venture capital investment, corporate restructuring investment, and basic facilities investment and so on.

Having studied the literature related to Industrial Investment Fund, we find that the research is mainly focused on general operational mechanism, such as capital resource, organizational structure and withdrawal approaches of Industrial Investment Funds. However, there are few scholars talking about the specific problem which is of vital importance, that is, how to make investment decisions for Industrial Investment Fund. In order to fill this blank, we tried to propose a design for the digital platform of Industrial Investment Fund, which helps to recognize appropriate industries and related companies whom to invest. In the process of this platform, Web Text

---

* Corresponding author.

H. Li et al. (Eds.): I3E 2014, IFIP AICT 445, pp. 204–212, 2014.

Extraction and Data Mining techniques are applied in order to help investors to make decisions, in a Big Data Analysis manner.

The rest of this paper is organized as follows. In section 2, we study the current situation of China's Industrial Investment Fund. Section 3 describes some related work relevant to our study. In section 4, we propose our design of digital platform to identify appropriate industries and related companies for the Industrial Investment Fund. Finally discussions and open issues for further research are given in section 5.

2   Current Situation of China's Industrial Investment Fund

Compared with other investment modes in the international market, China's Industrial Investment Fund is quite similar to Private Equity Fund. However, Industrial Investment Fund has its own characteristics, which are listed as follows.

Table 1. Characteristics of China's Industrial Investment Fund

Organizational Pattern		Two major types: Sino-foreign Joint Venture Fund and Large Chinese Fund
Financing Pattern	Sponsor	Generally some government departments, government policy banks or large state-owned enterprises
	Financing Size	Increasing greatly, less than one billion RMB before year 2005, and usually more than ten billion RMB after 2005.
	Financing Channel	Private
Investment Pattern	Investment Orientation	Usually some growing enterprises in the industries, which are booming or supported by the government.
	Investment Size	Quite large, usually millions RMB for a single investment project.
	Investment Period	Long-term investment, nearly ten years.
	Investment Tools	1. Equity investment 2. Quasi-equity investment 3. Investment on other funds
Exit Pattern		Usually through pre-IPO exit.

Since Chinese State Development Planning Commission (SDPC) has started making researches upon Industrial Investment Fund in 1995, China's Industrial Investment Fund has achieved great progress. According to the data published by Qingke Research Centre, till the end of 2012, there are nearly 560 Industrial Investment Funds in China, and the capital in this market reaches 30 billion US dollars. Since there are still many investors who would like to enter this market, the number mentioned above is estimated to keep growing. Taking 2013 market for example, there were 660 cases of investment in the market and the capital reaches 24.48 billion US dollars, with an increase of 23.7%.

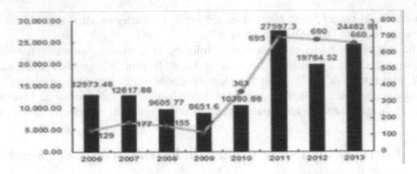

Fig. 1. Market Situation of 2013 China's Private Equity Investment
Source: Qingke Research Centre

Moreover, in order to support the development of Industrial Investment Fund, not only the central government but also many local governments, have made a series of policies and regulations. Till now a legal environment, which contains basic regulations, subsidiary assurances and specific policies, has been established.

3   Related Work

3.1   Foreign Industrial Investment Fund

In western countries, especially in the US, Industrial Investment Fund is considered as "Organized Private Equity Market". As the US is one of the earliest countries which have started to develop Industrial Investment Fund and it has the most influential financial market in the world, a vast number of related researches are about American's Industrial Investment Fund.

Many scholars are interested in the information transmission mechanism of Industrial Investment Fund. Hobbert proposed a theoretical model for the Industrial Investment Fund [7]. In this paper, Hobbert further pointed out that as a financial intermediary, the most important feature of Industrial Investment Fund is that it can effectively reduce the information asymmetry, investment risk and agency cost between the investors and entrepreneurs. Trester thought that the agreement between the investors and entrepreneurs can be made only when the information is symmetry between both sides. However, once the agreement has been made, the information may gradually become asymmetry [12].

Besides, many scholars have contributed to the risk assessment of Industrial Investment Fund. Reid came up with the idea that the biggest risk Industrial Investment Fund confronted was principal-agent problem [11]. Cornelli and Yosha made more detailed illustration about this case. They found that in a multi-stage investment program, since entrepreneurs always hope to get continuous investment, they are inclined to manipulate the short-term projects performance so as to get a next-stage investment [4]. Based upon these achievements, Kut *et al.* furthered the study and found that the principal-agent problem was caused by asymmetric information between the investors and entrepreneurs [8]. In order to deal with this challenge, Gompers pointed out that diversify investment portfolio is an effective approach to reduce adverse selection and moral risk [6].

3.2   China's Industrial Investment Fund

Different from foreign scholars, Chinese researchers are more concerned on practical operation of Industrial Investment Fund in China. Related work contributed by Chinese scholars is mainly in two fields:

1. Introduction about the operational mechanism of foreign Industrial Investment Fund. Ye and Li introduced organizational structure, operational process and management mechanism of both American and Japanese Industrial Investment Fund [13] [9]. Bao studied the partnership structure of foreign Industrial Investment Fund, and suggested it may also apply to Chinese market [2]. Cao compared operational pattern both at home and abroad, and suggested China need to develop different withdraw approaches for Industrial Investment Fund [5].
2. Suggestions about development of Chinese Industrial Investment Fund. Ai suggested that Chinese government should establish more regulations on foreign capital investment in order to make the market more efficient [1]. However, Bian hold a different view, and believed that government should make more encouraging regulations to help Industrial Investment Funds grow up rather than restrict them [3].

3.3   Our Contribution

Having reviewed literature work related to Industrial Investment Fund both at home and abroad, we found that the research is mainly focusing on general operational mechanism field, such as capital resource, organizational structure and withdraw approaches of Industrial Investment Funds. However, there are few scholars talking about the specific problem which is of vital importance: how to make investment decisions.

Through our investigation, we found that the major problem troubling Industrial Investment Fund investors is the recognitions of appropriate industries and related companies, which could be further illustrated as follows.

First of all, different from other investment modes, Industrial Investment Fund has a comparatively longer investment period. In light of current practice of China's Industrial Investment Fund, the investment period is approximately five to seven years. The longer investment period means the investors may face higher liquidity and credit risks.

Secondly, for Industrial Investment Fund investors, the total amount of invested capital is usually quite large. Given the possibility of investment loss constant, investors may have to suffer larger loss amount once they fail. Specifically, different from Private Equity investors in western countries, Industrial Investment Fund in China is more similar to the "Government Investment Pattern" that is popular in Japan, in which the investment are made by the large financial groups dominated by the government. Similarly, in China most Industrial Investment Funds are strongly influenced by government departments or state-owned policy banks, so the loss of investment will not only affect their financial status but also decrease the total social welfare.

Last but by no means the least, the investment targets of Industrial Investment Fund are usually unlisted enterprises. Compared with listed companies, they have much fewer information available to the public, and investors may suffer from serious asymmetric information problem.

Based upon the reasons above, we see that Industrial Investment Fund is facing great risk in defining the investment targets. An effective process is urgently needed to identify appropriate industries and related companies to invest in order to control the investment risk.

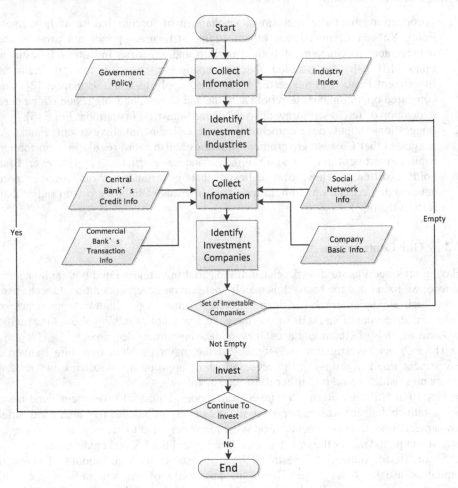

Fig. 2. Process Design for the Industrial Investment Fund Platform

4   Design of Digital Platform for Industrial Investment Fund

Having identified the main challenges of Industrial Investment Fund, we believe one of the most effective ways to solve these problems is to establish a digital platform for the Industrial Investment Fund and make information symmetric for both investors and enterprises. There are two functions of this platform: firstly, it helps to match the information between two sides effectively, thus reducing the investment risk; secondly, it could help to allocate the resources, so that the capital could be invested to the right enterprises. Hence, the goal of this digital platform is to achieve a win-win situation between investors and enterprises.

In order to achieve the functions mentioned above, the design of this platform could be divided into two stages: firstly, identify the specific industry (or industries) for the investment fund; secondly, identify the specific enterprises which have comparatively higher growth potential and lower risk. Our platform design is demonstrated as follows.

4.1   Identification of Investment Industries

Generally speaking, for Industry Investment Funds, there are strict restrictions on which industries to invest. So the first stage is to establish an effective investment industrial portfolio with limited choices.

4.1.1   Description of Data Source

When deciding which industries are worth investing, we believe two factors are of vital importance for consideration: one is the government's policy concerning different industries; the other is industry index, which is widely accepted as the indicator of development status for an industry.

First of all, investors should fully consider the influence that government policies have made on industries. Through our investigation, some source of government's policy concerning Industrial Investment Funds is listed as follows.

Table 2. Some Source of Chinese Government's Policy Concerning Industrial Investment Fund

	Regulation Name	Publishing Institution	Publishing Time
Central Government's Regulations	Interim Management Measures of China's Industrial Investment Fund	National Development and Reform Commission	2012
	Interim Management Measures on Emerging Industry Venture Capital	National Development and Reform Commission; Ministry of Finance	2011
	Notice on the Implementation of the Venture Capital Enterprise Income Tax Preferential Policies	Ministry of Finance; State Administration of Taxation	2009
	Acquisition management practices of listed companies	China Securities Regulatory Commission	2008
	Notice of the tax policy to promote the development of the venture capital business	State Administration of Taxation; Ministry of Finance	2007
	Interim Management Measures on Venture Investment Management	National Development and Reform Commission	2005
Local Government's Regulations	Notice on Approaches to promote equity investment fund industry development	Tianjin Local Government	2009
	Reply About agreed to support the construction of the Zhongguancun Science Park, National Innovation Demonstration Zone	Beijing Local Government	2009
	Promotion on Equity Investment Enterprises in Pudong	Shanghai Local Government	2009
	A number of provisions on the promotion of equity investment enterprise development fund	Shenzhen Local Government	2010

Secondly, as the key factor which indicating the potential and current development of the industry, industry index should also be taken into account. Taking energy industry for example, a number of index indicators are listed as follows.

Table 3. Some Source of Index Indicators for the Energy Industry

Index Name	Publishing Institution	Publishing Frequency
Power Industry Climate Index	China Economic Information Network	Quarterly
Petrochemical industry sentiment index	China Economic Information Network	Quarterly
Coal Price Index	National Coal Industry Network	Weekly
Coal industry prosperity index	China Economic Information Network	Quarterly

4.1.2 Technique Applied: Web Text Extraction

Having identified the source of data we need when choosing the investment industry, another challenge is how to extract and analyze the information effectively. One tool that has been generally used is the search engine. Though it's convenient and easy to use, its disadvantage is obvious. Firstly, people inference is needed during the whole searching process. Secondly, the aim of search engines is to cover the web as widely as possible, which may lead to irrelevant results returned by the search engine.

One effective way to solve the problem mentioned above is web crawler. Web crawler is an automatic program, which downloads web pages from World Wide Web. Begin with one or several URL from initial page, web crawler continues to extract new URL from the current page, and put them into a queue until the system meets certain stop conditions.

In our case, with the help with web crawlers, we could grab the government policy we simultaneously monitored. According to the information we got, a list of industries supported by the government may be summarized. Then, we capture the corresponding index of listed industries from the web site. Later, we rearrange the industry list, in a descending order, according to different industries' index. Finally, we choose the top K industries in the list as candidate investment industries.

4.2   Identification of Investment Companies

4.2.1 Description of Data Source

After we have chosen the industries to invest, the next stage is to identify the specific enterprises. Traditionally, investors' decisions are based upon mainly two kinds of resources: materials offered by the enterprise, and investors' on-site investigations. However, in the current "Big Data Society", only these two kinds of sources can hardly provide a solid basis for decision, we have to take more information into account and make comprehensive analysis. The information may include (but not limited to): enterprises' credit information from the central bank's credit system, enterprises' cash/deposit transaction information from the commercial banks, and comments or reputation of the enterprise captured from social networks.

For central bank's credit information, we need to reach an agreement with Chinese Central Bank, and get access to its credit system. So we can clearly see whether a company is "will" and "able" to pay for the loans. For enterprises' cash/deposit transaction information as well as their basic information, a company that is willing to be invested by our Industrial Investment Fund is supposed to submit its own information. For the reputation of companies, we could extract web texts concerning such comments - from the social networks through the Web Text Extraction technique mentioned above.

4.2.2 Technique Applied: Logistic Regression

Logistic Regression approach was first introduced in the 1970s; "*it became available in statistical packages in the early 1980s*" [10]. Logistic Regression is a type of regression analysis used for predicting the outcome of categorical dependent variables (i.e. "yes" vs. "no", or "high" vs. "low", etc.), based on independent variables (descriptive features). This technique attempts to model the probability of a "class/¬class" outcome using a linear function of the descriptive features, and then applying the log-odds of "class" (the *logit* of the probability) to fit the mentioned linear regression.

In our case, a data table is generated according to the information we got. The credit information of different companies, whether defaulted or not, is considered as the class-label. While other sources of data are included as data attributes of companies. Setting companies' past performance as training data, Logistic Regression model could offer us clues indicating what features may lead to a company's default. By putting into present performance of companies into this model, we are able to predict whether a certain company can afford to pay the loans or not.

5   Discussion

Since Industrial Investment Fund has become increasing important in Chinese financial market, effective investment decision mechanism is more and more important for both theoretical and practical aspects.

In this paper, we analyzed China's Industrial Investment Fund by clarifying its definition, summarized its characteristics and evaluated its current development situation. We then figured out two major challenges that investors confronted: the recognitions of appropriate industries and related companies, and analyzing the reasons causing the problem. In order to solve it, we came up with a process design of digital platform for the Industrial Investment Fund, and applied Web Text Extraction and Logistic Regression Classification techniques during the whole process.

In the current stage, we mainly focus on the theoretical design of the decision support process for the Industrial Investment Fund. Further research is suggested to make empirical verification of the study we proposed.

Acknowledgments. This work is sponsored by Discipline Construction Fund of University of International Business and Economics.

References

1. Ai, X.: Analysis on Foreign Mergers & Acquisitions and Development of Chinese Local Equity Fund. Special Zone Economy 3, 104–105 (2008)
2. Bao, Z.: System Innovation and Application Thinking of Limited Partnership in Venture Investment Organization. China Soft Science 7, 39–43 (2003)
3. Bian, H.: Solution about the Development of Chinese Equity Fund. Modern Enterprise Education 13, 42–43 (2007)
4. Cornelli, F., Yosha, O.: Stage Financing and the Role of Convertible Securities. Review of Economics Studies 70, 1–32 (2003)
5. Cao, W.: Understanding about the Development of Local Private Equity Fund. China Forex 11, 18 (2007)
6. Gompers, P.: Optimal Investment, Monitoring, and the Staging of Venture Capital. Journal of Finance 50, 1461–1490 (1995)
7. Hobbert, M.F.: Towards A Positive Theory of Venture Capital. University of Georgia, Georgia (1990)
8. Kut, C., Smolarshi, J.: Risk Management in Private Equity Funds: A Comparative Study of Indian and Franco-German Fund. Journal of Development Entrepreneurship 3(1), 35–55 (2006)
9. Li, L.: Analysis of Japanese Industrial Investment Fund. Foreign Economics & Management 12, 27–30 (1998)
10. Peng, C.-Y.J., So, T.-S.H.: Logistic Regression Analysis and Reporting: A Primer. Understanding Statistics 1(1), 31–70 (2002)
11. Reid, G.C.: The Application of Principal-agent Method to Investor-investee Relations in the UK Venture Capital Industry. Venture Capital 1(4), 285–302 (1999)
12. Trester, J.: Venture Capital Contracting Under Asymmetric Information. Journal of Banking & Finance 22, 675–699 (1998)
13. Ye, X.: American Industrial Investment Fund. Journal of Financial Research 10, 47–54 (1998)

Data Mining Challenges in the Management
of Aviation Safety

Olli Sjöblom

Turku University School of Economics, Turku, Finland
oljusj@utu.fi

Abstract. This paper introduces aviation safety data analysis as an important application area for data mining. Safety is a key strategic management concern for safety-critical industries and management needs new, more efficient tools and methods for more effective management routines. The aviation field is confronted with increasing challenges to provide safe and fluent services. Air travel has grown steadily during the last decades with a direct impact on the air traffic control. At the same time, the competition has become tougher because of increasing fuel prices and growing demand for air travel.

Keywords: Management, Flight Safety, Strategic Management, Data Mining, Text Mining, Analysis Method.

1   Introduction

Organisational decision making, especially in safety-critical systems, such as nuclear power and air traffic, is a complicated task. For successful operations, an acceptable air safety record has been required from the airline [1]. Air traffic has generally been forecasted to grow 5 – 6 % annually over the next two decades [2], or even over the next 10 – 15 years, the global air travel will probably double [3]. Consequently, the number of accidents will respectively increase if nothing were done to improve it, which development would, clearly, be unacceptable. This is why new and efficient ways for improving air safety need to be explored [4]. The conventional safety tools and methods based on data collection have reached their peak performance because of their inability to create new knowledge. Usually, data accumulates faster than it can be processed [5]. For further improvements new methods and tools are urgently needed [6].

2   Management in Safety-Related Context

Any system can be recognised to consist of elements, or factors, or parts that make up the whole [7]. Managing the organisation is exercised largely through management processes, in which the means of managerial communication inter-links with the environment. Johnsen (2002) defines the management process as *the interaction between people who want to attain mutual ends through mutual means.*" [8]. The strategy of

H. Li et al. (Eds.): I3E 2014, IFIP AICT 445, pp. 213–223, 2014.
© IFIP International Federation for Information Processing 2014

the corporation is according to Johnson et al. [9] to concern the organisation's mission, vision and objectives, developing plans and policies to use resources for enhancing the performance of the organisation.

Kettunen et al. [10] emphasise the managerial challenges in the safety-critical industries, which are typically related to finding a balance between diverging demands and expectations, like economy- and safety-related objects without forgetting the priorities-setting and maintaining focus on these components. The key action is a continuous balancing between taking risks and allocating resources for risk management. A scale with theoretical ends can be displayed, where at one end there is a situation where risks do not exist because the resources allocated are infinite; at the other end no resources are allocated because the risks are ignored and thus they are (practically) infinite. The reality is found somewhere in between, but no fixed location can be defined because all environments are somewhat unique and are also changing all the time. In daily operations perhaps existing hidden threats produce the need to maintain extra safety level naturally causing additional costs.

In studying risk management, the concept of tension cannot be ignored. It refers to the challenges of balancing conflicting objectives or expectations, like safety and other goals. These might exist for various reasons, even in the situation in which the executives of the organisation have set a high safety level as the priority official goal [11].In case warning signals appear, responding to those should happen without delay allocating safety resources to the critical area.

The safety decisions in an air traffic company follow the same pattern as other strategic decisions. Risk management should be carried out in parallel with safety management, referring to measures seeking to identify, assess and control risks on the organisational level having the goal to ensure the organisational and environmental safety. The executive management is responsible for recognising the safety significance of the ways the organisation is operated and maintained [12]. Managing risk and safety has been problematic in air transport: very high levels of safety are too costly – high levels of risk are unacceptable. Therefore, safety reports have been collected through decades to investigate and assess risks and to define risk standards, which are consistent with the value systems of the society [13, 14].

The value of safety cannot be estimated in any traditional way, because it has no determined price. Theoretically, limitless resources should be allocated to it, because one single failure may lead to significant losses in the form of missed business possibilities and claims for covering the damage caused to a third party. Kaplanski and Haim [15] have presented some estimates for the accident costs. A very large disaster with hundreds of casualties will cause a loss of about $1 billion for an airline company. However, the observed market effect has been found to be about 60 times larger; Kaplanski and Haim (2010) have found the evidence of a significant negative effect with an average market loss of more than $60 billion per aviation disaster. However, budget constraints set limits in practise and therefore a certain risk has to be accepted by achieving a sufficient safety level. There is never a 0-level risk. In case sufficient resources could not be allocated to achieve the required level of safety, the whole air traffic business would be critical. When confronting such a situation, the operations are to be adjusted by diminishing or changing them to correspond with the allocable safety resources so that a sufficient safety level is maintained.

Estimating the significance and importance of different alternatives in managing risks also needs tools, the exact definition of which is important for making strategic decisions. After the executive management has set goals as the thresholds of achievement, there must be methods and models to measure to what degree the achievements have been realised. In the decision process, there is always question about evaluating different alternatives. Any matter having significance enough to be taken into account in the evaluation process should be considered for evaluation [16]. Rumsfeld [17] has defined (simply expressed) three categories for knowledge: first, we know what we do know; second, we know what we do not know; and, finally, we do not know what we do not know. The hidden dangers belong to the last group, so in case we know what we are searching for, we obviously have means to reach it, but otherwise we need tools for finding something we do not know we are looking for. Thus, a deeper understanding is required for developing better methods and refining rules and practices that will contribute to higher levels of safety.

The unknown lethal factors brought into daylight could be eliminated; at least a significant part of them and a sufficient safety level could be reached with reduced investment allocation. For air traffic, there is theoretically no upper limit to allocate resources to safety in different forms. The relation between safety and cost efficiency could be illustrated explicitly comparing the costs between comprehensive maintenance programs and maintenance-induced accidents, the benefits that outweigh the accident costs [18]. The process for allocating extra resources to special projects might become even more troublesome in case there are interdependencies among the projects [16].

3   Flight Safety

According to the ICAO Safety Management manual [19], safety is defined as "*a state in which the risk of harm to persons or property damage is reduced to, and maintained at or below, an acceptable level through a continuing process of hazard identification and risk management*". Safety is not a matter-of-course, but the result of a rather complicated, carefully structured and comprehensive management process approaching to all airline safety aspects, particularly those of flight operations.

Air traffic is full of incidents and deviations that do not contain any hazard as such, but need to be reported and investigated to find out potential lethal trends. These undesirable, but very minor events are valuable investigation subjects for risk and safety specialists to build an understanding about their causes and to detect unsafe trends. Investigation also reveals whether countermeasures are warranted and how to reduce or eliminate potential accidents [20]. The appearance of similar recurring cases (a cluster, cf. Chapter 6) may indicate a hazardous trend that should be analysed very carefully to find out whether a real danger exists or not. The possibly existing lethal trends are trying to penetrate through the layers of defences, barriers and safeguards (cf. Figure 1) that, fortunately, usually stop them from proceeding. Because serious incidents and even accidents do happen, it can be presumed that after a certain amount of time they pass all the layers but the last one; then they will pass the last layer as well, which leads to accidents.

Finding trends from flight safety data, especially from narrative data has required significant human involvement. Thus, the analysis process and its possible results rely on the skill, memory and experience of the safety officers [21]. Watson [22] found that with conventional techniques it might take years to find meaningful relationships. Before text mining systems (one sub-class of data mining) were developed, there were no tools for analysing textual data with computers. Data mining provides a worthy analysis method in order to illustrate the safety indicators and to reveal undesired trends.

4   Safety Tools and Systems

Accident analysis as well as flight and operations modelling and simulation enhance the understanding of risk, but this is usually reactive and produces knowledge about causal factors potentially at the human and/or financial cost. Risk modelling typically collects knowledge resulting from flight safety analysis, human experience and theoretical and empirical studies. The goal of aviation risk assessment is to be comprehensive, timely and proactive, and this is why the analysis methods should be enhanced [23].

In aviation, the quantitative assessment of risk is particularly challenging, because the deviation events are extremely rare and the causal factors are non-linearly related to the events which makes them difficult to quantify [23]. The eventuality for the incident or accident occurring may be markedly reduced in case the risks can be efficiently diagnosed [24]. Then the question is: how to find and identify deviations leading to incidents and those leading to accidents? Reason [25] has modelled the process for the occurrence of accidents in his Swiss Cheese model, which is presented in Figure 1. The hazards appear from the right-hand side. Normally, their progress is stopped by successive layers of defences, barriers and lifeguards. If the process goes through all of these 'holes in the cheese slices', formally called the limited windows of accident opportunities, an accident will happen.

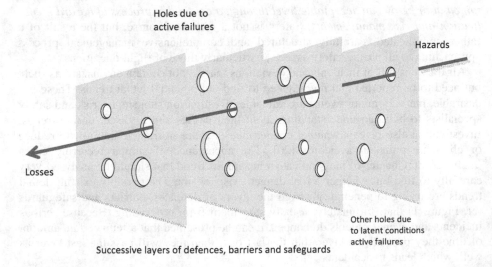

Fig. 1. The Swiss Cheese model (adapted from Reason 1997, 2000 [25, 26])

Kettunen et al. [10] regard redundancy as a method in improving safety by the duplication and overlap of critical factors like systems, functions and/or personnel. In general, redundancy can augment safety as such, but may also have counter-productive or unexpected effects, especially in case it is not managed properly. These unwanted effects can increase the complexity of the systems, which may hide individual failures and make them latent, so that they remain unnoticed and uncorrected and may even accumulate over time. Under these circumstances, a rather rare event might act as a trigger for an avalanche of unexpected events, which may be difficult to handle [10]. For situations of this kind, the Reason's Swiss Cheese model would work out excellently.

5   Data Mining in Flight Safety

Several different methods are recognised as data mining methods and a mining system can use the combinations of several of these methods. Parsaye [27] describes data mining as searching in the data for the patterns of information to guide a decision support process. These, often called "the nuggets of knowledge", are hidden in vast amounts of data and are practically undiscoverable with conventional techniques [22]. Using mining software, knowledge of data is combined by an analyst with advanced machine learning technologies to discover the relationships. In the discovery process to find hidden patterns, there are neither hypotheses nor any other predetermined model of the characteristics of the patterns. Obviously large databases, like those of aviation incidents and other deviations, contain a large number of patterns, so that the user of the discovery system can practically never ask the right question. The mining process acts as a decision support system that will not give straight answers to the questions, that is why skilled analytical and technical specialists are still required to interpret the created output [28]. The process contains several steps or phases (cf. Figure 2) that must be gone through to form knowledge from raw data. To be understandable the information must be presented with reports, graphs or in other suitable forms once found.

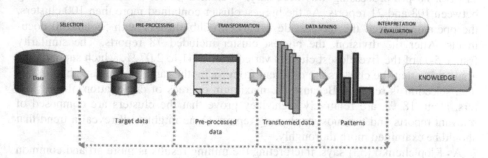

Fig. 2. The Knowledge Discovery in a Database Process (adapted from Fayyad et al. 1996 [29])

With structured data, the explanation of a case usually tells the truth to a certain extent, but completed with narrative data it can be close to 100 %, at least theoretically. Mining combined with other methods will give significant contributions to the decision processes. The idea to use text mining in the analysis of flight safety reports occurred along the need to analyse large amounts of narrative reports and when reports about successful text mining projects in the flight safety data analysis of English narratives were published [21, 30].

6   Testing Three Tools - Data Mining in Finnish

The basic idea of cluster analysis is that all the texts within each cluster have a high similarity in content [31]. This method was chosen for this study because it is an essential mining function in searching for similar documents, able to reveal a recurring hazard that might lead to an accident. It explores the data set and determines the structure of natural groupings without any preliminary assumptions. Another reason for its choice was the direct applicability to Reason's Swiss Cheese model presented in Figure 1. A third reason was that English literature gives several examples about using clustering in mining flight safety reports. These results have proved its better performance compared with more traditional statistical methods [32].

The beginning was finding text mining tools for processing Finnish. Three different systems seemed to be appropriate for benchmarking. The author was aware of one prototype (GILTA), one commercial product (TEMIS) with a Finnish module prototype, and one commercial system (PolyVista) with encouraging results mining Spanish, which seemed worth testing in Finnish. The Finnish Civil Aviation Authority granted the test data of 1240 cases (Target data on Figure 2), which created "a critical mass" for study.

The pre-processing produced filtered data containing 10572 word tokens, numbers and special characters, call signs, headings, the temperature, etc. The amount could be reduced to 8294 when parentheses and other similar characters without relevance were removed. The next procedure was preparing the lists of stop words (those to be ignored because of having no information) and synonyms. No transformation was needed because the data was extracted from one database.

The first round produced already promising results. Due to the Finnish module of TEMIS, no pre-processing was necessary. It created 26 clusters, their size varying between 108 and 21 reports. As the biggest cluster contained more than 100 clusters, the operator allowed the tool divide it into two sub-clusters with 58 and 50 documents. After the division, the biggest cluster included 78 reports. The similarity (range 5-1) of the five closest clusters varied from 3.41 to 2.07 %, which supports the assumption that the clusters are different from each other and thus this method in this data selection is reliable. Because the maximum degrees of explanation of the clusters, about 18 %, are relatively high, they prove that the clusters are composed of relevant reports and the most explaining reports alone might well reveal a trend that should be examined more thoroughly.

As Kloptchenko [33] says, interpreting the mining results is more art and common sense than science. The one single mining round of TEMIS made the direct comparison of the results challenging. Despite it, due to the high efficiency of the system with its in-built module for Finnish and because the mining results did not seem to require major changes, missing the second mining round was not considered a cause for losing significant information.

The smallest clusters began to produce some directly applicable information indicating that the sizes of the clusters play a significant role in the applicability of the results. This must, however, be scaled with the amount of production data. Additionally, a couple of similar cases found do not automatically create a dangerous trend; the way they occur and the reasons causing them can only be estimated by a thorough examination and investigation by human analysts. The results of TEMIS ought to be examined differently from the two other systems due to its interface and way of producing results which differ remarkably from the others. This, however, does not mean that these mining results would not be coherent with those of the other ones.

GILTA (manaGIng Large Text mAsses) divided the data on both rounds into 100 clusters (named classes) on the basis of the nine most significant words. Hence, on the first round 63 clusters contained less than 10 reports. These were easily analysable by a human analyst and could already be considered good mining results, proving clustering to be a useful method for this type of data. Some of the bigger classes could be interpreted as being real clusters, but according to experience the sizes should be reduced to less than 20. The results that were produced in Excel-form made it possible to carry out a comprehensible analysis and comparison of them with the results found with other tools. The system left out four reports beyond defined clusters.

PolyVista was originally built for using in English, but due to encouraging results with Spanish, its applicability for Finnish was tested, too. The system set score 100 for the most content describing word of the cluster and correspondent values to the others. The scores of the ten most important words of each cluster were only available, not the reports. The reports of the clusters could be 'guessed' by comparing the scores with the most important words in GILTA changing their relative weights for comparison. The data was processed determining the number of clusters first to be 6 and then raising it up to 20 in a second step. When there were 20 clusters, the smallest of them contained 10 reports and the biggest 232. In the case of 20 clusters, in eleven of them the scores of the three most important words were more than 50. In the last cluster containing 10 reports, the scores of the 10 most important words were 50 or more, which can be considered a good mining result.

As one result of the first mining round, the need for tuning, especially the definition of stop words and synonyms was discovered. Some pure mistakes, like some common stop words and synonyms forgotten from the list, were noticed. A more significant problem was the appearance of some frequently used "common" words (like 'plane' with its synonyms 'airplane' and 'aircraft') skewing the results. Their role in the data was carefully analysed [34], using an application called NVivo to get a deeper analysis. NVivo itself has no mining characteristics, but is used in analysing qualitative information, especially meeting the requirements of deep levels of analyses on different quantities of data, varying between a couple of sentences and thousands of text rows. In this context, the most important feature was cross-examining the mining results applying its search engine and query functions. Almost one hundred checking procedures were made with synonyms and stop words to prepare the data for the second round. After the careful estimation of the impact of possible changes, no major ones were made to keep the process unchanged but making the results more accurate.

After the second mining round with GILTA and PolyVista was performed, the results were studied carefully using the professional skills of a flight safety inspector.

The coherent clusters were taken into more detailed inspection. The progress as the change of distribution can be recognised through the increased percentage of 'sense making'[1] clusters, as for GILTA, illustrated in Table 1 displaying the minor, but perceptible change. First, the number of the relevant clusters increased from 9 to 11, and their average size diminished from 11.9 to 10.5 reports per cluster, shown in columns two and three. Further, the average weight of the nine most important words increased from 5.88 to 6.44 and the correspondent standard deviation diminished from 5.588 to 5.065, as shown in the two next columns. All these changes indicate the movement towards the aimed more homogenous clusters.

Table 1. Results illustration in GILTA rounds I and II

Round	Clusters	Average size	Average weight	Correspondent Standard Deviation
I	9	11.9	5.88	5.588
II	11	10.5	6.44	5.065

As already mentioned, the mining results of PolyVista must be analysed differently. Although comparing the weights of the most significant words is a cursory method, it was noticed to be relevant in this context. The results are illustrated in Table 2 showing an obvious progress between the two rounds. On the first round, 40.0 % of the clusters seemed to belong to the 'sense making' clusters, on the second 52.3 %. The size of the clusters did not seem to have any linear impact, but on both rounds those were found among the smallest ones. The average sizes changed from the first round being 37.9 compared with 62.5 of all clusters to the second, being then 20.6 compared with 28.2. These numbers illustrate that more information is achieved from the results of round II.

Table 2. Cluster distribution change between rounds 1 and 2 in PolyVista

Criteria / Round #	1	2
'Sense making' clusters of all content	40.0 %	52.3 %
Average size of all clusters (reports)	62.5	28.2
Average size of 'sense making' clusters	37.2	20.6

Proceeding with the same test and putting the results in a graphic presentation in Excel, the increased homogeneity was seen also from the 'centre of gravity' moving from the beginning of the rows rightwards as well as from the top downwards, when the clusters were sorted by the weight of the most significant words. It means that the number of clusters having more significant words increased. This occurred with both systems, indicating a slight improvement using this method, too.

Based on the professional skills and experience of the author, in case the safety personnel know what they are looking for, business intelligence (BI) methods could be applicable, allowing database queries using numerous keywords to search for known cases of a certain type or their combinations. BI could also be applied as a complementary method when mining is used to find something worth examining.

---

[1] Clusters, from which information can be seen clearly as such.

7   Results and Discussion

As already expressed before, the mining process does not give straight answers to the questions, but it acts as a support system for producing information for decision making. That is why experienced analytical and technical specialists are needed to interpret the created output. The testing process proved that data mining is neither an easy nor a fast method, but might be the only one for uncovering hidden information. All the results support the premise that it could reveal important safety information from fast accumulating, vast amounts of data, not accessible with other methods, to be used as an essential factor for strategic safety management. It is worth noticing that the test data was that contained no lethal trends, but in other case they could have been discovered and revealed using the method and tools as done in this study. An additional detail is worth noting - all the used tools left out almost the same reports as outliers.

The research process confirmed that text mining is a challenging task, especially in small language groups, where tools for text mining are scarcer than for big languages such as English which is an "easy" language for search technologies. Narrative text mining is generally demanding due to the multiplicity of languages spoken in the world. Especially languages with small user groups, such as Finnish, have to wait for efficient tools being developed much longer than the major languages. The search technologies are challenged by inflected forms and compounds. In Finnish, for example, the words may have thousands of inflected forms and in addition to that, they can be parts of compounds in almost countless combinations [35]. On average, every seventh word can be found in its basic form in fluent Finnish texts [36]. From the point of view of language processing, two significant results were achieved: first, Finnish texts were successfully mined with a tool originally to be used in an English environment. Secondly, the Finnish module for TEMIS was successfully production tested with real Finnish production data.

The number of clusters proved to be significant in the process: the more clusters, the better results. Mining is an iterative process although it makes no sense to increase the amount of rounds too much. Although this study has offered data mining as one solution to growing challenges, it is to be noticed that it is only one among several methods. Its special characteristic simply expressed is the ability to find something that is not known but expected to exist. Data mining has been used successfully for several years by a couple of airlines and other actors in the aviation industry. The process chain, beginning from the collection of safety data and ending in revised regulations for improving flight safety, going through several mining rounds and analyses to produce issued aviation rules and instructions, is rather long and demanding. Despite its complexity, it is worth going through, even for avoiding one single accident.

References

1. Liou, J.J.H., Yen, L., Tzeng, G.-H.: Building an effective safety management system for airlines. Journal of Air Transport Management 14(1), 20–26 (2008)

2. Netjasov, F., Janic, M.: A review of research on risk and safety modelling in civil aviation. Journal of Air Transport Management 14(4), 213–220 (2008)

3. Global Airline Industry Program. Analysis: The Airline Industry. Global Airline Industry Program [WWW-page] (2008),
http://web.mit.edu/airlines/analysis/analysis_airline_industry.html (cited September 5, 2011)

4. European Commission, Proposal for a DIRECTIVE OF THE EUROPEAN PARLIAMENT AND OF THE COUNCIL on occurrence reporting in civil aviation, Commission of the European Communities, Editor 2000: Brussels

5. Wang, X., Huang, S., Cao, L., Shi, D., Shu, P.: LSSVM with fuzzy pre-processing model based aero engine data mining technology. In: Alhajj, R., Gao, H., Li, X., Li, J., Zaïane, O.R., et al. (eds.) ADMA 2007. LNCS (LNAI), vol. 4632, pp. 100–109. Springer, Heidelberg (2007)

6. Evans, B., Glendon, A.I., Creed, P.A.: Development and initial validation of an Aviation Safety Climate Scale. Journal of Safety Research 38(6), 675–682 (2007)

7. Barnard, C.I.: The Functions of the Executive. Thirtieth Anniversary edn. Harvard University Press, Cambridge (1938)

8. Johnsen, E.: Managing the Managerial Process. A Participative Process. DJØF Publishing, Copenhagen (2002)

9. Johnson, G., Scholes, K., Whittington, R.: Exploring Corporate Strategy. Pearson Education Limited (2005)

10. Kettunen, J., Reiman, T., Wahlström, B.: Safety management challenges and tensions in the European nuclear power industry. Scandinavian Journal of Management 23(4), 424–444 (2007)

11. Sagan, S.D.: The limits of safety. Organizations, accidents, and nuclear weapons. Princeton University Press, Princeton (1993)

12. OECD/NEA, State-of-the-art report on systematic approaches to safety management, O.N.E. Agency, Editor 2006, OECD Nuclear Energy Agency: Issy-les-Moulineaux

13. Janic, M.: An assessment of risk and safety in civil aviation. Journal of Air Transport Management 6(1), 43–50 (2000)

14. Sage, A.P., White, E.B.: Methodologies for Risk and Hazard Assessment: A Survey and Status Report. IEEE Transaction on Systems, Man, and Cybernetics SMC-10(8), 425–446 (1980)

15. Kaplanski, G., Haim, L.: Sentiment and stock prices: The case of aviation disasters. Journal of Financial Economics 95(2), 174–201 (2010)

16. Kirkwood, C.W.: Strategic Decision Making. Wadsworth Publishing Company, Belmont (1997)

17. Rumsfeld, D.H.: News Transcript, U.S. Department of Defense Office of the Assistant Secretary of Defense, Public Affairs (2002)

18. Castro, R.: A Holistic Approach to Aviation Safety. In: Flight Safety Digest, pp. 1–12 (1988)

19. ICAO, Safety Management Manual, International Civil Aviation Organization: Montreal, Canada, p. 264 (2009)

20. Kirwan, B.: Incident reduction and risk migration. Safety Science 49(1), 11–20 (2011)

21. Nazeri, Z.: Application of Aviation Safety Data Mining Workbench at American Airlines. In: Proof-of-Concept Demonstration of Data and Text Mining, Center for Advanced Aviation Systems Development. MITRE Corporation Inc, McLean (2003)

22. Watson, R.T.: Data Management: Databases and Organizations, 2nd edn. John Wiley & Sons (1999)

23. Hadjimichael, M.: A fuzzy expert system for aviation risk assessment. Expert Systems with Applications 36(3), 6512–6519 (2009)
24. Lee, W.-K.: Risk assessment modeling in aviation safety management. Journal of Air Transport Management 12(5), 267–273 (2006)
25. Reason, J.T.: Managing the Risks of Organizational Accidents. Ashgate Publishing Limited, Aldershot (1997)
26. Reason, J.T.: Human error: models and management. British Medical Journal 320(7237), 768–770 (2000)
27. Parsaye, K.: A Characterization of Data Mining Technologies and Processes. Journal of Data Warehousing 2(3), 2–15 (1997)
28. Kutais, B.G. (ed.): Focus on the Internet. Nova Science Publishers, Inc. (2006)
29. Fayyad, U., Piatetsky-Shapiro, G., Smyth, P.: From Data Mining to Knowledge Discovery in Databases. AI Magazine 17(3), 18 (1996)
30. Megaputer Intelligence. Flight safety data analysis for Southwest Airlines (2004), http://www.megaputer.com/company/cases/southwest.php3 (cited December 17, 2004)
31. Rosell, M.: Text Clustering Exploration. Swedish Text Representation and Clustering Results Unraveled. In: School of Computer Science and Communication, p. 71. Kungliga Tekniska Högskolan, Stockholm (2009)
32. Saracoglu, R., Tütünkü, K., Allahverdi, N.: A new approach on search for similar documents with multiple categories using fuzzy clustering. Expert Systems with Applications: An International Journal 34(4), 2545–2554 (2008)
33. Kloptchenko, A.: Text Mining Based on the Prototype Matching Method, in Turku Centre for Computer Science, Åbo Akademi University: Turku. p. 117 plus additional pages including original papers (2003)
34. Lindén, K.: Word Sense Discovery and Disambiguation. In: General Linguistics 2005, p. 191. University of Helsinki, Helsinki (2005)
35. Karlsson, F.: Yleinen kielitiede1994. Helsinki, Yliopistopaino (1994)
36. Karlsson, F.: Finnish grammar 1987. WSOY, Porvoo (1987)

Function Inverse P-sets and the Hiding Information Generated by Function Inverse P-information Law Fusion

Kai-Quan Shi

School of Mathematics and System Sciences, Shandong University, Jinan, China
shikq@sdu.edu.cn

Abstract. Introducing the concept of function into inverse P-sets (inverse packet sets) and improving it, function inverse P-sets (function inverse packet sets) is obtained. Function inverse P-sets is the function set pair composed of function internal inverse P-set (function internal inverse packet set) $\overline{S}^F$ and function outer inverse P-set (function outer inverse packet set) $\overline{S}^{\overline{F}}$, or $(\overline{S}^F, \overline{S}^{\overline{F}})$ is function inverse P-sets. Function inverse P-sets, which have dynamic characteristic and law characteristic (or function characteristic), can be reduced to finite general function sets S under certain condition. Inverse P-sets is obtained by introducing dynamic characteristic to finite general element set X (Cantor set X) and improving it. Inverse P-sets is the element set pair composed of internal inverse P-set $\overline{X}^F$ (internal inverse packet set $\overline{X}^F$) and outer inverse P-set $\overline{X}^{\overline{F}}$ (outer inverse packet set $\overline{X}^{\overline{F}}$), or $(\overline{X}^F, \overline{X}^{\overline{F}})$ is inverse P-sets which has dynamic characteristic. In this paper, the structure of function inverse P-sets and its reduction, the inverse P-information law fusion generated by function inverse P-sets, and the attribute characteristics and attribute theorems of inverse P-information law are proposed. Using these theoretical results, the hiding image and its applications generated by inverse P-information law fusion are given, which is one of the important applications of function inverse P-sets.

Keywords: function inverse P-sets, inverse P-information law fusion, reduction theorem, attribute theorem, hiding information image, image camouflage, applications

1   Introduction

Shi (2008, 2009) indicated P-sets (packet sets), which has dynamic characteristic, are proposed by introducing dynamic characteristic to finite general element set X (Cantor set X) and improving it [1,2]. P-sets are a kind of mathematic structure using to research the information with dynamic characteristic. Function P-sets (function packet sets), which has dynamic characteristic and law (or function) characteristic, is put forward by introducing the concept of function to P-sets and improving it [3,4]. Function P-sets is a mathematic model used to research just the class of information law with dynamic characteristic. P-sets and function P-sets, are used in the theoretical and applicative research of dynamic information and dynamic information law

H. Li et al. (Eds.): I3E 2014, IFIP AICT 445, pp. 224–237, 2014.

respectively [1-10], and they have the same logic characteristic as following: If X is finite general element set, or S is finite general function set, α is the attribute set of X, or α is the attribute set of S, then $\forall x_i \in X$ whose attribute satisfies conjunctive normal form, where x_i has attribute $\wedge_{i=1}^{k}\alpha_i$ (or $\forall s_i \in S$ whose attribute satisfies conjunctive normal form , and s_i has attribute $\wedge_{i=1}^{k}\alpha_i$). Shi (2012) introduced dynamic characteristic into finite general element set X and improving it at the same time, inverse P-sets, which has dynamic characteristic, is put forward [12]. Inverse P-sets is the model to research the class of information with dynamic characteristic while it is a different class from that P-sets does, and inverse P-sets is also used in the theoretical and applicative research of a class of dynamic information. Shi (2013) introduced the concept of function to inverse P-sets and improving it at the same time, function inverse P-sets is proposed [13]. Function inverse P-sets, which have dynamic characteristic and law (or function) characteristic, is the mathematic model used to research the class of dynamic information law while it is a different class from function P-sets does. Inverse P-sets and function inverse P-sets have the same logical characteristic as following: If X is finite general element set, or S is finite general function set, α is the attribute set of X, or α is the attribute set of S, then $\forall x_i \in X$ whose attribute satisfies disjunctive normal form, where x_i has attribute $\vee_{i=1}^{k}\alpha_i$ (or $\forall s_i \in S$ whose attribute satisfies disjunctive normal form , and s_i has attribute $\vee_{i=1}^{k}\alpha_i$). In this paper, the structure and characteristic of function inverse P-sets, the inverse P-information law fusion of function inverse P-sets, the attribute characteristic and attribute theorems of inverse P-information law fusion, and the hiding information image generated by inverse P-sets and its applications are given.

In order to make readers accept the concept, structure and characteristic of function inverse P-sets easily, the characteristic and structure of inverse P-sets [12] are simple introduced to Appendix, where readers can compare function inverse P-sets with inverse P-sets. In Appendix, the existence fact of inverse P-sets and P-sets [1, 2, 4, 7, 8] and the proof are given respectively.

2   Function Inverse P-sets and Its Structure

Assumption. $U(x)$ is the finite function universe, $V(\alpha)$ is the finite attribute universe, and $S(x) = \{S(x)_1, S(x)_2, \cdots, S(x)_n\}$ is the finite general function set on $U(x)$, which is called function set for short. $\alpha = \{\alpha_1, \alpha_2, \cdots, \alpha_k\}$ is the finite attribute set on $V(\alpha)$, and $S(x)$ and $r(x)$ are both the function of x. $U(x)$, $V(\alpha)$, $S(x)$ and $r(x)$ are respectively written as U, V, S and r for short.

Definition 1. Given function set $S = \{s_1, s_2, \cdots, s_q\} \subset U$, if $\alpha = \{\alpha_1, \alpha_2, \cdots ,\alpha_k\} \subset V$ is the attribute set of S, and then $\overline{S}^F$ is called function internal inverse P-set (function internal inverse packet set) of S, moreover

$$\overline{S}^F = S \bigcup S^+ \tag{1}$$

While S^+ is called the F-function supplementary set of S, moreover

$$S^+ = \{r | r \in U, r \overline{\in} S, f(r) = s' \in S, f \in F\} \tag{2}$$

If $\overline{S}^F$ has the attribute set α^F, which satisfies

$$\alpha^F = \alpha \cup \{\alpha' | f(\beta) = \alpha' \in \alpha, f \in F\} \tag{3}$$

Where $\beta \in V, \beta \overline{\in} \alpha$, and $f \in F$ can change β into $f(\beta) = \alpha' \in \alpha$ in expression (3). $S = \{s_1, s_2, \cdots, s_r\}$, $q < r$, and $q, r \in N^+$ in expression (1).

Definition 2. Given function set $S = \{s_1, s_2, \cdots, s_q\} \subset U$, if $\alpha = \{\alpha_1, \alpha_2, \cdots, \alpha_k\} \subset V$ is the attribute set of S, then $\overline{S}^{\overline{F}}$ is called the function outer inverse P-set (function outer inverse packet set), moreover

$$\overline{S}^{\overline{F}} = S - S^- \tag{4}$$

While S^- is called the $\overline{F}$- function deleting set of S, moreover

$$S^- = \{s_i | s_i \in S, \overline{f}(s_i) = r_i \overline{\in} S, \overline{f} \in \overline{F}\} \tag{5}$$

If $\overline{S}^{\overline{F}}$ has the attribute set $\alpha^{\overline{F}}$, moreover

$$\alpha^{\overline{F}} = \alpha - \{\beta_i | \overline{f}(\alpha_i) = \beta_i \overline{\in} \alpha, \overline{f} \in \overline{F}\} \tag{6}$$

Where $\alpha_i \in \alpha$, $\overline{f} \in \overline{F}$ can change α_i into $\overline{f}(\alpha_i) = \beta_i \overline{\in} \alpha$ in expression (6); and $\overline{S}^{\overline{F}} \neq \phi$, $\alpha^{\overline{F}} \neq \phi$ in expression (4) while $\overline{S}^{\overline{F}} = \{s_1, s_2, \cdots, s_p\}$, $p < q$, and $p, q \in N^+$.

Definition 3. The function set pair composed of $\overline{S}^F$ and $\overline{S}^{\overline{F}}$, is called function inverse P-sets (function inverse packet sets) generated by function set S, moreover

$$(\overline{S}^F, \overline{S}^{\overline{F}}) \tag{7}$$

and finite function set S is called the ground set of function inverse P-sets $(\overline{S}^F, \overline{S}^{\overline{F}})$.

Using expression (3), we can get the following chain by adding attributes to α one after another,

$$\alpha_1^F \subseteq \alpha_2^F \subseteq \cdots \subseteq \alpha_{n-1}^F \subseteq \alpha_n^F \tag{8}$$

and function inter inverse P-set can be gotten from expression (8), moreover

$$\overline{S}_1^F \subseteq \overline{S}_2^F \subseteq \cdots \subseteq \overline{S}_{n-1}^F \subseteq \overline{S}_n^F \tag{9}$$

Using expression (6), we can get the following chain by deleting attributes from α one after another,

$$\alpha_n^{\overline{F}} \subseteq \alpha_{n-1}^{\overline{F}} \subseteq \cdots \subseteq \alpha_2^{\overline{F}} \subseteq \alpha_1^{\overline{F}} \tag{10}$$

and function outer inverse P-set can be gotten from expression (10), moreover

$$\overline{S}_n^F \subseteq \overline{S}_{n-1}^F \subseteq \cdots \subseteq \overline{S}_2^F \subseteq \overline{S}_1^F \tag{11}$$

Definition 4

$$\{(\overline{S}_i^F, \overline{S}_j^F) | i \in I, j \in J\} \tag{12}$$

is called function inverse P-sets family generated by function set S, and expression (12) is the general form of function inverse P-sets, if $(\overline{S}_\lambda^F, \overline{S}_k^F) \in \{(\overline{S}_i^F, \overline{S}_j^F) | i \in I, j \in J\}$ is function inverse P-sets.

Using expressions (1) to (12), the following can be gotten.

Theorem 1. (The first reduction theorem of function inverse P-sets) Function inverse P-sets $(\overline{S}^F, \overline{S}^F)$ and function set S can satisfy that

$$(\overline{S}^F, \overline{S}^F)_{F=\overline{F}=\phi} = S \tag{13}$$

Theorem 2. (The second reduction theorem of function inverse P-sets) Function inverse P-sets $\{(\overline{S}_i^F, \overline{S}_j^F) | i \in I, j \in J\}$ and function set S can satisfy that

$$\{(\overline{S}_i^F, \overline{S}_j^F) | i \in I, j \in J\}_{F=\overline{F}=\phi} = S \tag{14}$$

Using the expressions (1) to (16) in part 2, part 3 is given as following.

3   Data Disassembly-Synthesis and the Generation of Inverse P-information Law Fusion

In reference [15], the following is given.

The Principle of Data Disassembly-Synthesis
Given finite data set $Y = \{y_1, y_2, \cdots, y_n\}$, there are finite sub data sets $y_i = \{y_{i,1}, y_{i,2}, \cdots, y_{i,n}\}$ while y_i is a disassembly of Y, and Y and y_i fulfill $Y = \{y_1,$
$y_2, \cdots, y_n\} = \{\sum_{i=1}^{m} y_{i,1}, \sum_{i=1}^{m} y_{i,2}, \cdots, \sum_{i=1}^{m} y_{i,n}\}$, then Y is a synthesis of y_i. $\forall y_k, y_{k,i} \in R$,
R is real number set, $k=1, 2, \cdots, n, i=1, 2, \cdots, m$.

Using the principle of data disassembly-synthesis, the following can be gotten.

Definition 5. $w(x)$ is called the information law generated by function set $S = \{s_1, s_2, \cdots, s_q\}$, moreover

$$w(x) = \sum_{j=1}^{n} y_j \prod_{\substack{i,j=1 \\ i \neq j}}^{n} \frac{x - x_i}{x_j - x_i} = a_{n-1} x^{n-1} + a_{n-2} x^{n-2} + \cdots + a_1 x + a_0 \tag{15}$$

If $w(x)$ is generated by Lagrange interpolation depending on the data points (x_1, y_1), $(x_2, y_2), \cdots, (x_n, y_n)$ composed by the discrete data set $y = \{y_1, y_2, \cdots, y_n\} = \{\sum_{i=1}^{q} y_{i,1}, \sum_{i=1}^{q} y_{i,2}, \cdots, \sum_{i=1}^{q} y_{i,n}\}$ of S, and $y_i = \{y_{i,1}, y_{i,2}, \cdots y_{i,n}\}$ is the discrete data set of $s_i \in S$, $i = 1, 2, \cdots, q$.

Definition 6. $\overline{w}(x)^F$ is called the inter inverse P-information law fusion of $w(x)$ generated by $\overline{S}^F$, moreover

$$\overline{w}(x)^F = b_{n-1}x^{n-1} + b_{n-2}x^{n-2} + \cdots + b_1 x + b_0 \tag{16}$$

If $\overline{w}(x)^F$ is generated by expression (16) depending on the data points $(x_1, y_1^f), (x_2, y_2^f), \cdots, (x_n, y_n^f)$ composed by the discrete data set $y^F = \{y_1^f, y_2^f, \cdots, y_n^f\} = \{\sum_{i=1}^{r} y_{i,1}, \sum_{i=1}^{r} y_{i,2}, \cdots, \sum_{i=1}^{r} y_{i,n}\}$ of $\overline{S}^F$.

Definition 7. $\overline{w}(x)^{\overline{F}}$ is called the outer inverse P-information law fusion of $w(x)$ generated by $\overline{S}^{\overline{F}}$, moreover

$$\overline{w}(x)^{\overline{F}} = c_{n-1}x^{n-1} + c_{n-2}x^{n-2} + \cdots + c_1 x + c_0 \tag{17}$$

If $\overline{w}(x)^{\overline{F}}$ is generated by expression (17) depending on the data points $(x_1, y_1^{\overline{f}})$, $(x_2, y_2^{\overline{f}}), \cdots, (x_n, y_n^{\overline{f}})$ composed by the discrete data set $y^{\overline{F}} = \{y_1^{\overline{f}}, y_2^{\overline{f}}, \cdots, y_n^{\overline{f}}\} = \{\sum_{i=1}^{p} y_{i,1}, \sum_{i=1}^{p} y_{i,2}, \cdots, \sum_{i=1}^{p} y_{i,n}\}$ of $\overline{S}^{\overline{F}}$.

Where p, q and r fulfill $p < q < r$ and $p, q, r \in N^+$ in definitions 5 to 7.

Definition 8. The information law fusion pair composed of $\overline{w}(x)^F$ and $\overline{w}(x)^{\overline{F}}$, is called the inverse P-information law fusion of $w(x)$ generated by function inverse P-sets $(\overline{S}^F, \overline{S}^{\overline{F}})$, and is called the inverse P-information law fusion for short, moreover

$$(\overline{w}(x)^F, \overline{w}(x)^{\overline{F}}) \tag{18}$$

Theorem 3. (The relation theorem between inter inverse P-information law fusion and information law) If there is a difference information law $\Delta w(x) \neq 0$, inter inverse P-information law fusion $\overline{w}(x)^F$ and information law $w(x)$ satisfy that

$$\overline{w}(x)^F - \Delta w(x) = w(x) \tag{19}$$

Theorem 4. (The relation theorem between outer inverse P-information law fusion and information law) If there is a difference information law $\nabla w(x) \neq 0$, outer inverse P-information law fusion $\overline{w}(x)^{\overline{F}}$ and information law $w(x)$ satisfy that

$$\overline{w}(x)^{\overline{F}} + \nabla w(x) = w(x) \tag{20}$$

Theorem 5. (The relation theorem between inverse P-information law fusion and information law) If there is a difference information law $(\Delta w(x), \nabla w(x))$, $\Delta w(x)$ $\neq 0$, $\nabla w(x) \neq 0$, inverse P-information law fusion $(\overline{w}(x)^F, \overline{w}(x)^{\overline{F}})$ and informa-tion law $w(x)$ satisfy that

$$(\overline{w}(x)^F, \overline{w}(x)^{\overline{F}}) = (w(x) + \Delta w(x), w(x) - \nabla w(x)) \tag{21}$$

There are $\overline{w}(x)^F = w(x) + \Delta w(x)$ and $\overline{w}(x)^{\overline{F}} = w(x) - \nabla w(x)$ in expression (21).

It should be pointed out that the generation of information law $w(x)$ can use piecewise interpolation method, linear regression method and other methods, and the discussions are omitted.

The Engineering Background and Engineering Significance of Law Fusion

$f(t)$ is a rectangular wave or rectangular function, and $f(t)$ can be decomposed to several $\sin k\omega t$, $k=1,2,\cdots,m$; or there is another saying, $f(t)$ = $\sin\omega t +$ $\sin 2\omega t + \cdots + \sin\lambda\omega t + \sin m\omega t$. If $f(t)$ and $\sin k\omega t$ are defined as laws, it is obvious that $f(t)$ is gotten by the fusion of $\sin\omega t$, $\sin 2\omega t$, $\cdots$, $\sin m\omega t$. Another saying, law $f(t)$ is gotten by the fusion of $\sin\omega t$, $\sin 2\omega t$, $\cdots$, $\sin m\omega t$. Conversely, $\sin\omega t$, $\sin 2\omega t$, $\cdots$, $\sin m\omega t$ are the law fusion of $f(t)$. In the general mathematics, Fourier's sinc series of $f(t)$ are $f(t) = \sum_{n=1}^{\infty} b_n \sin n\omega t = \quad b_1 \sin \omega t + b_2 \sin 2\omega t + \cdots + b_\lambda \sin \lambda\omega t + \cdots$, under certain conditions, where $f(t)$ and $b_k \sin\lambda\omega t$ are defined as laws. Apparently, law $f(t)$ is the law fusion of $b_1 \sin \omega t + \quad b_2 \sin 2\omega t + \cdots + b_\lambda \sin \lambda\omega t$. $\sin\omega t$ is called as fundamental wave in electric engineering, and $\sin 2\omega t$ and $\sin 3\omega t, \cdots$, are called as "second harmonic", "third harmonic", and so on.

4   The Reduction Theorem of Inverse P-information Law Fusion

Theorem 6. (The reduction theorem of inter inverse P-information law fusion) If $F = \phi$, inter inverse P-information law fusion $\overline{w}(x)^F$ and information law $w(x)$ fulfill

$$\overline{w}(x)^F_{F=\phi} = w(x) \tag{22}$$

Theorem 7. (The reduction theorem of outer inverse P-information law fusion) If $\overline{F} = \phi$, outer inverse P-information law fusion $\overline{w}(x)^{\overline{F}}$ and information law $w(x)$ fulfill

$$\overline{w}(x)^{\overline{F}}_{\overline{F}=\phi} = w(x) \tag{23}$$

Theorem 8. (The reduction theorem of inverse P-information law fusion) If $F = \overline{F} = \phi$, inverse P-information law fusion $(\overline{w}(x)^F, \overline{w}(x)^{\overline{F}})$ and information law $w(x)$ fulfill

$$(\overline{w}(x)^F, \overline{w}(x)^{\overline{F}})_{F=\overline{F}=\phi} = w(x) \tag{24}$$

Corollary 1. Inverse P-information law fusion families satisfy

$$\{(\overline{w}(x)_i^F, \overline{w}(x)_j^{\overline{F}}) \mid i \in \mathrm{I}, j \in \mathrm{J}\}_{F=\overline{F}=\phi} = w(x) \tag{25}$$

5   The Attribute Characteristic of Inverse P-information Law Fusion

Theorem 9. (The attribute theorem of inter inverse P-information law fusion) $\overline{w}(x)^F$ is the inter inverse P-information law fusion of $w(x)$ if and only if there is attribute set $\Delta\alpha \neq \phi$, and the attribute set α^F of $\overline{w}(x)^F$ and the attribute set α of $w(x)$ fulfill

$$\alpha^F - (\alpha \cup \Delta\alpha) = \phi \tag{26}$$

Theorem 10. (The attribute theorem of outer inverse P-information law fusion) $\overline{w}(x)^{\overline{F}}$ is the outer inverse P-information law fusion of $w(x)$ if and only if there is attribute set $\nabla\alpha \neq \phi$, and the attribute set $\alpha^{\overline{F}}$ of $\overline{w}(x)^{\overline{F}}$ and the attribute set α of $w(x)$ fulfill

$$\alpha^{\overline{F}} - (\alpha - \nabla\alpha) = \phi \tag{27}$$

Theorem 11. (The attribute theorem of inverse P-information law fusion) $(\overline{w}(x)^F, \overline{w}(x)^{\overline{F}})$ is the inverse P-information law fusion of $w(x)$ if and only if there is attribute sets $\Delta\alpha \neq \phi$, $\nabla\alpha \neq \phi$, and the attribute sets $(\alpha^F, \alpha^{\overline{F}})$ of $(\overline{w}(x)^F, \overline{w}(x)^{\overline{F}})$ and the attribute set α of $w(x)$ fulfill

$$(\alpha^F, \alpha^{\overline{F}}) - ((\alpha \cup \Delta\alpha), (\alpha - \nabla\alpha)) = \phi \tag{28}$$

There are $\alpha^F - \alpha \cup \Delta\alpha = \phi$ and $\alpha^{\overline{F}} - (\alpha - \nabla\alpha) = \phi$ in expression (28).

Using the structure of function inverse P-set in part 2 and part 3 to 5, part 6 is given as following.

6   The Hiding Information Image Generated by Inverse P-information Law Fusion and Its Application

1. The Generation of Hiding Information Image and Its Structure

Definition 9. $O(a, \overline{w}(x)_0^{\overline{F}}, b, \overline{w}(x)_0^F)$, which is called the information image with two boundary, is generated by function inverse P-sets $(\overline{S}_0^F, \overline{S}_0^{\overline{F}})$, while $\overline{w}(x)_0^F$ and $\overline{w}(x)_0^F$ are respectively called the lower-boundary and upper-boundary of $O(a, \overline{w}(x)_0^{\overline{F}}, b, \overline{w}(x)_0^F)$.

Where a and b are the common points of $\overline{w}(x)_0^F$ and $\overline{w}(x)_0^{\overline{F}}$, $a \neq b$; $a, b \in R^+$; $\overline{w}(x)_0^F$ is the inter inverse P-information law fusion, and it is generated by $\overline{S}_0^F$; $\overline{w}(x)_0^{\overline{F}}$ is the outer inverse P-information law fusion, and it is generated by $\overline{S}_0^{\overline{F}}$.

Definition 10. $O^*(a, \overline{w}(x)_k^{\overline{F}}, b, \overline{w}(x)_\lambda^F)$ is called the hiding information image of $O(a, \overline{w}(x)_0^{\overline{F}}, b, \overline{w}(x)_0^F)$, if its lower-boundary and upper-boundary respectively satisfy

$$\overline{w}(x)_0^{\overline{F}} - \overline{w}(x)_k^{\overline{F}} \geq 0 \tag{29}$$

$$\overline{w}(x)_0^F - \overline{w}(x)_\lambda^F \leq 0 \tag{30}$$

Figure 1 shows $O(a, \overline{w}(x)_0^{\overline{F}}, b, \overline{w}(x)_0^F)$ and $O^*(a, \overline{w}(x)_k^{\overline{F}}, b, \overline{w}(x)_\lambda^F)$ visually in the form of folder line.

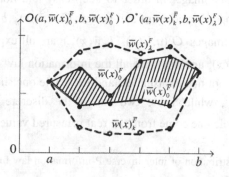

Fig. 1. $O(u, \overline{w}(x)_0^{\overline{F}}, b, w(x)_0^\Gamma)$ is information law with two boundary, $O(a, \overline{w}(x)_0^{\overline{F}}, b, \overline{w}(x)_0^F)$ is shown in real line; $O^*(a, \overline{w}(x)_k^F, b, \overline{w}(x)_\lambda^F)$ is the hiding information image of $O(a, \overline{w}(x)_0^{\overline{F}}, b, \overline{w}(x)_0^F)$, and $O^*(a, \overline{w}(x)_k^{\overline{F}}, b, \overline{w}(x)_\lambda^F)$ is shown in broken line; a, b are common points, and $a \neq b$; $O(a, \overline{w}(x)_0^{\overline{F}}, b, \overline{w}(x)_0^F)$ is shown in shade.

Theorem 12. (The non-unique existence theorem of hiding information image) If $O(a, \overline{w}(x)_0^{\overline{F}}, b, \overline{w}(x)_0^F)$ is the information image with two boundary, then there are some $O^*(a, \overline{w}(x)_n^{\overline{F}}, b, \overline{w}(x)_m^F)$, and any $O^*(a, \overline{w}(x)_k^{\overline{F}}, b, \overline{w}(x)_\lambda^F)$ in them is one of the hiding information images of $O(a, \overline{w}(x)_0^{\overline{F}}, b, \overline{w}(x)_0^F)$, $k \in (1, 2, \cdots, n)$, $\lambda \in (1, 2, \cdots, m)$.

The proof can be gotten by definitions 9 and 10, theorems 2 to 5, and corollaries 2 to 4, and it is omitted.

2. The Application of Hiding Information Image in the Information Image Camouflage

Definition 11. $O^*(a, \overline{w}(x)_i^{\overline{F}}, b, \overline{w}(x)_j^F)$ is called an information image camouflage of $O(a, \overline{w}(x)_0^{\overline{F}}, b, \overline{w}(x)_0^F)$, if $O^*(a, \overline{w}(x)_i^{\overline{F}}, b, \overline{w}(x)_j^F)$ is a hiding information image of $O(a, \overline{w}(x)_0^{\overline{F}}, b, \overline{w}(x)_0^F)$.

Using definitions 9 to 11, the following can be gotten.

Stealth Camouflage Principle of Information Image

Any hiding information image $O^*(a,\overline{w}(x)_p^{\overline{F}},b,\overline{w}(x)_q^F)$ is the stealth camouflage of real information image $O(a,\overline{w}(x)_0^{\overline{F}},b,\overline{w}(x)_0^F)$, and $O(a,\overline{w}(x)_0^{\overline{F}},b,\overline{w}(x)_0^F)$ is hidden in $O^*(a,\overline{w}(x)_p^{\overline{F}},b,\overline{w}(x)_q^F)$, or

$$O(a,\overline{w}(x)_0^{\overline{F}},b,\overline{w}(x)_0^F) \subset O^*(a,\overline{w}(x)_p^{\overline{F}},b,\overline{w}(x)_q^F) \tag{31}$$

In expression (31)," $\subset$ "expresses that $O(a,\overline{w}(x)_0^{\overline{F}},b,\overline{w}(x)_0^F)$ is surrounding by $O^*(a,\overline{w}(x)_p^{\overline{F}},b,\overline{w}(x)_q^F)$, and $p \in (1,2,\cdots,n)$, $q \in (1,2,\cdots,m)$.

The example in this part is from a sub-image of an important information image, which is a two-boundary image. In order to keep easy and not lose generality, the lower-boundary and upper-boundary of sub-image $O(a,\overline{w}(x)_0^{\overline{F}},b,\overline{w}(x)_0^F)$ and those of its hiding information images $O^*(a,\overline{w}(x)_k^{\overline{F}},b,\overline{w}(x)_\lambda^F)$ are all expressed in folder line (here $\overline{w}(x)_0^{\overline{F}},\overline{w}(x)_0^F$, $\overline{w}(x)_k^{\overline{F}}$ and $\overline{w}(x)_\lambda^F$ are all the information law fusion in the form of folder line), which can not make misunderstanding. Table one shows the discrete data of $\overline{w}(x)_0^{\overline{F}}$ and $\overline{w}(x)_0^F$, while Table two shows the discrete data of $\overline{w}(x)_k^{\overline{F}}$ and $\overline{w}(x)_\lambda^F$. The data in Table one come from the real measured value of the sub-image.

Table 1. The discrete distribution of inter inverse P-information law fusion $\overline{w}(x)_0^F$ and outer inverse P- information law fusion $\overline{w}(x)_0^{\overline{F}}$

k	1	2	3	4	5	6
$\overline{w}(x)_0^F$	1.20	1.35	0.94	1.55	1.63	1.38
$\overline{w}(x)_0^{\overline{F}}$	1.20	0.76	0.83	0.92	0.80	1.38

Table 2. The discrete distribution of inter inverse P-information law fusion $\overline{w}(x)_k^F$ and outer inverse P- information law fusion $\overline{w}(x)_\lambda^{\overline{F}}$

k	1	2	3	4	5	6
$\overline{w}(x)_k^F$	1.20	1.70	1.83	1.69	1.76	1.38
$\overline{w}(x)_\lambda^{\overline{F}}$	1.20	0.46	0.35	0.37	0.44	1.38

It should be pointed out that the data in Table 2 is gotten depending on the principle of data disassembly-synthesis and the disassembly-synthesis rule given (the coefficient of data extension and contraction is the random number on $(0,1)$). The real values of the coefficient of data extension and contraction are omitted for some reason.

Basing on table one and two, it can be gotten that the stealth camouflages $O^*(a, \overline{w}(x)_k^{\overline{F}}, b, \overline{w}(x)_\lambda^F)$ and $O(a, \overline{w}(x)_0^{\overline{F}}, b, \overline{w}(x)_0^F)$ of sub-image $O(a, \overline{w}(x)_0^{\overline{F}}, b, \overline{w}(x)_0^F)$ fulfill expression (31). It is difficult to finding out $O(a, \overline{w}(x)_0^{\overline{F}}, b, \overline{w}(x)_0^F)$ from hiding information image $O^*(a, \overline{w}(x)_k^{\overline{F}}, b, \overline{w}(x)_\lambda^F)$, and it is difficult to steal $O(a, \overline{w}(x)_0^{\overline{F}}, b, \overline{w}(x)_0^F)$ from $O^*(a, \overline{w}(x)_k^{\overline{F}}, b, \overline{w}(x)_\lambda^F)$, too. But it is easy for the image transmission to reduce $O^*(a, \overline{w}(x)_k^{\overline{F}}, b, \overline{w}(x)_\lambda^F)$ to $O(a, \overline{w}(x)_0^{\overline{F}}, b, \overline{w}(x)_0^F)$ by using the rule of data disassembly-synthesis. Figure 1 gives out $O(a, \overline{w}(x)_0^{\overline{F}}, b, \overline{w}(x)_0^F)$ generated by table one and $O^*(a, \overline{w}(x)_k^{\overline{F}}, b, \overline{w}(x)_\lambda^F)$ generated by table two. In order to keep simple, the lower-boundary $\overline{w}(x)_0^{\overline{F}}$ and upper-boundary $\overline{w}(x)_0^F$ of $O(a, \overline{w}(x)_0^{\overline{F}}, b, \overline{w}(x)_0^F)$, and the lower-boundary $\overline{w}(x)_k^{\overline{F}}$ and upper-boundary $\overline{w}(x)_\lambda^F$ of $O^*(a, \overline{w}(x)_k^{\overline{F}}, b, \overline{w}(x)_\lambda^F)$ are all shown in the form of folder line.

7   Discussion

Function inverse P-sets are gotten by introducing dynamic characteristic into finite general function set S and improving it. In other words, introducing the concept of function into inverse P-sets and improving it, function inverse P-sets is gotten. Function inverse P-sets has dynamic characteristic and law characteristic, and it is a new model to research the characteristic and application of a class of dynamic information laws which is different from that of function P-sets does. Function inverse P-sets have the dynamic characteristic and law characteristic, which are contrary to that of function P-sets [3-4]. And function P-sets are also a new model to research the characteristic and applications of some one class of dynamic information laws, which is a different class from the one function inverse P-sets does. Function inverse P-sets and function P-sets are two separate dynamic models, which can't be replaced by each other and can only be used separately.

In order to understand the characteristic of function inverse P-sets and compare it with inverse P-sets, the structure of inverse P-sets in expressions (1*)-(14*) are given in appendix.

Acknowledgements. This work was supported by National Natural Science Foundation of China (61273277).

References

1. Shi, K.Q.: P-sets. Journal of Shandong University (Natural Science) 43(11), 77–84 (2008) (in Chinese)
2. Shi, K.Q.: P-sets and its applications. An International Journal Advances in Systems Science and Application 2, 209–219 (2009)

3. Shi, K.Q.: Function P-sets. Journal of Shandong University (Natural Science) 2, 62–69 (2011)
4. Shi, K.Q.: P-sets and its applied characteristics. Computer Science 8, 1–8 (2010)
5. Shi, K.Q.: P-reasoning and P-reasoning discovery-identification of information. Computer Science 7, 1–9 (2011)
6. Lin, H.K., Fan, C.X., Shi, K.Q.: Backward P-reasoning and attribution residual discovery-application. Computer Science 10, 189–198 (2011)
7. Fan, C.X., Lin, H.K.: P-sets and the reasoning-identification of disaster information. International Journal of Convergence Information Technology 1, 337–345 (2012)
8. Lin, H.K., Fan, C.X.: The dual form of P-reasoning and identification of unknown attribute. International Journal of Digital Content Technology and its Applications 1, 121–131 (2012)
9. Shi, K.Q., Zhang, L.: Internal P-sets and data outer-recovery. Journal of Shandong University (Natural Science) 4, 8–14 (2009)
10. Fan, C.X., Huang, S.L.: Inverse P-reasoning discovery identification of inverse P-information. International Journal of Digital Content Technology and its Applications 20, 735–744 (2012)
11. Zhang, L., Xiu, M., Shi, K.Q.: P-sets and applications of internal-outer data circle. Quantitative Logic and Soft Computing 2, 581–592 (2010)
12. Shi, K.Q.: Inverse P-sets. Journal of Shandong University (Natural Science) 1, 98–109 (2012)
13. Shi, K.Q.: Function inverse P-sets and information law fusion. Journal of Shandong University (Natural Science) 8, 73–80 (2012)
14. Shi, K.Q.: P-sets, inverse P-sets and the intelligent fusion-filter identification of information. Computer Science 4, 1–13 (2012)
15. Shi, K.Q.: Function S-rough sets, function rough sets and separation-composition of law for information systems. Computer Science 10, 1–10 (2010)
16. Robinson, G.D., Harry, N.: Evaluation of two application of spectral mixing models to image fusion. Remote Sens Environ. 71, 272–281 (2000)
17. Varshney, P.K.: Multisensor data fusion. Journal of Electronics Computer Engineering 6, 245–253 (1999)

Appendix

Inverse P-sets and Its Structure

Given $X = \{x_1, x_2, \cdots, x_q\} \subset U$ while $\alpha = \{\alpha_1, \alpha_2, \cdots, \alpha_k\} \subset V$ is the attribute set of X, $\overline{X}^F$ is called the inter inverse P-sets generated by X, and called $\overline{X}^F$ is inter inverse P-sets for short, moreover

$$\overline{X}^F = X \cup X^+ \tag{1*}$$

X^+ is called F-element supplementary set of X, moreover

$$X^+ = \{u \mid u \in U, u \,\overline{\in}\, X, f(u) = x' \in X, f \in F\} \tag{2*}$$

If $\overline{X}^F$ has attribute set α^F, moreover

$$\alpha^F = \alpha \cup \{\alpha' \mid \beta \in V, \beta \,\overline{\in}\, \alpha, f(\beta) = \alpha' \in \alpha, f \in F\} \tag{3*}$$

Here in expression (1*), $\bar{X}^F = \{x_1, x_2, \cdots, x_r\}$, $q, r \in N^+$, $q \le r$.

Given $X = \{x_1, x_2, \cdots, x_q\} \subset U$, $\alpha = \{\alpha_1, \alpha_2, \cdots, \alpha_k\} \subset V$ is the attribute set of X, and $\bar{X}^{\bar{F}}$ is the outer inverse P-sets of X, and $\bar{X}^{\bar{F}}$ is called outer inverse P-sets for short, moreover

$$\bar{X}^{\bar{F}} = X - X^-$$ (4*)

X^- is called the $\bar{F}$-element deleting set of X, moreover

$$X^- = \{x \mid x \in X, \bar{f}(x) = u \bar{\in} X, \bar{f} \in \bar{F}\}$$ (5*)

If $\bar{X}^{\bar{F}}$ has attribute set $\alpha^{\bar{F}}$, moreover

$$\alpha^{\bar{F}} = \alpha - \{\beta_i \mid \alpha_i \in \alpha, \bar{f}(\alpha_i) = \beta_i \bar{\in} \alpha, \bar{f} \in \bar{F}\}$$ (6*)

Here in expression (4*), $\bar{X}^{\bar{F}} = \{x_1, x_2, \cdots, x_p\}$, $p, q \in N^+$, $p \le q$, $\bar{X}^{\bar{F}} \ne \phi$, and in expression (6*), $\alpha^{\bar{F}} \ne \phi$.

The element set pair composed by inters inverse P-sets $\bar{X}^F$ and outer inverse P-sets $\bar{X}^{\bar{F}}$, is called inverse P-sets generated by X, and called inverse P-sets for short, moreover

$$(\bar{X}^F, \bar{X}^{\bar{F}})$$ (7*)

and finite general element set X is called the ground set of inverse P-sets $(\bar{X}^F, \bar{X}^{\bar{F}})$.

By using expression (3*), the following can be gotten

$$\alpha_1^F \subseteq \alpha_2^F \subseteq \cdots \subseteq \alpha_{n-1}^F \subseteq \alpha_n^F$$ (8*)

According to expression (8*), inter inverse P-sets $\bar{X}^F$ fulfill

$$\bar{X}_1^F \subseteq \bar{X}_2^F \subseteq \cdots \subseteq \bar{X}_{n-1}^F \subseteq \bar{X}_n^F$$ (9*)

By using expression (6*), the following can be gotten

$$\alpha_n^{\bar{F}} \subseteq \alpha_{n-1}^{\bar{F}} \subseteq \cdots \subseteq \alpha_2^{\bar{F}} \subseteq \alpha_1^{\bar{F}}$$ (10*)

According to expression (10*), outer inverse P-sets $\bar{X}^{\bar{F}}$ fulfill

$$\bar{X}_n^{\bar{F}} \subseteq \bar{X}_{n-1}^{\bar{F}} \subseteq \cdots \subseteq \bar{X}_2^{\bar{F}} \subseteq \bar{X}_1^{\bar{F}}$$ (11*)

By using expressions (9*) and (11*), the following can be gotten:

$$\{(\bar{X}_i^F, \bar{X}_j^{\bar{F}}) \mid i \in I, j \in J\}$$ (12*)

Expression (12*) is called inverse P-sets family, and it is the general form of inverse P-sets.

Using expressions (1*) to (12*), the following can be gotten:

Theorem 1*. If $F = \bar{F} = \phi$, then inverse P-sets $(\bar{X}^F, \bar{X}^{\bar{F}})$ and finite general element set X fulfill

$$(\bar{X}^F, \bar{X}^{\bar{F}})_{F=\bar{F}=\phi} = X \qquad (13*)$$

Theorem 2*. If $F = \bar{F} = \phi$, then inverse P-sets $\{(\bar{X}_i^F, \bar{X}_j^{\bar{F}}) | i \in I, j \in J\}$ and finite general element set X fulfill

$$\{(\bar{X}_i^F, \bar{X}_j^{\bar{F}}) | i \in I, j \in J\}_{F=\bar{F}=\phi} = X \qquad (14*)$$

Figure 2 shows inverse P-sets $(\bar{X}^F, \bar{X}^{\bar{F}})$ directly.

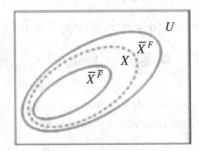

Fig. 2. U is finite element universe, X is the finite general element set on U, which has the attribute set α, and X can be shown in broken line. $\bar{X}^F$ is inter inverse P-sets, it has the attribute set α^F, and $\bar{X}^F$ can be shown in real line. $\bar{X}^{\bar{F}}$ is outer inverse P-sets, it has the attribute set $\alpha^{\bar{F}}$, and $\bar{X}^{\bar{F}}$ can be shown in real line. Inverse P-sets $(\bar{X}^F, \bar{X}^{\bar{F}})$ is composed of $\bar{X}^F$ and $\bar{X}^{\bar{F}}$.

The Existence Fact of Inverse P-sets and Its Proof

A company can produce m kinds of productions which can be named $x_1, x_2, \cdots, x_m$; $x_1, x_2, \cdots, x_m$ can compose to production universe U; $x_1, x_2, \cdots, x_q$ have the purchase contract (written as "Contract") $\alpha_1, \alpha_2, \cdots, \alpha_q$ respectively; and if $x_i \neq x_j$, then $\alpha_i \neq \alpha_j$. If $x_1, x_2, \cdots, x_q$ are defined to element set $X = \{x_1, x_2, \cdots, x_q\} \subset U$, and "Contract" $\alpha_1, \alpha_2, \cdots, \alpha_q$ are defined to the attribute sets of $x_1, x_2, \cdots, x_q$ respectively, apparently, set $X = \{x_1, x_2, \cdots, x_q\}$ has the attribute set $\alpha = \{\alpha_1, \alpha_2, \cdots, \alpha_q\}$ where $q < m, q, m \in \mathbb{N}^+$. If new attributes written as $\alpha_q, \alpha_{q+1}, \cdots, \alpha_r$ are adding to α, and α can be changed to $\alpha^F = \alpha \cup \{\alpha_q, \alpha_{q+1}, \cdots, \alpha_r\} = \{\alpha_1, \alpha_2, \cdots, \alpha_r\}$, then set $X = \{x_1, x_2, \cdots, x_q\}$ can be change to inter inverse P-sets $\bar{X}^F = X \cup \{x_{q+1}, x_{q+2}, \cdots, x_r\} = \{x_1, x_2, \cdots, x_r\}$. If $\alpha_{p+1}, \alpha_{p+2}, \cdots, \alpha_q$ are deleting from α, and α can be changed to $\alpha^{\bar{F}} = \alpha - \{\alpha_{p+1}, \alpha_{p+2}, \cdots, \alpha_q\} = \{\alpha_1, \alpha_2, \cdots, \alpha_p\}$, then set $X = \{x_1, x_2, \cdots, x_q\}$ can be changed to outer inverse P-sets $\overline{X}^{\overline{F}} = X - \{x_{p+1}, x_{p+2}, \cdots, x_q\} = \{x_1, x_2, \cdots, x_p\}$; $p < q < r$, $p, q, r \in N^+$. By promoting this fact, we can get that if some new attributes are adding to α while other attributes are deleting from α, then set X can be changed to set pair $(\overline{X}^F, \overline{X}^{\overline{F}})$, and $(\overline{X}^F, \overline{X}^{\overline{F}})$ is inverse P-sets. It is easy to get that in inverse P-sets, element x_i and attribute α_i satisfy (disjunctive normal form), and x_i has the attribute set $\vee_{i=1}^k \alpha_i$.

The Existence Fact of P-sets and Its Proof

x_1, x_2, x_3, x_4, x_5 have the same attributes α_1 = red and α_2 = sweet apple. If x_1, x_2, x_3, x_4, x_5 are defined to element set $X = \{x_1, x_2, x_3, x_4, x_5\}$ and both α_1 and α_2 are defined to the attributes of x_1, x_2, x_3, x_4, x_5, it is apparent that set $X = \{x_1, x_2, x_3, x_4, x_5\}$ has the attribute set $\alpha = \{\alpha_1, \alpha_2\}$. If the new attribute α_3 = weight 150g are adding to α, and α can be changed to $\alpha^F = \alpha \cup \{\alpha_3\} = \{\alpha_1, \alpha_2, \alpha_3\}$, then set X can be changed to inter P-sets $X^{\overline{F}} = X - \{x_1, x_2, x_4\} = \{x_3, x_5\}$. If α_2 is deleting from α and α is changed to $\alpha^{\overline{F}} = \alpha - \{\alpha_2\} = \{\alpha_1\}$, then set X is changed to outer P-sets $X^F = X \cup \{x_6, x_7, x_8\} = \{x_1, x_2, x_3, x_4, x_5, x_6, x_7, x_8\}$. By promoting this fact, we can get that if some new attributes are adding to α while other attributes are deleting from α, then set X can be changed to set pair $(X^{\overline{F}}, X^F)$, and $(X^{\overline{F}}, X^F)$ is P-sets [1,2]. It is easy to get that element x_i and attribute α_i satisfy conjunctive normal form in P-sets, and x_i has the attribute set $\wedge_{i=1}^k \alpha_i$.

From the upper facts and proofs, we can get that inverse P-sets and P-sets [1,2,5,8,9] are two dynamic mathematics models which has the oppose dynamic characteristic and logical relation from each other, inverse P-sets and P-sets are two separate model, and they can not be replaced by each other.

Classification of Customer Satisfaction Attributes: An Application of Online Hotel Review Analysis

Jian Dong[1], Hongxiu Li[2], and Xianfeng Zhang[3,*]

[1] School of Economics and Finance, Xi'an Jiaotong Univesrity, Xi'an, China
aolang-y@163.com
[2] Information Systems Science, Department of Management and Entrepreneurship, Turku
School of Economics, University of Turku, Turku, Finland
hongli@utu.fi
[3] School of Information Science and Technology, Hainan Normal Univesrity, Haikou, China
xfzhangchina@gmail.com

Abstract. With the wide penetration of Internet, online hotel reviews have become popular among travellers. Online hotel reviews also reflect customer satisfaction with hotel services. In this study we use online hotel reviews to classify the attributes of customer satisfaction with hotel services. The empirical data was collected via Daodao.com, the Chinese affiliated brand of online travel opinion website tripadvisor.com. Based on text mining and content analysis, we found that the following seven dimensions are important attributes generating customer satisfaction with hotels: hotel, location, service, room, value, food and dinging, and facility availabilities. Finally we concluded on the research findings and also highlight the research limitations and future research directions.

Keywords: e-Service, e-Commerce, eWOM, Online reviews.

1   Introduction

The wide penetration of Internet applications in the hospitality and tourism field has greatly changed the way travelers retrieving tourism information, managing their trips, and booking flights or hotels. As Li and Liu (2014) indicated that individuals yield large amount of user generated contents (UGC) via Internet, and the UGCs spread to others via various online media, such as email, chat rooms, personal Web pages, bulletin boards, newsgroups, discussion forums, blogs, social networks, and virtual communities [1, 2, 3]. Consumers would like to rely on UGC to support their consuming decisions [1]. Since travelers can easily get access to the Internet, where they may freely express their subjective assessments, sentimental thoughts or even the emotional feelings towards the hotels, destinations, or the trip process, and share their travel experience with others, and the most important, individuals are also trying to use these travel-related UGC to support their travel decisions, such as hotels or destinations. According to Gretzel and Yoo (2008), three-quarters of travelers take online consumer reviews as the main information

---

* Corresponding author.

H. Li et al. (Eds.): I3E 2014, IFIP AICT 445, pp. 238–250, 2014.

source when planning their trips [4]. The study by Gretzel and Yoo (2007) confirms the vital role (77.9%) of online reviews in making decision "where to stay" [5]. The similar trends also show in China. According to the recent report released by China Internet Network Research Center in 2013, 69.5% of the travelers in China considered online travel-related reviews as the essential factor supporting their hotel booking decision [6]. Information asymmetry has been dramatically eliminated [7] by referring to the UGC, such as online reviews (also named as eWOM), based on travelers' previous experience.

This growing reliance on the Internet as the information source has inversely led to the prosperous development of different online opinion facilitated websites, and inspired travelers' enthusiasm in sharing their experience. In the hospitality and tourism industry, the professional social network websites normally employ numerical rating (overall rating or rating components) and text comments with or without numerical ratings [8], such as tripadvisor.com, booking.com. The highlighted summary of the travel-related services performs like a powerful illustration of an Internet-mediated abstract system [9], or the reputation feedback system [10]. The popularity of online review makes the related websites gradually develop to be a popular intermediary in travel industry, or even a more trustful third party for individuals, when comparing to the traditional travel agents [11].

The rating system in hospitality and tourism industry, composed of rich numerical ratings and text contents from experienced travelers, can not only offers other individuals reliable information to support their travel decision, but also delivers to travel service providers, i.e. hotels, valuable feedbacks about their service quality, and even as a good marketing channel [10]. Ye et al. (2009, 2011) found that the relationship between travelers' hotel rating behavior and the room reservation performance is significant [12, 13]. A 10 percent increase in the travelers' ratings, will boost online bookings by more than five percent. Some researchers worked on the components of user satisfaction in order to better understand the factors determining hotels' service quality and user satisfaction. In practice, different online rating systems are also adopted to measure hotel's service quality, such as Tripadvisor.com, Booking.com.

Though online reviews have been argued to be important information source for both individuals and hotels, research on hotel customer satisfaction mainly focus on the attributes of service quality based on the perceptions from hotel customers, and little research has attempted to examine the attributes of hotel customer satisfaction from the perspective of online hotel review -- the real feedback of hotel customers. In addition, few studies have explored to explain the importance level of various attributes generating customer satisfaction based on online hotel reviews, which might shed light on how hotel can satisfy their customers and retain customers, as well as how to recruit customers as marketing channel for their products or services via the WOM simultaneously. Thus, it is meaningful to investigate the attributes of hotel customer satisfaction based on the online reviews from real hotel customers.

The rest of this paper is organized as follows. A literature review on hotel customer satisfaction research is presented in the next section. Section three describes the research instrument development and the collection of the data. The results are presented in section four with a discussion on the findings as well. Section five presents the conclusions in this research and followed by the discussion on research limitations and future research directions.

2   Research Background

2.1   Hotel Customer Satisfaction

Customer satisfaction has been an important research topic in the marketing field, such as to explore customer loyalty, and purchasing intention. According to Oliver (1980), customer satisfaction refers to an attitude [14]. It is an evaluation formed by customers based on the expectations of what customers would receive from a product or service and on their perceptions on the performance of a product or service they actually received. Prior literature has attempted to identify the attributes which generate customer satisfaction as well as to identify the importance levels of different attributes in enabling user satisfaction [15]. Much of the literature researches on customer satisfaction from the lens of service quality [16]. The research on user satisfaction of hotels also follows the research tradition of customer satisfaction and investigates customer satisfaction based on service quality of hotels, mainly employing survey, interview and case study research methods [17,18]. The research stream mainly identifies the attributes of service quality. More details are presented in Table 1. Deng (2008) identified the importance level of different indicators in generating customer satisfaction based on the SERVQUAL scale [17], proposed by Parasuraman et al. (1985, 1988) [19,20]. Choi and Chu (2001) conducted empirical study in the context of hotels in Hongkong and found that staff service quality, room quality, value, general amenities, business service, IDD and security are important attributes in generating user satisfaction with hotels [18].

Table 1. Research on hotel customer satisfaction from the lens of service quality

Literature	Scales
Deng 2008 [17] Case study, and survey Hotels in Taiwan	Responsibility and empathy: Personal warm care given by staff, Have customers' best interest at heart, Easy to get staff's attention & help. Readiness to respond to customer's requests, Knowledgeable to answer customers' request, Courtesy and friendliness of staff, Individual attention for customer, Willingness to help customers, Understand the specific needs of customers, Provision of safe environment and equipment. Reliability and assurance: Provision of services as promised, Dependability in handling customers' service problem, Reasonable price, Prompt reply to customers, Perform service right at the first time. Tangibility: The physical facilities are visually appealing, Multiple hot spring facilities, Availability of adequate fire & first aids facilities and instructions, Cleanness of hot spring facilities, Convenient hotel location.
Choi & Chu 2001 [18] Survey Hotels in Hongkong	Staff service quality (efficiency in check-in/out, helpfulness of staff, politeness/friendliness of staff, neat appearance of staff, efficiency of staff), Room quality (cleanliness, comfort of bed/mattress/pillow, in-room temperature control and quietness of room) Value (room value for money, hotel food & beverage value for money, comfortable ambiance of the hotel and hotel being part of reputable chain). General amenities (availability of mini bar, variety of food and beverage facilities, quality of food and beverage facilities, reliability of wake-up call, efficiency of valet/laundry service, efficiency of room service and availability of information desk)

Table 1. *(Continued)*

Literature	Scales
	Business services (availability of secretarial service, availability of business-related meeting rooms, availability of business-related facilities). IDD (International direct dial) facilities Security (responsibility of security personnel, reliability of loud fire alarms and availability of safe box)
Mey et al. 2006 [21]; Survey; Malaysia	Reliability (5 items), responsiveness (3 items), assurance (3 items), empathy (3 items), tangibles (11items)
Shanka & Taylor 2008 [15] Survey Hotels in Perth of AU	Physical facilities: personal safe, safe deposit boxes, IDD, long bath, in-house movies Service experienced: friendly front office staff, efficient check-in/check-out, 24 hour reception Service provision: taxi booking, lobby ambience, tour bookings, on-site parking
Ramanathan 2012 [22]; Regression; Hotels in UK	Physical-product management: cleanliness, room quality and family friendliness People and process management: customer service Marketing management: Value for money

2.2   Research on Online Hotel Reviews

Online travel reviews have become popular among travelers. Traveler would like to make comments on the travel services they have experienced and share travel experience with others. The popularity of online hotel reviews offers researchers the possibility to get access to online hotel reviews generated by hotel customers, and to get the real data reflecting travelers' satisfaction, but not based on their perceptions about travel services as in the traditional research on user satisfaction.

Recently, quite much research on hotel customer satisfaction has used online hotel reviews in their research to explore user satisfaction. Though these researches are also trying to identify the indicators of customer satisfaction, the same as the traditional research, its research context and research methods are different. Normally text mining and content analysis are employed in these researches. Chaves et al. (2012) conducted content analysis of online hotel reviews on the SME hotels, and found customer satisfaction with hotel services are mainly determined by the indicators related to room, staff, location and neighborhood of hotels [23]. Zhou et al. (2014) made analysis of the online hotel reviews in a city in China and found that physical setting, (including room, hotel, and food), value, location and staff are the most important attributes generating hotel user satisfaction [24]. More detailed research based on online hotel reviews is presented in Table 2.

Table 2. Research on online hotel reviews

Literature	Scales
Chaves et al. (2012) [23] Content analysis SME hotels	Room: cleanliness, size, silence and bed. Cleanliness, confort, scenery and the Internet (positive), air-conditioner, bed and sound proofing (negative). Staff: friendliness and helpfulness. Knowledge and indication of sights and landmarks, foreign language skills. Professionalism (negative). Location: near the city center, proximity to access points for public transport, proximity to the beach (positive). Neighborhood: safe surroundings.
Zhou et al. (2014) [24] Content analysis Hotels in Hangzhou city, China	Physical setting –room: Amenities in the room/bathroom; size and layout of the room; cleanliness of the room; welcoming extras Physical setting – hotel: Availability of Wifi; public facilities (lounge, lobby, pool and fitting centre); dated level (old/new); noise level; entertainment facilities Physical setting – food: A variety of food (including western food); the quality of food; the dining environment; availability of special food service (room service; vegetarian and gluten free options), Value: Room price; food and beverage price, and other prices Location: Close to attractions; close to city centre; close to the airport/railway station; and accessibility Staff: Friendliness of the staff; language skills of the staff; efficiency of the staff in solving problems
Li et al. (2013) [25] Content analysis Hotels in Beijing city, China	Logistics: Transportation convenience, convenience to tourist destination Facility: room, air conditioning, network, sound insulation, TV, parking Reception services: luggage, check in/out, morning call, lobby Food and beverage management: breakfast, lunch, dinner, room service Cleanliness and maintenance: Room cleaning, bathroom, bedding replacement and bed Value: Value for money
Magnini et al. (2011) [26]; Content analysis Blogs	Customer service, cleanliness, location, value, facility, guestroom size and decor, renovation/newness, food, amenities, quietness
Lu and Stepchenkova (2012) [27] Content analysis Online review of hotels in USA	Ecology settings (grounds/surroundings, lodge amenities, ambiance, noise, ecofriendliness, other guests) Room (room/bathroom decor and layout, room amenities, room/bathroom facilities, insect problem) Nature (nature-based activities, natural attractions, weather) Services (customer service, tour/tour guider service, extra service, restaurant service, reservation service, management policy) Food quality Location (closeness to town, accessibility, closeness to attraction) Value for money (food/drink price, room rates, other price)
Stringam and Gerdes (2010) [28] Content analysis Online review of hotels in USA	Clean, staff, breakfast, bed, price, restaurant, pool, bathroom, airport, downtown, view, shopping, shower, dirt, coffee, towel, noise, smell, courteous, pillow, buffet, noisy, attractions, smoking, accommodating, pay, expensive, mall, cheap, employee, eat, rate, decor, dine, sheet, valet, Disney, toilet, clerk, refrigerator, stain, broken, bath, bus, affordable, store, cost, shops, sink dinner

3   Research Methodology

3.1   Data Collection

The source selection of hotel review websites is crucial to the research. The platform should be capable of assembling larger population and generating higher external validity. Based on the rule, daodao.com was finally opted for the online hotel review/eWOM data source. Daodao.com is fully affiliated to the world's largest travel site - Tripadvisor, which operates worldwide in 42 countries, and generates more than 150 million reviews and opinions covering over 4 million accommodations, restaurants and attractions. Since launched in 2009, Daodao.com has been providing valuable insights to Chinese travelers in trip planning, for example, over 3 million reviews in Chinese has been accumulated at the website of Daodao.com by February 2014. Currently Daodao is ranked as the top one tourism-related UGC website among the Chinese community. According to the iResearch statistics, the monthly visits of Daodao.com have reached 7.62 million in August 2013.

Only 4 and 5 star hotels are considered in this research.Sanya, a coastal city deemed as the oriental Hawaii in the far south part of China is selected. Sanya is the second largest city of Hainan which is an isolated island in the south sea, and is also the designated international tourism island by Chinese government in December 2009. Sanya has been a popular leisure travele destination for its tropical attractions and the yearly wide low PM2.5, when comparing with other bigger tourism cities like Beijing or Xi'an. A number of 100 top star hotels located in Sanya are registered in daodao.com, consisting 49 five star hotels and 51 four star hotels. The 100 hotels generated a total volume of 24051 reviews, with online reviews on individual hotel ranging from 2 to 1744, and averaging in 240.

All reviews are crawled from the website by using a spider program. The two-week data crawling process started from March 23, 2014, and ended in April 5, 2014. Numbers are also revised and re-crawled when there are new reviews during the two weeks.

Taking in sight that the reviewers of Tripadvisor are from 42 countries and can write reviews in 25 languages, Daodao.com merged international reviews in the general opinion pool for each hotel. Correspondingly, there are numbers of reviews written in other languages, such as English, Russian, Japanese, and so on. This portion of opinions is not considered in the research so as to neatly focus on the context of Chinese eWOM. Daodao.com also cooperates with similar review or booking brands, through borrowing travelers' opinions left in other branded websites like Expedia and Agoda. These segments are also deleted, since differed reviewing regulations in diversified platforms may deliver varied thinking. The online hotel reviews are finally purified to 19,659 items, with individual hotel reviews ranging from 1 to 1632.

3.2   Data Content Mining Procedure

Due to the unique feature of Chinese language, one native Chinese software called ROST CM6.0 is chosen to conduct the research. ROST content miner is designed by a research group from Wuhan University in China, for the purpose of solving the problems that other non-Chinese originated content mining software met in parsing and merging Chinese words. ROST CM6.0 is capable of splitting, filtering, merging Chinese words, and researchers can also automatically count frequencies, cluster and classify the words, construct the sentiment networks or social networks, and display the co-occurrence matrix. It has been downloaded by researchers worldwide from more than 100 universities to deal with Chinese qualitative data in social sciences.

The word parsing and initial word frequency counting is operated first, followed by the filtering stage. Though ROST CM6.0 has its own filtering dataset, the manually selected filtering lists are still needed for each individual research under its unique context. The full list of the parsed words is therefore carefully reviewed, while those words that either do not conforms to the common Chinese usage or contains no meanings under this research context are selected and added in the filtering database. When the filtering word list is done, the parsed word dataset is then re-input in the software to generate the new word frequency counts.

3.3   Data Coding

The filtered output displays a number of 1899 items of words with the frequency of 1 or above. The highest frequency climbs up to 35089, whereas a list of 812 words counts less than 10. The number of words with frequencies up from 57 takes the top 499 lists, and those ranging from 11 to 49 surmounts 588. Among the 1899 word lists, the word hotel ranks the highest, being mentioned by 35089 times. Room and service follow with the frequency of 18482 and 15111. Surroundings, amenities, convenience, breakfast, seashore, sandy beach and sea view are also the frequently used words in the top 10 lists.

Though the review portraits are sketched out by referring to the 1899 word lists, it is still far from reaching the fundamental essence. The coding process is therefore continued so as to stress out the important word groups and accordingly pinpoint the key dimensions that travelers care. The top 500 words are picked out and reviewed carefully. Through aggregating the synonyms or similar words, the 500 vocabulary is then reduced to 100 groups. The frequency of the word groups range from 56 to 37342, and average in 3300. The denomination related with hotel still ranks the first, including hotel, resort, villa and etc. Room, beach, service, location, facilities, environment, cleanness, swimming pool and price also enlist in top 10 lists. Generally speaking, the groups of words outlining the varied facets of each dimension still take the majority. Take service for instance, vocabulary pools specifically describing empathy, politeness, professionalism each take the frequency of 7670, 3735 and 2590.

In order to reveal the diversified dimensions and explain their respective significance, attribute coding is conducted according to the scales proposed by prior literature. A small number of word lists are deleted, for either they are the general denomination, such as hotel name related lists (hotel, resort, villa and etc.), or the ambiguous words that cannot unveil the real context, like lovely, delicious, tasty.

4   Results and Discussions

Based on the literature review and careful coding, the 100 word groups are categorized into seven attributes, each containing detailed indexes, as shown in Table 3. Among the seven attributes, hotel, location and room attracts the majority of the stress, respectively taking the percentage of 25.9%, 21.7% and 21.4%, which clearly reflect the travelers' willingness to stay in a conveniently located and well equipped hotel, and to enjoy the comfort of the room fully supported by the well furnished amenities. Service and food are also frequently mentioned, sharing another proportion of 22.4% in altogether. This finding also conforms to the prior research finding of Li et al. (2013) that service and food are important attributes generating customer satisfaction with hotel [1]. Surprisingly, only no more than 5% of the review words have connections with the elements of hotel value. The finding reflects the true thoughts of travelers. Travellers consider more about quality of travel service, and less on the hotel value considering the popularity of Sanya as a leisure travel tourism destination and aiming for relaxation when traveling to Sanya. The denomination regarding the general facility availabilities, ambiguous in the underneath context of either hotel or in-rooms, for instance, facilities, equipments, deployments, hardware, equipped, and etc., also counts for 3.8%, though it cannot unveil much significant findings.

In the attributes related to hotels, the index of beach and sea view far surmount the other indexes, standing over one third of review frequency in its category. This is crucial to hotels in the tropical cities like Sanya, which is also why most top star hotels locate nearby the sea to avail beautiful sea view in-rooms possible, and to run their own beach with seashore supports for enjoyment. Room rates of those hotels can even triple the price of the similar hotels which are not near the sea, since they follow the fashion and pinpoint the unique feature. The element of decoration turns to be the second important attribute related to hotels, sharing a percentage of 19.0%. This finding reveals that customers care very much about the hotel decoration. The decoration, furnishing, ornamentation of hotels can help customers to get the feeling of luxury, modern, fashionable, dated and etc. The public facilities and fitness and entertainment facilities are the two commonly mentioned indexes, individually standing for 12.5% and 15.5%. Lobbies, elevators, corridors, and facilities for children and elder people sketch out the public facilities travelers concern, whereas fitness and entertainment facilities mainly refer to fitness centers, sonar and message, bars and clubs, and swimming pools. In addition, reviewers also care about the chains or star ratings of the hotel (7.9%), revealing that endorsement based reputation can truly make some influence towards the travelers. Some reviewers also mentioned about the beautiful surroundings and ambience (7.5%) inside the hotel, such as gardens, fresh air, quietness and etc. The two non-significant yet indispensible elements, parking and in-hotel commuters and Internet facilities are also highlighted, though respectively with only 2%.

The general denomination of room has been intensely remarked in the room attribute, constituting over one third of the proportion. Nonetheless, the detailed hints regarding room facilities still draw a panoramic view. Cleanness (24.5%) and comfort of the room (22.5%) are the two vital variables to measure the perceptions customers hold towards the room, standing more than half of the detailed descriptions. To well explain the comfort of the room, noise & quietness (16.4%) and sleep quality (9.5%) are proposed multiple times to reflect the two influencing facets. Toilet & bathroom (13.9%), furniture & electronics (12.5%) sharing the portions of 26.4% altogether, primarily emphasize the internal facilities either in rooms or in the bathroom, which can provide the solid supports to customers. Interestingly, a great number of reviewers stress out the room size and layout (23.1%), particularly referring to size, stories, balcony, casement and floor windows, space, scenery, airiness, etc. A small number of reviews comment on beddings (3.5%), including pillows, beds, sheets, and mattress. Of course, well beddings can normally bring comfortable and tight sleep.

Same as the prior literature, travelers discuss the hotel location in regard to its proximity to public transport, attractions and city centers, and its general environment. Surprisingly, travelers intently headline the surrounding environment of the hotel (51.2%), through adopting words like landscape, scenery, beautiful, tropical, greens, nature, and etc. Expectancy and conformity theory might help explain this finding. Travelers who choose the most famous tropical city of China mainly expect to experience the exotic views. Accessibility of the public transportation (22.3%) draw some heeds, while vicinity to the attractions (15.9%) follows. The proximity to city center (10.6%) is comparatively the least highlighted. This fits in the research context, since Sanya is more a tourism spot than a gigantic city with prosperous shopping malls.

The word service in single has been mentioned for 15111 times, standing nearly 38% of the service attribute, though there are many facets of service that have been highlighted. Friendliness and helpfulness (44%), and staff and professionalism (41%) are the two equally important index, stating either the consideration of the service people, i.e. warm, polite, greeting, kind, patient, attentive, smile, help and etc., or the professionalism that displays through efficient, speedy and timely service, and their neat appearance and nice manage. Front desk service and reception (11%) can also deeply impress the customers, which implies in the process of check-in/out, inquiry, complaint, room change, welcome etiquette, and etc. Pick-up service (4%) attracts the least concern, yet it is still vital for hotel.

Dining and food is also frequently mentioned. The general description regarding food, foodstuff, eating, dining, feast etc. takes some proportion (18%), and portrait the overall profile. The two elements depicting dining and food are the dining place and the contents. Food and beverage takes the transcending advantage with the percentage of 87.4%, including the variety of food (western/oriental food, buffet, breakfast/lunch/supper, seafood, cakes/fruits/ barbecue etc.), beverage (drinks, liquors, juice, coffee, etc.), and the quality of food (delicious, flavor, etc.). Breakfast attracts the highest remarks (36% out of the category), since most hotels put breakfast in the package bundle. The dining place (12.6%), namely Chinese/western restaurant and room service, also counts. Value is mainly reflected by means of perceived value (23.1%) and price (76.9%), with the former revealing the customers' perception towards value (value for money, economic, cheap, worthy, etc.), and the latter being price of room, food, service, or the discounts and freebies.

Table 3. High Frequency Review Words Coding Profile

Index	Ratio	Notes and Examples
Hotel (25.9%)		
- beach & sea view	33.5	beach, seashore, sea view, sand, sea, etc.
- decoration	19.0	renovation, furnishing, ornamentation, luxury, modern, etc.
- entertainment facilities	15.5	fitness, sonar & message, bars & clubs, swimming pools, etc.
- public facilities	12.5	lobbies, elevators, corridors, fountains, slides, toys, etc.
- star ratings	7.9	star ratings, reputation, chain, brand, 1st class, etc.
- ambience	7.5	natural ambience like gardens, fresh air, quietness, etc.
- Internet facilities	2.1	computers, Internet, cable, Wi-Fi, etc.
- parking & commuters	2.0	parking lot, in-hotel storage battery car, etc.
Location (21.7%)	36.6/-	general description, e.g. location, district, distance, etc.
- environment	32.4/51.2	landscape, scenery, beautiful, tropical, greens, nature, etc.
- close to transportation	14.2/22.3	bus, airport, railway station, walking, taxi, etc.
- close to attractions	10.1/15.9	famous scenery spots and attractions, and place for diving, etc
- close to city center	6.7/10.6	city center, shopping, supermarket, downtown, etc.
Room (21.4%)	34.5/-	general descriptions, e.g. room, guestroom, indoors, etc.
- cleanness	16.0/24.5	cleanness, cleaning, disinfection, dust,rubbish, etc.
- room size and layout	15.2/23.1	size, stories, balcony, windows, space, scenery, airiness, etc.
- comfort	14.7/22.5	quietness, sound proofing, noise, comfort
- toilet & bathroom	9.1/13.9	bathroom facilities, e.g. shower, bathtub, towel, slippers etc
- furniture & electronics	8.2/12.5	curtain, carpet, sofa & chairs, telephone, TV, cooler, etc.
- bedding	2.3/3.5	bed, mattress, pillow etc
Service (14.3%)	38/-	the single word of service for general description
- friendliness & helpfulness	28/44	warm, polite, greeting, kind, patient, attentive, smile, and etc.
- staff & professionalism	26/41	efficient, speed, timely, manage, neat appearance, and etc
- front service & reception	7/11	check-in/out, inquiry, complaint, welcome etiquette, and etc.
- pick-up service	2/4	pick-up service including shuttles, drivers, pick-up, and see-off
Dining & Food (8.1%)	18.1/-	general description, e.g. food, foodstuff, eating, etc.
- foods & beverages	71.5/87.4	variety, quality, and taste of food, drinks and liquors, etc.
- restaurants	10.4/12.6	Chinese/western restaurant and room service
Value (4.8%)		
- price	76.9	room/food charges, service fee, discounts, freebies, etc.
- perceived value	23.1	value for money, economic, cheap, worthy, etc.
Facility availabilities (3.8%)		facilities denomination ambiguous in the underneath context, such as facilities, equipments, deployments, hardware, equipped, and etc.

5   Conclusion

This research attempts to identify the attributes generating customer satisfaction with hotels based on the online hotel reviews from hotel customers. Based on text mining and content analysis, we found that the following seven dimensions are important attributes generating customer satisfaction with hotels: hotel, location, room, service, food and dinging, value, and facilities. Hotel related attributes is the most important factors travelers care about and influence customer satisfaction strongly, followed by location, room, service, food and dining, value and facility availabilities. The research finding implies that physical settings of hotel (such as hotel, room), services and location play more important role in generating customer satisfaction, whereas dining and food, value and facility availabilities are not as so important as the role of them.

The research findings also offer some practical guidelines to hotels on how to improve customer satisfaction. Providing good service and satisfying customers are always important for hotel.

6   Limitations and Future Research

The same as other research, this study involves some limitations that need to be acknowledged. First, this research was conducted in context of hotels in Sanya in China. This gives a possible avenue for future studies to research on hotels in different cities or different countries to see whether there is difference among varied cultural background of hotels. Second, to compare the findings of research on online hotel reviews and on traditional customer perceptions, such as survey, might also be interesting in this field.

Acknowledgement. This research was supported by the National Natural Science Foundation of China (Grant No.71362027), and MOE Humanities and Social Sciences Project of China (No.13YJC630228).

References

1. Li, H.-X., Liu, Y.: Understanding the Post-Adoption Behaviors of E-Service Users in the Context of Online Travel Services. Information & Management (2014)
2. Litvin, S.W., Goldsmith, R.E., Pan, B.: Electronic word-of-mouth in the Hospitality and Tourism Management. Tourism Management 29(3), 458–468 (2008)
3. Reichheld, F.F., Markey, J. R.G., Hopton, C.: E-customer Loyalty—Applying the Traditional Rules of Business for Online Success. European Business Journal 12(4), 173–179 (2000)
4. Gretzel, U., Yoo, K.: Use and impact of online travel reviews. In: O'Connor, P., Hopken, W., Gretzel, U. (eds.) Information and Communication Technologies in Tourism 2008, pp. 35–46. Springer-Verlag, Wien, New York (2008)
5. Gretzel, U., Yoo, K.H.: Online Travel Review Study: Role & Impact of Online Travel Reviews. Laboratory for Intelligent Systems in Tourism, Texas A&M University (2007)

6. Online Travel Booking Development in China: 2012-2013, http://www.cnnic.cn/hlwfzyj/hlwxzbg/lxgb/201310/t20131022_41722.htm

7. Akerlof, G.: The Market for Lemons: Quality under Uncertainty and the Market Mechanism. Quarterly Journal of Economics 84, 488–500 (1970)

8. Bronner, F., Hoog, R.: Vacationers and eWOM: Who Posts, and Why, Where, and What? Journal of Travel Research 50(1), 15–26 (2011)

9. Giddens, A.: Modernity and self-identity: Self and society in the late modern age. Polity Press, Cambridge (1991)

10. Resnick, P., Kuwabara, K., Zeckhauser, R., Friedman, E.: Reputation systems: Facilitating Trust in the Internet Interactions. Communications of the ACM 43(12), 45–48 (2000)

11. Jeacle, C.: In TripAdvisor we trust: Rankings, calculative regimes and abstract systems. Accounting, Organizations and Society 36, 293–309 (2011)

12. Ye, Q., Law, R., Gu, B.: The impact of online user reviews on hotel room sales. International Journal of Hospitality Management 28, 180–182 (2009)

13. Ye, Q., Law, R., Gu, B., Chen, W.: The influence of user-generated content on traveler behavior: An empirical investigation on the effects of e-word-of-mouth to hotel online bookings. Computers in Human Behavior 27, 634–639 (2011)

14. Oliver, R.L.: A Cognitive Model for the Antecedents and Consequences of Satisfaction. Journal of Marketing Research 17(4), 460–469 (1980)

15. Shanka, T., Taylor, R.: An Investigation into the Perceived Importance of Service and Facility Atributes to Hotel Satisfaction. Journal of Quality Assurance in Hospitality & Tourism 4(3-4), 119–134 (2004)

16. Kandampully, J.: Service Management: The New Paradigm in Hospitality. Pearson Education, Australia (2002)

17. Deng, W.J.: Fuzzy Importance-Performance Analysis for Eetermining Critical Service Attributes. International Journal of Service Industry Management 19(2), 252–270 (2008)

18. Choi, T.Y., Chu, R.: Determinants of Hotel Guests' Satisfaction and Repeat Patronage in the Hong Kong Hotel Industry. International Journal of Hospitality Management 20(3), 277–297 (2001)

19. Parasuraman, A., Zeithaml, V.A., Berry, L.L.: A conceptual model of service quality and its implications for future research. Journal of Marketing 49(4), 41–50 (1985)

20. Parasuraman, A., Zeithaml, V.A., Berry, L.L.: SERVQUAL: A multiple-item scale for measuring consumer perceptions of service quality. Journal of Retailing 64(1), 12–40 (1988)

21. Mey, L.P., Akbar, A.K., Fie, D.: Measuring Service Quality and Customer Satisfaction of the Hotels in Malaysia. Journal of Hospitality and Tourism Management 13(2), 387–394 (2006)

22. Ramanathan, R.: An Exploratory Study of Marketing, Physical and People Related Performance Criteria in Hotels. International Journal of Contemporary Hospitality Management 24(1), 44–61 (2012)

23. Chaves, M.S., Gomes, R., Pedron, C.: Analysing Reviews in the Web 2.0: Small and Medium Hotels in Portugal. Tourism Management 33(5), 1286–1287 (2012)

24. Zhou, L.Q., Ye, S., Pearce, P.L., Wu, M.Y.: Refreshing Hotel Satisfaction Studies by Reconfiguring Customer Review Data. International Journal of Hospitality Management 38, 1–10 (2014)

25. Li, X., Ye, Q., Law, R.: Determinants of Customer Satisfaction in the Hotel Industry: An Application of Online Review Analysis. Asia Pacific Journal of Tourism Research 18(7), 784–802 (2013)

26. Magnini, V.P., Crotts, J.C., Zehrer, A.: Understanding Customer Delight: An Application of Travel Blog Analysis. Journal of Travel Research 50(5), 535–545 (2011)
27. Lu, W., Stepchenkova, S.: Ecotourism Experiences Reported Online: Classificiation of Satisfaction Attributes. Tourism Management 33(3), 702–712 (2012)
28. Stringam, B.B., Gerdes Jr., J.: An Analysis of Word-of-Mouth Ratings and Guest Comments of Online Hotel Distribution Sites. Journal of Hospitality Marketing & Management 19(7), 773–796 (2010)

Research on Non-verbal Graphic Symbol Communication of Cross-Border e-Commerce

Yuhui Feng* and Meiyun Hua

Business School, Yunnan University of Finance and Economics, China
fengyu9@hotmail.com
huameiyun@126.com

Abstract. Promoting e-commerce in Southeast Asia and South Asia has language barriers. There are many languages in the region that have a very limited number of speakers. And not many ordinary people use English for their communication purposes. Online translation software does not include such languages like Burmese and Laos. Therefore, it is necessary to explore alternative methods for instant network communication. This paper explores the use of non-verbal graphic symbols for instant communication in the cross-border e-commerce context by borrowing theories and methods from linguistics, psychology, semiotics, graphic design and computer science to enrich this approach.

This paper designed and implemented three experiments: (1) the cognitive effect of non-verbal symbols and network graphical symbols experiment, (2) the cognitive efficiency experiment of non-verbal graphic symbols and a global language such as English and (3) a simulated communication experiment using graphical symbols. Experimental results show that designed non-verbal graphical symbols are recognizable can be used as media for simple communication purposes in e-commerce after some training.

This study has the potential to contribute to cross-border e-commerce by less educated groups of small regional language speakers. It may also contribute to special purpose communications as well as providing an embodiment of the need to use graphic expressions.

Keywords: Non-verbal graphic communication symbols, online instant communication, cross-border e-commerce, e-Business.

1   Introduction

Current cross-border e-commerce websites provide three language solutions: (1) web pages are in different language; (2) translation software is available for instant web page translation;(3) human interpreters are used to offer synchronous or asynchronous translation services. The first method is used for news release in static web pages but not very applicable to interactive e-commerce activities. The second method has two main problems. First, software translation has limited accuracy of about 70%. Second, most of the translation software does not provide translation for languages with small

---

* Corresponding author.

H. Li et al. (Eds.): I3E 2014, IFIP AICT 445, pp. 251–263, 2014.

number of speakers such as Khmer, Laotian and Burmese and other ethnic minority languages. The third method is restrained by the high cost of human translators, the delay in response and the difficulty to maintain long-term high quality services. In addition to that, the Europe approach of using English as the communicative media does not apply in other cultures or regions.

Given that e-commerce operates in a virtual environment, users' level of education is unpredictable; the limitation of English as the communicative media for e-commerce and translation software does not support small languages, exploring alternative methods of communication has practical significance.

So, our research question is: Can we conduct simple and preliminary communications using non-verbal graphic symbols in cross-border e-commerce?

If it is any hint, the application of graphical symbols in Internet communications may have practical implications.

On September 19, 1982 at 11:44 am, Carnegie Mellon university professor Scott Fahlman set the precedent in the BBS of the Internet by typing a string ": -)", emoticons to express feelings. The emoticons spread quickly in the BBS, and developed. Later the Martian symbols appeared in online social networks to convey specific information. Examples are 'Orz' (bowing down in hieroglyphic method, like a man three letters fell to his knees on the ground) and '@ @' (feeling dizzy). Along with the development of Web instant communications technology, in the instant communication network applications such as MSN, QQ and WeChat, people developed graphic emoticons to replace the character emoticons that are not easy to remember. After that animated graphic symbols were adopted thus adding business significance to these symbols (Fig. 1). Feng Cheng (2005) found that "in the SCMC (based synchronous computer - mediated communication) context, people used the non-verbal symbols for entertainment mentality and emotional needs hence giving these non-verbal symbols the inherent function of communicating entertainment and expression."[1]. Since then, graphic symbols in addition to expressing emotions have gained added meanings.

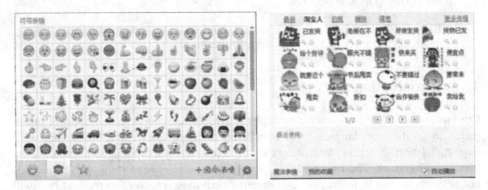

Fig. 1. Emoticons graphic symbols, animated graphic symbols and business graphic symbols

In addition to expressing emotions, graphic symbols have gained new uses. With the development of social media, people began to upload large numbers of pictures in forums such as Baidu Post Bar and via communication software to express richer meanings and meanings that are hard to convey in language. This then led to the development of a visual language.

There are also attempted complete narratives by using graphical symbols. For example, a painter named Xu Bing (2012) depicted the life of a white collar worker in a day using graphic symbols in his book "To book: from point to point"[2]. Figure 2 is a paragraph from the book demonstrating Bing Xu's words-graphic symbol translation system. This is an experiment using graphic symbols to convey meaning.

Fig. 2. Bing Xu's narrative in graphic symbols

2   Conceptual Background

No research paper has been documented studying the use of non-verbal graphic symbols in business, especially in the context of the multi-lingual environment in Southeast Asia. But that does not exclude the following relevant studies.

Nonverbal graphic symbols communication has become a way for people to communicate. Cheng (2005) argues that "computer-mediated communication (CMC) has become an important means of human communication." "In online socializing computer users have developed a set of nonverbal cues popular among netizens to compensate for the lack of nonverbal cues in CMC context. Non-verbal communication in SCMS is classified into three types: graphic accents, electronic paralanguage and emoting. Graphic accents fall into two kinds: emotions and pictures. Emotions usually include keyboard symbols and ASCII codes to indicate users' mood or states of mind. Pictures are content-rich picture icons or animated pictures originated from the Internet. Electronic paralanguage is a set of innovative nonverbal cues created by communicators mainly through the strategic use of typography for emphasis or effect enhancement. Action description is also categorized as a method of nonverbal communication. Action description (Emotion?) refers to the narrative descriptions of the user's current emotional or physical state in the third person"[1].

Studies show that nonverbal symbols can be directly felt, do not require a lot of thinking and analysis. It is also found that clear and systematic nonverbal symbols can improve cognitive efficiency and reliability. The most representative study was conducted by Standing in 1973 [3]. In this study, 2500 slides were shown to the subjects in 10 seconds. Then the slides were shown again in pairs for the subjects to identify. The results showed that even if the new slides were shown only for 1 second, or when the original slides were mirrored, 85% - 95% of the slides were correctly identified.

King (1971) did an experiment comparing the cognitive difference between nonverbal graphical symbols and text. His founding was that people were able to match nonverbal graphic symbols more quickly than the text. Also, 65% of the subjects believed that it was easier to match the nonverbal graphic symbols than the text [4]. Based on another study regarding the correct recognition of texts and symbols, Walker (1965) came to a similar conclusion that nonverbal graphic symbols have a higher recognition rate than texts [5]. Thus, Horton (1994) agreed that non-verbal graphic symbols were visually easier to recognize than texts and easier to remember too. Nonverbal graphic symbols have both verbal memory and visual memory, and texts have only verbal memory. Therefore, the human visual perception system has powerful cognitive and identifying ability for symbols. For well-designed graphical nonverbal symbols, people's cognition is even more rapid and accurate [6].

As can be seen from the above studies in other fields, nonverbal graphic symbols have better cognitive recognition and higher cognitive efficiency. Thus we propose the following hypotheses.

H1: In the context of e-commerce, people can effectively recognize graphic symbols designed by following the rules of cognitive psychology and linguistics.

H2: The cognitive effect of graphic symbols is acceptable relative to that of the verbal signs.

H3: Preliminary communication in graphic symbols is possible after some training in them.

In order to verify the above hypothesis, we have designed three experiments.

E1. The cognitive effect experiment of non-verbal symbols and network graphic symbols, designed to assess whether the designed nonverbal graphic symbol system is reliable as a system and whether it is easy to recognize or easy to use.

E2. The cognitive efficiency experiment of non-verbal symbols and such global language symbols as English. The experiment was designed to evaluate the cognitive effect of designed graphic symbols by means of a cognitive test of both designed nonverbal graphic symbols and linguistic symbols.

E3. A simulated communication experiment using nonverbal graphic symbols. The purpose of this experiment was to test whether two-way communication using nonverbal graphic symbols in the e-commerce context can be achieved after both parties of the communication were properly trained.

3   Methodology

3.1   Experimental Material Preparation

We prepared two sets of non-verbal material and a set of graphical symbols for English learning material, called material 1, material 2 and material 3.

Material 1: picture collected from Internet, show as Table 1. These graphical symbols have the same style, more commonly used. To avoid the influence of the color, all graphic symbols are processed into black and white figure.

Material 2: icons designed the nonverbal graphic symbols as shown in table 2. It is designed based on some knowledge of cognitive psychology, cognitive linguistics and graphic design.

Material 3: learning materials adopted 'New Cambridge Business English (primary)' in business consulting part of school textbooks.

Table 1. Sample picture part of material 1 **Table 2.** Sample graphic symbol part of material 2

3.2   Participants

In this study, subjects with diverse gender, age, education, respectively, from Thailand, Laos, Vietnam, China, undergraduate, graduate students and exhibitors ASEAN Expo, A total of 94 people. Intelligence and vision of all the subjects are normal.

3.3   Experimental Design

Experiment 1

For 94 subjects, due to the Likert scale questionnaire survey results has independent of features with personnel level, experience, and knowledge and so on, and therefore don't need to consider when designing the experimental program the subject components affect experimental results generated.

Using questionnaire survey, the symbol design results are subjective evaluation. Experiment principle using Likert (1932) scale of psychology experiment method. It belongs to the one of the rating aggregation type scale that most commonly used. The scale consists of a set of statements consisting. Each group statement has five kinds of answers, such as "very", "less", "general", "more" and "very". Subjects are asked to express the views for each group of statements and icon related to the experiment, and t and clearly the extent of its holdings attitudes [7]. In making Likert scale questionnaire, this paper adopted Gittins (1986) put forward to evaluate symbols of the six factors, namely, can associate, can be recognized, meaning clear, concise design, attractive and symbolic [8]. The researchers randomly selected 10 symbols from material 1 and 2, and made the questionnaire. Each symbol is set up six factors scale, each scale are set five decision values. Table 3 is an example of the questionnaire.

Table 3. The sample of questionnaire

			can associate	can be recognized	meaning clear	concise design	attractive	symbolic
1-1								
2-1		money						

Experiment 2

24 invited participants were not with graphic symbols or English learning experience. Therefore, under the same conditions that reflects characteristics of two language systems in learning and cognitive efficiency. In order to make the experiment with a more intuitive contrast, this test used two different language systems, material 2 and material 3. Depending on the material, we grouped subjects. On the use of material 2 symbol system team labeled A, using English language system team labeled B. Where group A and group B, each of 12 people.

Experimental 2 consists of two processes: learning process and testing process. Learning process: By reading training to learn the designed Non-verbal graphic symbols and common language (English), then let the subjects get some understanding about Non-verbal graphic symbols and common language. Based on the learning, through identification tests set out by the experiment, we can analysis level of these subjects reorganization between Non-verbal graphic symbols and international common language, and take the result as judgment to evaluate these symbols. Training and testing process should be alternately, as learning, testing, learning…. At the same time, each learning time specified in Table 4.This time lamination is determined by some subjects' experimentation (Note: They don't participate in formal experimental test).

Table 4. Cumulative learning time of six experiments

Cycle experiment	1st	2nd	3rd	4th	5th	6th
Cumulative time	3min	3min	6min	6min	9min	9min

This experiment adopts filling in the blanks type rules for identification test. I give some symbols to subjects, that subjects can describe and explain graphic symbols on the basis of learning symbol's system. This test can reflect whether the graphic symbol system is easy to understand recognize, remember and use. Developed a total of

12 sets of experimental test papers, numbered A-1, A-2, ..., A-6 (randomly selected from the material 2) and B-1, B-2, ..., B-6 (random selected from material 3). Each set of test papers total of ten symbols, need the subjects to answer symbolic representation of meaning after learning. Experiment 2 test papers are shown in Table 5 and 6.

Table 5. Experiment 2 Test paper (Part)

Sequence Number	Graphic symbol	Please give its characterization based on the significance of graphic symbols
1	$	
2	⚖	
...	......	......

Table 6. Test paper (Part)

Sequence Number	English word	Please give its characterization based on the English word
1	Money	
2	Weight	
...	......	......

From cognitive experimental variables analysis, Experiment 2 mainly affected by "the length of training time" the independent variable. Therefore, in this experimental program designed to deal with the independent variables investigated. Working staff were asked to try to make rigorous, reducing experimental noise accidental errors.

Experiment 3

Experiment 3 is based on Experiment 2. The first group of members in experiment 2 received training in non-verbal graphic symbols designed for this study so that they had some basic understanding of the graphs thus satisfying the requirement for research participants. The experiment verify whether can use nonverbal symbols to two-way communication on both sides under the e-commerce context. This experiment uses the symbols system of material 2 and English language of material 3. When developed, we retain a number of symbols that are not used in experiment 2 to be used in experiment 3.

Members were in Experiment 2. Then these participants were divided into groups X and Y with group X organizing the symbols into graphs using provided conceptual ideas and group Y attempting to translate these organized graphs. The questionnaire is shown in Table 7.

3.4   Procedures

Experiment 1

Grant questionnaire to subjects, and clarify the relevant matters needing attention when answer the questionnaire, ensure that all participants personnel seriously think and evaluate symbols. After subjects answer the questionnaire, researchers receive questionnaire.

Table 7. Sample volume

Experiment 3 : **Communication process simulation I (first cycle of group X)**	
1. Basic information of group X members（Omission）	
2. Answer team Y symbol combinations	
Sequence Number	Semantic
1	The price can't be any cheap.
......	......
5	Free shipping

_ Cutting line _ _ _ _ _ _ _ _ _ _ _ _ _ _ _ _ _ _

Experiment 3 : **Communication process simulation I (first cycle of group Y)**		
1. Basic information of group Y members（Omission）		
2. Answer team X symbol combinations		
Sequence Number	The result of graphic symbol made up by group X	The result of graphic symbol Answere by group Y
1		
......		......
5		

Experiment 2

Experiment 2 were arrangement of six "learning - recognition" cycle. For each learning cycle are made of different test papers. Each piece of paper is randomly drawn from the use of two language systems. The purpose of doing so is to investigate the extent of the subjects to understand and grasp the whole symbols system, which reflects on the cognitive performance in two sets of language system.

Experiment 3

(1) A and B teams in experiment 2 were regrouped (each team having the same number of participants) to simulate X and Y in figure 3.

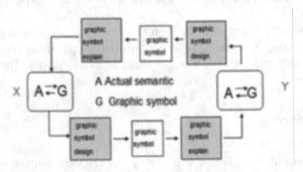

Fig. 3. Communication model using graphical symbols

(2) Following the simulation process of "X (understanding of reality) -> symbol ->Y (understanding of reality)", experiment 3 gave the conceptual ideas of several new symbols which had not been used before (to be called the "source semantics"). After that, group X organized the symbols into communicative sentences in graphs. Then these organized graphs were given to group Y to translate into sentences (to be called "target semantic"). In each cycle of these experiments, symbols were first distributed to group X (or Y) to present in graphs. When the graphs were ready, they were given to group Y (or X) to translate.

(3) Recycling semantics "source semantics" and "target semantic" test, carries on the comparison and evaluate the degree of similarity between them.

4   Analyses and Results

4.1   Measures

Experiment 1

In this paper, fuzzy comprehensive evaluation method (Fang Ke, etc., 2011) conducted a questionnaire survey data processing and analysis.

  a. handling method

Establishing factor set and evaluation set. The factors set are expressed as:

$$U=\{U_1, U_2, U_3, U_4, U_5, U_6\} \tag{4.1}$$

Which, U_1: can associate, U_2: can be recognized, U_3: meaning clear, U_4: concise design, U_5: attractive, U_6: symbolic.

Evaluation sets vector as follows:

$$V=\{v_1, v_2, ..., v_5\} \tag{4.2}$$

This experiment evaluation set as shown in table 8.

Table 8. Credibility evaluation set

evaluation	very significant v_1	more significant v_2	general significant v_3	generally not significant v_4	not very significant v_5
value	83	72	65	60	20

Calculate fuzzy evaluation matrix. To statistics of the questionnaire survey, establish fuzzy relations from factors set to comment, calculating the membership of each.

For a symbol evaluation factor U_i ($i = 1, 2, .., 6$), the evaluation statistic is obtained through the questionnaire (x_1, x_2, x_3, x_4, x_5), the membership function is defined:

$$U_{ij}= x_i/\sum x_j \ (i=1, 2,..., 6; j=1, 2,...,5) \tag{4.3}$$

Can get fuzzy matrix:

$$R=\begin{bmatrix} u_{11} & u_{12} & ... & u_{15} \\ u_{21} & u_{22} & ... & u_{25} \\ ... & ... & ... & ... \\ u_{61} & u_{62} & ... & u_{65} \end{bmatrix} \tag{4.4}$$

Using the method of contrast sort, the factor of factors set and calculates the weight of each factor was sorted. Questionnaire, subjects were asked to sort the 6 evaluation factors on importance, endowed the scores for sorting (i.e. ranked first 6 points, second place five points... ranked sixth 0 points), calculate the total score for each factor U_i, and thus can get indexes weight vector:

$$\omega = (\omega_1, \omega_2, \omega_3, \omega_4, \omega_5, \omega_6) \quad (4.5)$$

Among them,

$$\omega_i = \sum_{j=1}^{6} k_{ji} / \sum_{j=1}^{6} \sum_{i=1}^{6} k_{ji} \quad (4.6)$$

i =1,2,...,6, j=1,2,...,6 and $\sum \omega_i = 1$.

Experiment 2
Using sememe analysis (Feng, 2010) score, while scoring hits by analysis of the semanteme of the subjects answer hit on all the proportion of the semanteme[9]. The answer on subjects sememe analysis conducted with the answer set semantics and comparison, hit 1 point, the semantic meaning of the answer appears non-prime minus -1 point (because it wrong), no negative points, the lowest score is 0 point.

Experiment 3
Using the same "plain meaning analysis" from experiment 2, an analysis was done of the recycled "source semantics" and "target semantic" to identify the similarity and difference of "source semantics" and "target semantics". Below is the equation used for similarity calculation:

$$W=[SS \cap TS] / [SS \cap TS] \quad (4.7)$$

Which, SS= Source Semantic, TS= Target Semantic.

4.2   Results

Experiment 1
The statistics and weight calculation results of questionnaire are shown in table 9.

Table 9. Statistics and weights and calculation result of the questionnaire

no	efs	Rank Occurrence Statistics k_{ji}						tsf	weight ω_i
		s1	s2	s3	s4	s5	s6		
1	U_1	8	44	30	2	0	10	404	0.205
2	U_2	14	22	38	6	0	14	378	0.192
3	U_3	18	14	10	34	4	14	342	0.173
4	U_4	28	6	12	30	8	10	362	0.184
5	U_5	10	6	2	18	34	24	244	0.124
6	U_6	16	2	0	6	48	22	242	0.122

Note: no=number, efs=evaluation factors set, tsf=total score of factor, s1= section 1, s2=section 2,...

Calculate the evaluation results. According to the principle of fuzzy comprehensive evaluation, each symbol fuzzy comprehensive evaluation results can be calculated by the following formula:

$$f = \omega\, RV^T$$

Experimental Evaluation of a test groups participate in the survey results of fuzzy symbols shown in Table 10, where I-i represents symbols from 10 randomly selected material 1, II-i represents symbols from 10 randomly selected material 2.

Table 10. Each symbol fuzzy comprehensive evaluation results

I-1	I-2	I-3	I-4	I-5	I-6	I-7	I-8	I-9	I-10	v	sd
3033.06	3301.63	3288.93	3351.75	3022.67	3277.08	3166.97	3255.63	3124.10	3091.23	3191.31	98.12

II-1	II-2	II-3	II-4	II-5	II-6	II-7	II-8	II-9	II-10	v	sd
3542.87	3546.23	3390.23	3887.70	3509.96	3468.34	3301.07	3586.05	3562.79	3606.45	3560.17	92.01

Note: v= variance, sd=standard deviation

Table 9 is the sorting result of subjects to male the evaluation factors. Table 10 shows the fuzzy comprehensive evaluation score of material 1 and material 2 samples. Assuming random samples can reflect the characteristics of the whole symbol system, and then you can get the following conclusion: As can be seen from Table 5 weights, subjects generally considered that six Gittins evaluation factors are importance for symbolic design, and the importance is more average. So the evaluation set U is effective.

Material 1 and material 2 symbol system got the average scores was 3191.31 and 3560.17, respectively. Therefore, within the scope of the evaluation set U, material 2 symbol system is superior to the overall material 1 symbol system, which reflects the desirability of this symbol design method.

Material 1 and material 2 symbol system got the standard deviation was 98.12 and 92.01. The experiment shows that slightly of designed nonverbal graphic symbols are better than the picture of selected from Internet.

Experiment 2
After scoring and statistical analysis, the experiment 2 obtained the results shown in Table 11, the table shows the cognitive efficiency comparison between two systems.

Table 11. Statistics data of non-verbal symbol systems and common language (such as English) cognitive efficiency comparison test

Cycle experiment		1st	2nd	3rd	4th	5th	6th
Graphic symbol A	Accuracy100%	60.82	70.65	78.92	82.46	86.38	86.92
	st. dev	0.18224	0.17622	0.16157	0.10225	0.07559	0.03268
English language B	Accuracy100%	30.46	42.21	52.24	60.68	72.65	78.26
	st. dev	0.30889	0.28323	0.20034	0.12769	0.04643	0.03234

Speaking on the overall circumstance, two sets of experimental languages' cognition accuracy improved along with learning time. And the standard deviation of two tests' language is decreasing with learning time increasing. This test reflects the subjects were able to improve the ability to master the language and graphic symbols through learning.

Six experiments, material 2 graphic symbols system accuracy was higher than the material 3 in English learning system. The average correct rate of graphical symbols is 77.69%, while the correct rate of English language system is 56.08%. Can be considered after a short learning, the designed symbol has certain efficiency. And after a short learning, English with little effect.

To view the change of accuracy standard deviation from six experiments, except for the last almost standard deviation, the material 2 of standard deviation is lower than the material 3 of standard deviation. This shows that material 2-designed symbol system has good stability in the statistical sense.

Experiment 3
The similarity data is obtained through analysis and comparison and is shown in Table 12.

Table 12. Similarity Ratio Data

Experiment cycle		1st	2nd	3rd	4th	5th	6th
Graphic symbol	Similarity	0.6036	0.7318	0.7032	0.7782	0.7622	0.8375
	St. dev.	0.2781	0.2011	0.2983	0.2133	0.1015	0.0621
English language	Similarity	0.2492	0.3274	0.4265	0.5728	0.4872	0.6852
	St. dev.	0.4625	0.3010	0.2801	0.2379	0.1583	0.0798

Overall, all of the six experiments indicate that the material graphic symbols system is overwhelmingly superior to the material global language of English in terms of "source semantic" and "target semantic", the graphic symbols demonstrate an overall tendency of decreased standard deviation in similarity, superior to that of the global language of English. In this simulated communication experiment, graphic symbols demonstrate a relatively stable high transmission efficiency.

It is also found that the material graphic symbols can reach a similarity standard of 60% and above after some training. If the training time is increased, this similarity standard will increase and remain stable after reaching as high as 83.7%. This suggests that the symbols used in the experiments are not perfectly designed. For future study, further adjustments and modifications of the rules are needed for the symbols to most effective in communication and exchange.

5   Discussion and Conclusion

From the above experiments it can be concluded that:

(1) Relative to the graphic symbols obtained from the Internet, the designed nonverbal graphic symbols are better thus validating hypothesis 1.

(2) After some brief training, nonverbal graphic symbols designed for this study are proved to be superior in cognitive effect than the English language. This suggests that graphic symbols are easier to acquire than language thus validating hypothesis 2 as well.

(3) After several rounds of training, simple two-way communication using graphic symbols was achieved hence validating hypothesis 3 too.

In summary, the experiments demonstrate that designed non-verbal graphic symbols are recognizable and can be used for simple communications in the e-commerce context after some training in them.

6   Further Research

The experiments designed in this research were rather rudimentary. The content of information to be communicated is obvious. And the designed nonverbal graphic symbols adopted in the experiments are easy to understand too. For future studies, abstract concepts can be incorporated to increase the level of content difficulty and more easy-to-understand rules can be explored to enhance the cognitive effect of nonverbal graphic symbols.

At the same time, efforts should also be made in launching online experiments on the web so that research participants could be expanded to include a more diverse population in Southeast Asian countries.

Acknowledgements. This research was sponsored by the Humanities and Social Sciences General Project (Research on methods of nonverbal symbol instant communication in GMS, grant 11XJA870001) commissioned by China's Ministry of Education.

References

1. Cheng, F.: A Tentative Study of Nonverbal Communicative Modes in Text-based SCMC. Central China Normal University (2005)
2. Xu, B.: To Book: from point to point. Guangxi Normal University Press (2012)
3. Standing, L.: Learning 10,000 pictures. Quarterly. Quarterly. Journal of Experimental Psychology 25(2), 207–222 (1973)
4. King, L.E.: A laboratory comparison of symbol and word roadway signs. Traffic Engineering and Control 12(10), 518–520 (1971)
5. Walker, R.E., Nicolay, R.C., Stearns, C.R.: Comparative accuracy of recognizing American and international road signs. Journal of Applied Psychology 49(5), 322–375 (1965)
6. Horton, W.K.: The icon book: Visual Symbols for computer Systems and Documentation. John Wiley & Sons, New York (1994)
7. Likert, R.: A Technique for the Measurement of Attitudes. Archives of Psychology (1), 1–55 (1932)
8. Gittens, D.: Icon-based Human-computer interaction. International Journal of Man Machine Studies 24 (1968, 1986)
9. Feng, Z.: Formal models of natural language processing. University of Science and Technology of China Press (2010)

How to Set and Manage Your Network Password: A Multidimensional Scheme of Password Reuse

Yang Cheng[1,2] and Zhao Qi[2,*]

[1] China's Research Centre for Payment System
[2] School of Economic Information Engineering, Southwestern University of Finance and Economics, Chengdu 611130, China
Yangcheng@swufe.edu.cn

Abstract. The rapid development of the Internet and highly decentralized network services, prompting the majority of Internet users continue to register more accounts, and cause a high incidence of password reuse, which makes the user information leakage risks facing the domino-style. Based on the data of Internet password leak door at the end of 2011 as well as the college students' online survey, the paper analyzed the structural characteristics and reuse behavior of netizen passwords in detail, and thus designed a multidimensional password scheme that infused into the information dimensions and classified management. This scheme, based on the structure of "seed - reuse code", includes three dimensions: the content dimension contains multi-independent "information factor", which constitutes the main part of the password; the formal dimension is responsible for conversion formatting, in order to enhance the complexity and security of the password; and space-time dimension is targeted designed to protect the password timeliness and reusability. Through comparative analysis and quantitative analysis, the new password scheme not only has good memorability and convenience, and can effectively resist the violent attacks and acquaintances attacks.

Keywords: Password Security, Password Reuse, Seed Password, Difference code, Multidimensional Password Scheme.

1 Introduction

With the rapid expansion of the Internet, the network has been closely together with people's life, become an indispensable part of life. In order to fully enjoy the convenience of the Internet, such as business transactions, information access, communication and network entertainment, users need to register account more and more. At present, the most common site authentication mechanism is still the username and password combination model (ID-PWD). Although the mode has the obvious disadvantage of [1][2][3] in terms of safety compared with biometrics, smart cards and other methods, but it is easy to be accepted by users, and low cost, thus no one mature authentication mechanism can replace it so far, because of the convenience and practicality. Network certification will be mainly ID-PWD model [4] for a long period.

---

[*] Corresponding author.

H. Li et al. (Eds.): I3E 2014, IFIP AICT 445, pp. 264–276, 2014.
© IFIP International Federation for Information Processing 2014

As the ID-PWD model is not perfect in the protection of network information security, so users often have to face the problem that account password was stolen and personal information been leaked, or distress. According to the CNNIC "twenty-eighth times Chinese Internet development statistics report" [5] data shows, the number of Internet users who had account or password stolen reached 121000000 at the first half of 2011, account for 24.9 percent of the total number of Internet users. At the end of this year, Chinese Internet outbreak a large-scale user data leaks. From the 6400000 user data leakage of the programmer website (CSDN.net), to the 30000000 user password leakage of Tianya (Tianya.cn), and Renren (renren.com), duowan (duowan.com), 7K7K network (7K7K.com), baihewang (baihe.com), mop (maopu.com) and other famous sites have been stolen. So far, a large number of the big and famous website in China has been involved, about hundreds of millions of users. Not only the amount is amazing, and the leaking data is a plaintext password, non-encrypted storage, so the "secret door" caused a panic of the password security in china. In the user information leak incident, password reuse behavior of Internet users play an important role, causing huge losses to the majority of Internet users.

The so-called "password reuse" refers to the user's behaviors that choose the same password between multiple different accounts. Cognitive psychology indicate that this behavior is rooted on human memory limitations [6]: ordinary users can remember a meaningless, random, high-intensity password composed of numbers, letters and special characters after training, but after multiplied by 10 or more human mind will reach the physiological memory limit. Therefore, the design of ID-PWD combination would face trade-offs between memory and security. Based on the specificity and sensitivity of password data, there has little research about this kind of behavior, also not deep enough, and the domestic is almost a blank, but the hidden danger lead to network security is a consensus: using a small amount of password at numerous sites repeatedly would face security risk, if a combination of ID-PWD leaked it may cause the user to lose many other accounts [3] [6] [7] [8] [9]. For example, the protection of the user account information is different as different websites; especially some small and medium-sized website may become a short board password protection, which leads to the large websites' pay went up in smoke, because of funding, technology limitations. So, in order to avoid further spread of password reuse phenomenon, put forward a reasonable imminent, efficient password design and management are strategy, based on a large amount of empirical research we attempt to do some exploration in this respect.

2   Analysis the Structure of Passwords

The serious user data leak which occurred at the end of 2011 is the sorrow of the entire Internet industry. On the other hand, it also provides researchers priceless data with which to study the password security and password feature. At present, similar studies are often limited by the objectivity and representative of data, because the data before are got either from a small scale questionnaire survey or from the little leak data of a single website, thus there has never have a study that involves so many sites, cross category, large scope user password data. More importantly, most of the existing literature on the user password feature is about the English speaking world, but these user password data are usually related with the culture and language of the country, for example, we have found a lot of unique characteristics of Chinese in our research, such as Chinese Pinyin. In addition, as the password data involve a number of different categories of representative sites, we can get the value sequence of these kinds of websites and the empirical research of reusable data through the correlation analysis. Therefore, in this section we will analyze some of the leak data and the students password reuse behavior, to study the universal ways users design the password, including structure features, application habits and reuse mode, so as to bring forward the password design scheme and management strategies.

2.1 Structure

In the analysis of password structure in this section, we select the most representative leaked data of Tianya community as the main analysis object, while other data for comparative analysis and verification, this is because the Tianya community uses blog, micro-blog as a basic way to exchange, human emotion as the characteristics of the integrated virtual community and large social networking platform, it is the most influential global Chinese online home, while the user groups are widely distributed, including different age, different class, different occupation of Internet users. At the same time, Tianya is also the website who has been stolen the most data in this incident, thus, it is suitable to select it as the analysis object.

This leaked data of Tianya is backup data for 2009, a total of 29865731 account records, each record contain the account ID, password and E-mail information, so we can quickly get some basic structure characteristics of the Internet users password using some simple SQL statements.

For example, length, two thirds of the passwords range from 6 to 8, where the average is 7.94, the ratio of number and letter is about 3:1. About structure, 63.8 percent of the accounts are pure digital password; while 10.3 percent are pure alphabetic password. On the other hand only 24 percent are mixed, and the choice of the special character for the password is lower than 1.9 percent (the last two numbers in the Myspace users are respectively 81% and 8.3%) [10]. Visible, China users prefer digital password; password security awareness is generally low.

To analyze the commonly used password, we found a lot of difference from the western caused by the unique characteristics of Chinese culture. In addition to the highest frequency of 1, 2, and0, the number eight ranked fourth (the pronunciation is similar to "death "); while the number four is the least frequently used number (the pronunciation is similar to "making fortune ").

In addition, due to the differences in culture and modes of thinking, Western Internet users are different from their Chinese partner in the choice of 26 letters (case insensitive) as the password character. Through the statistics the frequency of each letter used in four different application environment (English text, Western password, Pinyin text and Chinese password), and calculate the space cosine angle between the four groups of data, we found that the correlation of China users' password and Chinese characters Pinyin text is 0.928, far higher than the English one 0.841, showing that people in the password design used to reference the pinyin. [11] This point is also reflected in these commonly used Chinese Pinyin characters: Ang, Jia, Hao, Wan and Xiao with a frequency of 0.5 percent, while the frequency of password, baby, ball, boy and other strings are often appeared in the Western passwords are below 0.005 percent, only ABC and love have a frequency close to 0.5 percent. Further studies revealed that the frequency of consonant characters are higher in Pinyin. This reflects that people like use Mnemonic Phrase-based Passwords, for example, with reference to the idiom " Man proposes, God disposes ", the password may be set to "m4zrc4zt".

2.2 Reusability

In this section, we select the leak data of four websites Tianya, CSDN, 7K7K and Renren as the research object, to establish the association table across the sites account for a common mailbox, and study the phenomenon of password reuse. These four sites belong to different subject categories, in addition to the previously mentioned Tianya, CSDN is the largest (6428631 records) technology forum for programmers, 7K7K is one of the most professional casual games website (19138451 records), Renren is a famous social networking sites (4768600 records). They are the leader in their categories, has a huge and representative registered users.

Table 1. Analysis of Password reuse between any two of the four sites

	full password				root password			
	Tianya	CSDN	7K7K	Renren	Tianya	CSDN	7K7K	Renren
Tianya	-	34.10%	91.60%	51.70%	-	19.80%	3.30%	6.90%
CSDN	34.10%	-	36.00%	29.30%	19.80%	-	21.90%	22.30%
7K7K	91.60%	36.00%	-	57.50%	3.30%	21.90%	-	5.30%
Renren	51.70%	29.30%	57.50%	-	6.90%	22.30%	5.30%	-
Average Val	59.10%	33.10%	61.70%	46.20%	10.00%	21.33%	10.17%	11.50%

Account association table "Relation (Email, Tianya_PWD, CSDN_PWD, 7K7K_PWD, renren_PWD)" has a total of 4718269 records; each of them is at least email registered at two different sites. Through the table we found that, in addition to password reuse, there are a large number of multiple password corresponding with the same email exist only minor differences, they are from the same root password (PWD-Seed) and derived, this should be regarded as an extension of the password reuse. In order to accurately measure the multiple correlation relationship, we use the standard LCS algorithm. The standard LCS algorithm was adopted here to calculate password similarity from the same users on different websites:

$$sim(PWD1, PWD2) = \frac{2 \times len(lcs(PWD1, PWD2))}{len(PWD1) + len(PWD2)} \quad [1] \quad (1)$$

$len(x)$ returns the length of the string X; $lcs(X, Y)$ returns the longest common substring of X and Y, such as, $lcs('12345678', 'a13458') = '345'$. In the formula, "Sim=1" indicates that users use the same passwords on multiple websites and "Sim=0" indicates that the passwords used in different websites are totally different. Therefore, "0.75< Sim<1" indicates that passwords used in unique websites are

---

[1] Function Description: *len* Returns the length of the string, *lcs* Returns the longest common substring, for example, $lcs('12345678', 'a13458')='345'$.

basically the same. Table 1 shows the average Sim between the four sites of two values, namely password reuse rate of different websites.

Table 1 indicates that in these four sites more than half of the public user pass-word existing reuse phenomena between any two sites (the sum of two reuse patterns between any two sites are larger than 50 percent), where Tianya and 7K7K have a proportion of more than 90 percent are completely password reuse, with password basically the same is more close to 95 percent. In the two reuse mode, ordinary users tend to use fully pass-word reuse is simple, but has the background of IT users (CSDN) tend to relatively complex password reuse. Further analysis showed that, in the multi-ply site registered users, nearly 80% (79.1%) users existing a behavior of password reuse (4 password corresponding to the same Email are at least 2 identical), including 4 password identical up to 33.9 percent, which is more than one to three users.

Usually, password reuse rate between websites are related to many components, the value of website and its user component structure are the two most important. For example, in four sites in Table 1, CSDN has the highest value, followed by Renren, the two passwords are unique compared with others, to prevent theft; while Tianya and 7K7K account value is relatively low, more likely to share the password, to facilitate the application. As the user structure, network safety consciousness of the four in descending order: CSDN user has the background of IT technology, Renren to college students, while Tianya and 7K7K are relatively popular. The above two rea-sons, jointly determined the password reuse of four big websites and its representative categories with a sort: CSDN (IT) < Renren (SN) < Tianya (BBS) < 7K7K (GAME).

Finally, the generalized password reuse should also include different users use the same password. It is embodied in the high-frequency password. In the case of Tianya, there have an average 1.5 people sharing a same password per 1000000 people, of which there are more than 4% accounts use "123456" as the password, the top 20 high-frequency password account proportion reached 8.42 percent, 100 reached 11.38 percent, while the first 10 percent password should cover more than forty percent (41.11). Obviously, the security of these passwords is relatively low, when faced with force attacks; it is very likely that they are the first to be compromised.

2.3   Behaviors Study for Students

In order to analyze users' password reusing habits, we developed an online re-search of 123 students majoring information manager, which received 118 effective questionnaires (including 25 boy-students, 93 girl-students). These participants having better knowledge of computer science and network safety realization, and we tried to find some valuable devise thought through their reusing behavior.

By the previous interview, we surveyed and sifted 29 website which always visited by college students (or website software), including QQ, Alipay, the ABC online bank (blank cooperated with college), 163 mailbox, Sina Web, Worry-free future, Baidu Library and so on, which can roughly divide into 6 kinds: online classes, communication kind, forum kind, job wanted kind, datum kind and entertainment kind, and add to an daily "starting up password". Subjects involved were asked to visit these website from the first one, and fill out one typical figure one/two/three..., and the website which having reusing password use one public figure to mark, unregistered website use zero to mark. For the survey using anonymous way, and didn't involve material password data, or privacy reveal, so the data of this survey are relative real and objective.

By analysis, we found that almost all 118 students interviewed having reusing password behavior (94.9%), and among them 56.8% used the root password mode. The average account number is 10.6 for each student (between 3 and 28), however the independent password number is only 4.6 (between 2 and 8). Using the number of registered website and independent password of each student, we calculated all inter-viewers' average rate of reusing password is 2.6(between 1.0 and 6.5). This meaning, when the password of a reusing website leaked, an average of 2.6 websites facing a potential threat. Compared with the previous survey, we found that some data is coin-cide. For example, Florencio and Herlry analysis 500,000 users found that the average number of password is 6.5, the reusing rate of password is 3.9; Brown and others researched 218 students from American universities found similar result: each student had the average number of password is 4.45, and the reusing rate of pass-word is 1.84[13].

Researchers always use reusing rate to measure the potential threat level of pass-word system, but Youngsok and some others thought this data ignored users' bias of password use, that is, it reflected the unbalance of reusing password on the account. They construct passwords on the base of graph theory, and put forward a new kind of Vulnerability Index [14],

$$VI = \sum_{i=1}^{m} (\frac{n_i}{N})(\frac{n_i - 1}{N - 1}) \tag{1}$$

N is the total number of register website, m is the number of independent pass-word(that is, the number of reusing subnet, $m \leq N$), n_i is the number of websites in-cluded in reusing subnet.

Using this formula, we calculate 118 students' average value of VI is 0.33, that is, when one website leaked, averagely 1/3 of remaining website facing threat, com-pared with 2.6 before is much higher. Figure 1 shows the reusing rate and VI value of all students interviewed, from which we can found that, the VI value is very different with same reusing rate. For example, the reusing rate of student A and student B is 3.0, but the VI value of them is 0.67 and 0.20. It's thus clear that, using reusing rate only to measure Vulnerability Index is not enough, it also need independent password distribution in reusing network to evaluate it.

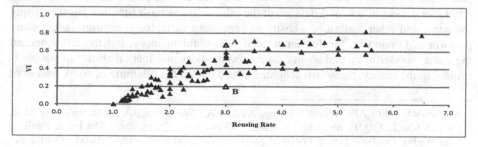

Fig. 1. Password reuse rate and VI value of interviewed students

However even so, we recognized VI value is still not exact enough, it ignored the value difference among websites: the same leak information, the loss of different value account is very different. Although the registered websites number of users is large, but most of them have low value, and the high value accounts only a little part. The value of websites is also observed "two-eight principle", this phenomenon reflected by our survey data.

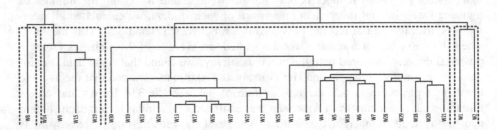

Fig. 2. Clustering pedigree chart of 30 surveyed sites

According to the data of all users' passwords, Clustering of 30 websites: firstly, numbered all websites with w1.w30[2], secondly, make a table [30*30] by the original data, the data in the table is the number of users having reusing password between two websites, and last, using "the longest distance way" in R software make cluster analysis, figure 2 is hierarchical diagram. The figure showed, the interviewers' reusing password having obvious classified phenomenon: the password similarity of w1(the ABC online bank)and w2(alipay) is higher; w8(QQ)is an independent class; however the passwords of w9(Fetion), w14(Renren), w15(Sina Weibo) and w19(Baidu Library) are similar; and the remaining 23 websites is a class. Obviously, the two classes previously belong to high useful website, and the number is very small; the number of latter class more than 2/3 of the total number, but for college students, they belong to small useful website. This classification can meet the real need of large password use, is a worthy mirror in our password design system.

3   Multi-dimensional Password

For the phenomenon of reusing the same password for multiple websites, there are two better solutions usually: one is using "one website one password" pattern, setting a random password for each password; the other is using hash function pat-tern, using the name and other feature of website as dependent variable, generating independent passwords. Obviously, the former needed wonderful memory, but the latter needed the same wonderful mental ability. Since the memory and mental ability of people is limited in the reality, these two schemes would entrap the dilemma of weak password

---

[2] 30 sites(w1..w30): ABC online banking, Alipay, Ali Wangwang, Amazon, Jingdong Mall, Ctrip, VANCL, QQ, Fetion,Gmail mailbox, 163 mailbox, 126 mailbox, The Farm, RenRen, Sina Weibo, Douban, Tianya Forum, CSDN, Baidu library, Douding network, DaoKe Baba,51job,ChinaHR,Street site, The recruitment of Chile, Youku, Tudou, The most liked online games, The most liked game sites, power-on password.

or with the aid of tools for storage/calculation. So, on the problem of multiple website password design, lots of scholar considered pessimistic cannot meet the memory and security at the same time. But there are some scholar considered that memory and security are opposite actually, so finding suitable solution meeting the real demand is possible, and they presented principle thought: (i) the quality of initial passwords; (ii) the memory of passwords; (iii) the complexity of passwords [15].

The short-term memory survey of American psychologist Miller indicated that, the capacity of short-term memory had inner relationship with the quality of memory material and individual material processing, the memory limitation of general people is around 7±2 chunk. Here the chunk is the unit of short-term memory, it's the process of uniting several separate stimulation into meaningful, larger information unit, that is stimulation of information recoding, this can effectively improve the capacity and effect of short-term memory. And short-term memory through repeated col-lection, changed into long-term memory. [16] For this, we imagined a password en-coding way: changing a long password string into the suitable combination of several small "information factor", and each information factor is a sub-string through the designer considered deeply, containing rich personal information, and having closed relation-ship with the experience of designers. In this way, it remained the complex and mul-tiple form, at the same time improved the durability of memory.

Concretely, we referenced "Dimension of Information" concept in Information Theory, extended it and then proposed" Multiple Dimensions Password System" designed way. In Information Theory, dimension of information is a measure to esti-mate the effective value of information, including dimension of content, dimension of form and dimension of time, to measure the relationship of content, the rationality of form and the effective of time separately. Expanded into multiple dimension system, dimension of content corresponded several independent information factors, consti-tuted of the memurial main part of password; dimension of form responded to the form change, improving the complexity and security of pass-words; and dimension of time reflected specific design, ensuring the effect and importance of passwords. The following is a detailed instruction for each latitude.

Dimension of Content: It's the base of the entire password system, and it can be any personal information of users, such as name, date, address, telephone number, motto, pet phrase, or liked poetry, proverb and so on.

Designing the dimension of password, we should avoid the case of single "in-formation factor". Such as, analyses the Tianya passwords, we found lots of simple mnemonic passwords, they often from a poetry or proverb. Designed in this way could solve formal violence attack, but couldn't avoid friend attack. When content dimension contained several information factors, it could be recoded, changed it into a entire, meaningful information unit, improving the effect of memory further. For example, a dog named "diandian", and its mother was born at 62, and it wanted to go to "Vienna", so the content could be designed as"Dot-62-Vienna", memory point: diandian went to Vienna at 62.

Dimension of Form: It was effective guarantee of password security; it included both password type diversity, and its changed complexity.

The Dimension of Space-time: it was the key of multiple password system reusing, including the content of time and space.

By selecting the content dimension and form processing, we got the password having excellent performance in memory and security. Besides that, compared with the independent password problems, the design of multiple websites password system should consider the ease problem. In section 2.2, mentioned password reuse phenomenon except for complete reuse patterns, also had root mode, which is especially obvious in Internet population with higher security awareness. It was not hard to find that, the root password reuse is actually compromise of "one network one password" and "multiple websites one password", it enabled users to meet certain safety while took the least amount of time and energy in the design, memory and application of the password. This also conformed to the American Harvard professor G.K.Z put forward "Principle of least effort", namely, all people have the nature of streamline save and pursuit of benefits, always want to obtain the maximum benefit with the minimum cost (including current and expected). But studied the leaked data in-depth found that, the reuse of code (that is, the different part of the outside of the root password) design were too simple, they were more attached to the root password in the form of a prefix or suffix, and on content used multiple websites pinyin initials and digits as simple encoding. This design might be meet the security requirements of a single password, but as a password system, this faced the threat of collapse once a point breakthrough. For once the interpreter got two or more passwords of the same account, he would easily analysis the root password, and easily guess the user' password for other websites.

To avoid the password system domino collapsed, the reuse of code must be based on the personalized design, it should be diversified, no rules, and should not be stereotypes and unified. Formally, although the multiple information factors' order of root password relatively fixed, but the reuse of code can be inserted into the information factors, it could be integrated and also could be spread; and on content, it was not only related to the public information of the websites, but also related to the user's personal experience of the website, which was joint coding based on the two parts of information (Net-Public, Net-Private). Such as alipay website, Net - the Public can be zfb (alipay), ali (alibaba), ww (ali wangwang), 82 (taobao TB phone stretchable latex pinyin) and so on, even $, EC (electronic commerce) and characters contacted with alipay, such as Net private could be the time of the user shopping online, one of the most expensive online, or the nickname or avatar of ali wangwang,or...any private information associated with "alipay" . For example, the same root password ".62Vie", its alipay password could be designed for "67.62$Vie" (67, the user usually shopping on weekend), and sina blog password could be designed as "eCat. 62 Vie" (e with sina logo similar big eyes; the Cat, the user's blog, nicknamed "big Cat") or "M24.62 Vie" (24, sina's top two vowels; M, with big Cat logo shape).

At last, drew lessons from the results of the survey in 2.3, we need to divide websites into 3-4classes by the value degree, designed different root password, and constituted different password system for classified management. This classification not only reflected the value difference among websites, more was that it could en-hence the security of the entire password system further, especially be conducive to protect small amounts of high value accounts. This was because, on subjective the user had stronger sense of security for high value accounts, would pay more attention to the environmental security, reduced the possibility of self-leak; On objective, ac-count of the high value often correspond to some large websites, which could provide more capital and technology for the client data protection, isolated such accounts could avoid safe hidden trouble caused by the short board effect. Therefore, balance d the convenience and safety, users classified managed the Internet accounts ac-cording to the value, for online banking, alipay, QQ and other first or second high value should be focused on protected, let the root password contained more information factors, and carefully constructed to reuse code; And for the last class of account with low value or zero value, should focus on convenience, using relatively simple root pass-word and reuse code, and even didn't need reuse code.

4 Security Analysis

Compared with the traditional password, multi-dimensional password system has obvious advantages in memory and convenience, and in this section we will analyze the quantity of safety. Usually, password attack was mainly divided into two kinds, hackers and acquaintances hack. The former attack is more widely, while the latter has a higher success rate. In the system scheme of this paper, the hacker attack is still decided by the password length and alphabet size, while acquaintance attack is decided by the number and form of the root password information factors and the privacy levels of the reused string in the password.

As Shirley and Edward [3] conducted a survey in 2006; their purpose is to find who poses a biggest threat to their password in the eyes of the users. They considered the population in risk and the ability of hackers, on the basis of computer ability and the relationship with the attacker, the population in risk was divided into 6 categories. Similarly, we have classified test on the security of password system, also divided into 6 types: SN^0, SN^1, SN^2 and AN^0, AN^1 and AN^2, according to the revealed pass-word number (not leak, one leakage as N^0, N^1, revealing 2 and above as N^2) and the degree of closeness (strangers and acquaintances remembered as S and A).

We propose that the length of multidimensional password is at least 10 or more, because it will be unable to contain enough information factors and enough number of reusable codes if it is shorter than that, thus affecting the security of the password. Table 2 take a password with 3 information factors and the minimum length of 10 characters as an example, to analyze the 6 cases safety. At this time, the multidimen-sional password usually has 3 kinds of structure mode: 6-4 (6 root passwords, 4 reuse codes), 7-3 and 8-2. Table3 estimated the number of possible combinations in a con-servative way, to reveal the worst safety case: assuming that risk groups are "Acquaintances" and the hacker is very understanding, knowing that the possible 100 information factors (10 classes, each class contains 10 items of information), and 10 common formal transformation, namely each factor about 1000 value. In the reuse of code, Net-Private is highly personal, similar to another information factor, while the Net-Public is limited to specific sites, equivalent to a large class of "weak" informa-tion factor, set the value as 100 (10 items of information *10 forms transformation). In addition, Table2 also provides a reference system: simple mnemonic type reuse passwords. This kind of password generated by a mnemonic phrase and simple reuse code, its essence is a simple combination of a single factor and Net-Public.

As shown in Table2, multidimensional password was significantly better than the simple mnemonic type password in safety; in any case it has several orders of magnitude higher than the latter. If the simple mnemonic type password was good for violent attack, then it is almost completely transparent when encounters an acquaintance attacks, especially under the condition that the attacker got multiple passwords. For multidimensional password system, not only it can effectively resist all kinds of pure technology attacks, but also close to 8 bit random password when faced "Acquaintances". Even the highest AN^2, the most conservative estimate, it also had a combined million species. The current online login mechanism, such as incorrect password restrictions and additional parity check code, can effectively ensure the security.

Table 2. Multidimensional password system security quantify table

type	combination	simple mnemonic	description
SN^0	$94^{10}, O(10)$		Available password characters 96.
SN^1	$C_{10}^6 \times 7^6 \times 94^4 + C_{10}^7 \times 8^3 \times 94^3 + C_{10}^8 \times 9^2 \times 94^2$ $O(6.9)$	$2^2 \times (94^4 + 94^3 + 94^2)$ $O(4.3)$	Randomly selected 6-8 root password, reuse code can appear in any position in the multidimensional system, and only in the simple mnemonic is a former suffix.
SN^2	$7^4 \times \dfrac{94^4}{8^3} \times \dfrac{94^3}{9^2} \times 94^2$	$2 \times \dfrac{94^2}{2} \times \dfrac{94^3}{2} \times 94^2$	Known the root password, three Numbers corresponding structure in turn 6-4/7-3/8-2
AN^0	$1000^2 \times 1000 \times 100 \times 4^2$ $O(7.7)$	$2 \times 1000 \times 100$ $O(2.7)$	Information factors (including Net - Public and Net - Private) is the basic unit of the constructing multi-dimensional password, usually does not make the split
AN^1	$4^2 \times 1000 \times 100$	2×100	Known the root password and every information of the factor (the worst possible)
AN^2	$4^2 \times 1000$	2	Net - Public encoding be cracked (worst possible)

Note: $O(x)$ represent password combinations of equivalent to the length of x random password

In addition, Table2 also shows that when faced the high risk acquaintance attack, the security of multidimensional password system is mainly determined by the number of information factor and private levels, and has no direct relationship with its characteristic length. Therefore, suggests that high value sites contain at least more than 3 information factors, while the reusing code should be not less than 3 characters, in order to ensure system security after particular individual password leaked.

5   Discussion

Based on the survey data and the leaked data, we designed and created the multidimensional password system, "root password - reuse code" as the structure, involved the information dimension and the classification thought. After comparison and quantized analysis, which not only has a good memory and convenient, but also can effectively resist brute force attack and acquaintance attack. Although the complete website authentication mechanism contains two parts: ID and PWD, many researchers have suggested using different ID in secure password system, but we think that the essence of ID is the password extension, it is not recommended to do too much change in the account. Because most of the formal sites were required to pro-vide email when the users register, or even mobile phone number, for confirmation of registration and retrieve password and other services, and users often fill in. And it is often the key to judge whether it is the same user, rather than ID itself. Therefore, too much account transform can do nothing but increase the burden of memory, meaningles to reduce the risk of reuse.

Notable is, password scheme in this paper is only for network user. And a good password security system needs three aspects to cooperation: users, Internet companies and policy makers. Especially for the enterprise, safe storage and encryption transmission problems need to be solved for user data. If these data were stored in plain text, once encounter "drag library", and then the password was also in vain even if it designed perfectly.

References

1. de Rodrigo, L.G., Carlos, A.L., Atman, A., et al.: Biometric identification systems. Signal Processing 83(12), 2539–2557 (2003)
2. Halderman, J.A., Waters, B., Felten, E.W.: A convenient method for securely man-aging passwords. In: Proceedings of the 14th International Conference on World Wide Web, pp. 471–479. ACM (2005)
3. Shirley, G., Edword, W.F.: Password management strategies for online ac-counts. In: Proceedings of the Second Symposium on Usable Privacy and Security, pp. 44–55. ACM (2006)
4. Pinkas, B., Sander, T.: Securing passwords against dictionary attacks. In: Proceedings of the 9th ACM Conference on Computer and Communications Security, pp. 161–170. ACM (2002)
5. China Internet Network Information Center (CNNIC). The 28th Statistical Report on Internet Development in China [EB/OL], (July 19, 2011), http://www.cnnic.net.cn
6. Zhang, J., Luo, X., Akkaladevi, S., et al.: Improving multiple-password recall: an empirical study. European Journal of Information Systems 18(2), 165–176 (2009)

7. Notoatmodjo, G., Thomborson, C.: Passwords and perceptions. In: Proceedings of the 7th Australasian Conference on Information Security, vol. 98, pp. 71–78. Australian Computer Society, Inc. (2009)

8. Devi, S.M., Geetha, M.: OPass: Attractive presentation of user authentication protocol with resist to password reuse attacks. International Journal of Computer Science and Mobile Computing 8(2), 174–180 (2013)

9. Ives, B., Walsh, K.R., Schneider, H.: The domino effect of password reuse. Communications of the ACM 47(4), 75–78 (2004)

10. Schneier.B. Real-world passwords [EB/OL]. Schneier on Security 12 (2006)

11. http://www.schneier.com/blog/archives/2006/12/realworld_pass w.html

12. Yang, C., Hung, J.-L., Lin, Z.: An analysis view on password pat-terns of Chi-nese internet users. Nankai Business Review International 4(1), 66–77 (2013)

13. Florencio, D., Herley, C.: A large-scale study of web password habits. In: Proceedings of the 16th International Conference on World Wide Web, pp. 657–666. ACM (2007)

14. Brown, A.S., Bracken, E., Zoccoli, S., et al.: Generating and remembering pass-words. Applied Cognitive Psychology 18(6), 641–651 (2004)

15. Bang, Y., Lee, D.J., Bae, Y.S., et al.: Improving information security management: An analysis of ID–password usage and a new login vulnerability measure. International Journal of Information Management 32(5), 409–418 (2012)

16. Sasse, M.A., Brostoff, S., Weirich, D.: Transforming the 'weakest link' — a human/computer interaction approach to usable and effective security. BT Technology Journal 3(19), 122–131 (2001)

17. Lisman, J.E., Idiart, M.A.: Storage of 7+/-2 short-term memories in oscillatory subcycles. Science 267(5203), 1512–1515 (1995)

The Analysis of Advertising Pricing Based on the Two-Sided Markets Theory in Social Network

Qiongwei Ye[1,*], Zhang Qian[2,*], and GuangXing Song[1]

[1] Business School, Yunnan University of Finance and Economics, Kunming, China
ye.qiongwei@gmail.com
[2] School of Finance and Economics, Yunnan University of Finance and Economics,
Kunming, China

Abstract. The mushroom development of social networks has brought opportunity to the analysis of social ad pricing. On the one hand, compare with traditional ad pricing, social networks advertising pricing (SNAP) enables greater consumer surplus and profits to social network companies; On the other hand, reasonable SNAP can provide guidance to network users and advertisers and coordinate the interests between bilateral participants to maximize their behavior. In this regard, using the methodology of bilateral market, this paper firstly analyzed the conduct of bilateral participants to maximize the benefits of social network companies. Secondly, the paper investigates the characteristics of bilateral markets and social networks comprehensively and proposes the Relation-Intensity Model (R-I model) to measure the strength of social relation to optimal ad asking price. Finally, the paper draws a conclusion that the SNAP increases along with the growth of the number of users at first and performs a downward trend after the number of users comes to a certain value (threshold). Thus, the paper explains that after exceeding certain amount of users (a higher network clustering coefficient), the price elasticity of demand of advertising is relatively large, lower price for the enterprise can realize higher profits, i.e. the scale effect of advertising exceeds its price effect.

Keywords: Bilateral market; social network; social relationships intensity; online advertising pricing; cross-network externality.

1   Introduction

On basis of the 33th "Statistical Report on Internet Development in China", the scale of China's Internet users has reached 618 million, the Internet penetration rate has been 45.8% relatively, and social networking users in the overall utilization rate has come to 45.0% by December 2013 [1], and according to the latest report of iResearch, the scale of Online Advertising in China reaches 110 billion Yuan with an increase rate of 46.1% [2]. In addition, a consuming psychology test of U.S. shows that the differences of the impact power between online advertising and friends' recommendation is 12 times. DCCI also shows that 75% of people are willing to buy

---

* Corresponding authors.

H. Li et al. (Eds.): I3E 2014, IFIP AICT 445, pp. 277–287, 2014.

products from a friend's recommendation [3]. Obviously, the value of social commerce basis of the social relationship is that making the transactions among strangers turns to be the market of acquaintances, so as to strength the confidence and improve the efficiency. Therefore, taking the background above as a starting point to analyze the issues of SNAP is reasonable and necessary both in theory and in practice.

The analysis of social network advertising is mainly reflected from its profit model and the way of ad pricing. On the one hand, online advertising is the main profit model of Chinese social network, social ad pricing contains mainly brand advertising, product placement and precision marketing advertising [4]. Brand advertising in social network is China's main social network advertising presently. However, compared with the portal sites, social brand advertising is not dominant. On the other hand, online advertising pricing has remained mostly on the traditional way of ad pricing, such as flat-rate model, the cost per mille (CPM), cost per click (CPC) and so on[5]. Facing with the demand analysis of SNAP, they are difficult to form a unified, flexible and efficient pricing way. Thus, based on the environment of big data, this paper try to seek a more convincing pattern of ad pricing accordingly.

According to relative data of iResearch, the consuming behavior of social network users in China is still conservative in 2011: the proportion of social network users that paid fees is only 47.3 percent, and most of them consume very less [4]. This suggests that the spread of advertising is still relatively low among social users, analysis of user behavior or content services are not in place for social network platform. Thus, the SNS focusing on enhancing the experience of social network users is necessary to improve the value of social network advertising, which is mainly reflected from the marketing value and the path of promotion [4].

2   Literature Review

2.1   Summary of Research on Social Network

Social networks (Social Network Service) refers to online relationship net that is based on the real social interpersonal relationships, which comes into being from social users' friends of friends (Friend of a friend) [6] [7]. Lu [8] points out that social network is a huge network system that is woven of a large number of interrelated user nodes, which can be described with a network diagram indicated data sets of heterogeneous relationship [8]. Watts and Strogatz [9] believe that, the increase randomness among users will make the social network topology tend to be random network [9]. After studying the impact of network structure on the spread behavior, Centola [10]. reckons that a larger cluster of (strong ties) network topology will impact great effects on the spread of behavior, comparing to random network (weak ties) [10]. Borgatti et al. [11] also suggests that, more centralized network structure (such as star structure) is more excellent than decentralized structure (such as a circle) both in the rate and efficiency [11].

Vaughn [12] creates a FCB grid model to describe the behavior characteristics of consumer purchase decisions by quantifying the user's perception. Lee et al. [13] exploits a theoretical model of online brand community to analyze the impact of brand community to users' behavior. Meng and Cui [7] [14] measure the ad price of social network by using the linear fitting of several traditional ad prices. These methodological analyses of advertising still does not walk out of the plight of traditional online advertising, which consider the effects of social users on ad pricing sufficiently.

2.2   Summary of Research on Bilateral Market

Through the middle layer or platform, two kinds of participants conduct a transaction, and the benefits of a group of participants that joint the platform depends on the number of participants in another group, this kind of market is called bilateral market [15] [16] [17]. Bilateral market involves two distinct types of users, each of which obtains value by interacting with another through the common platform [17]. Mark Armstrong notes that bilateral prices are affected by three factors: the strength of cross externalities, pricing method and single home or multi-home [15]. Roson [18] believes that the distribution of bilateral price affects the market participation and overall demand scale. Therefore, the determination of price relies on the price transfer to some extent.   Kaiser and Wright [19] advocates that advertisers pay much more attentions to the users than versa, the growth of users' demand will lead to higher advertising rates, while increased demand of advertising brings a decline price of the layout. Cheng [20] [21] divides social users into ad-averse users and no-difference users, suggests that the ad pricing performs differently periodically for a distinct "effects of relative value ratio", and then appears unilateral pricing, bilateral pricing and so on.

3   R-I Model Framework

The paper focuses on the analysis of ad pricing on a single monopolistic social platform (the choices of participants restrict to be "access" or "no access"). Based on the existing theoretical analysis, assumptions of R-I model is made firstly:

Hypothesis 1: The number of users and unit users' utility are relevant to the number of advertisers

For the issue of ad pricing belongs to the scope of the bilateral market, it mainly investigates the impact of the number of users to the ad pricing [15] [17]. Therefore, this paper mainly concerns the effects of social users' (fixed network structure) interaction on the SNAP (i.e. the cross-network externality). In this case, the bilateral market theory requires quantifying the impact of the cross-network externality [15].

Hypothesis 2: Social Advertising brings social users disutility

Cheng [20] [21] divides social users into ad-averse users and no-difference users. With the starting of the interactions between users and advertisers, the users limit to be ad-averse users effectively [21].

Hypothesis 3: Social networking platforms seek to maximize their own welfare

Given the failure of measuring the impact of users' behavior on the interests of social platform in the traditional environment, the paper cares more of the welfare of social network platform, when it comes to the social network environment.

3.1   Constructing R-I Model

Enterprises of monopolistic social network platform can change the bilateral price to maximize their behavior. Based on the assumptions above, we are able to quantify these effects.

With social network topology, the paper quantifies the social network externality. In this case, we pay attention to the network structure within fully connected diagram [22] (Figure 1), thus network externality can be measured by the permutation of nodes (A_n^2), each of which represents a social network user. As shown below:

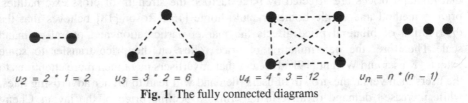

$$u_2 = 2 * 1 = 2 \qquad u_3 = 3 * 2 = 6 \qquad u_4 = 4 * 3 = 12 \qquad u_n = n * (n - 1)$$

Fig. 1. The fully connected diagrams

In general, the social network externality can be linearized as $U = bN * (N - 1)$ (b is the strength of social relationships). Nitzan [23] uses joint strength, homogeneity, connection intensity to measure the social effects within social network. While Wu [22] divides the social relationship structure into two sides: the relationship between knowledge-acquired instrumental relations and friends-interacted expressive relations, and analyses these two relations. Here, in order to survey social network topology and the service level social platform comprehensively, the paper selects the clustering coefficient of all nodes [11] [23] [24] and users' online time length [10] to indicate the intensity of social relationships.

$b = f(C, T)$, b measures the monetary utility where unit user obtains from others within effective time; C represents network clustering coefficient, whose object is confined to be the inherent or spontaneous social circles; T measures users' average online length, which reflects the service level of social network platform. Centola [10] believes that spread behavior decays exponentially with time increases. Thus users' online time length can also expose users' preference for social platforms.

Clustering coefficient of a node represents the ratio of the total number between the most connections it may be connected to its neighboring nodes and all those close to the node [11] [23] [24], that is:

$$CG = \frac{n}{k_i(k_i - 1)/2} \qquad C = \sum_{i=1}^{N_1} G/N_1$$

Where ki represents the degree of node i, which involves the number of edges that connects to the node. Moreover, clustering coefficient of a network is the average of all nodes' clustering coefficient within the network. Where N1 is the number of nodes. b is the social relationship intensity. Clearly, strong ties impacts more influence to its relative users than weak ties, which indicates that on the relationship between joint users is higher than the weak joint, indicating that users are more susceptible to the impact of a friend instead of a friend of a friend, and this has been proven to be true [10].

Adding the utility model of social network externality and fixed-proportion price transformation into Armstrong's two-sided market theoretical model [15], we can derive the Relationship Intensity model. Here, users' (represented by u) utility is impacted by the cross-network externality, social network effects and the price, while advertisers' (represented by a) utility is derived from the cross-network externality and advertising prices. Then the effects of unit bilateral participant can be expressed as:

$$u_u = a_u n_a + br_u(n_u^0 - 1) - p_u, \quad (a_u < 0) \quad u_a = a_a n_u - p_a \tag{1}$$

p_u and p_a represents monopoly platform for users and advertisers initial asking price separately; α_u is the strength of the cross-network externality that advertisers acts to users, and α_a is the strength of the cross-network externality that users act to advertisers; While b still represents the social relationship intensity. n_u^0 is the number of social users within a certain social circle contained in the whole social web, and if the social network has one social circle, the number of social users and that of social circle will be equal. In this way, the paper will mainly pay attention to the number of effective social users, which connects the amount of social users and the topology structure of social network, and we are pleasure to make it simplified. According to the theoretical bilateral market model [15], The participants in the utility function is expressed as the number of participants, and assuming that the unit cost of the participants were bilateral image And image . The profits of social network platform is:

$$\pi = n_u(p_u - f_u) + n_a(p_a - f_a) \qquad V - \pi(u_u, u_a) + v_u(u_u) + v_a(u_a) \tag{2}$$

Where f is the cost of unit participants (user and advertiser). In this case, the benefits of the platform are added by the profits and bilateral participants' surplus (Vu and Va).

Take the equations above into consideration, we have:

$$p_u = f_u - a_a n_a - bn_u(2n_u - 1) \qquad p_a = f_a - a_u n_u$$

After calculating the initial asking price of social platforms, we need to draw into the fixed-proportion price transformation (ε is the proportion ratio).

$$P_a = p_a + \varepsilon p_u$$

Furthermore, the relationship intensity model (R-I model) is:

$$P_a = \varepsilon f_u + f_a - \varepsilon a_a n_a - 2\varepsilon bn_u^2 + (\varepsilon b - a_u)n_u \qquad a_u \le 0, a_a \gg 0, \varepsilon > 0$$

3.2   Analysis of R-I Model

The strength of social relationship affects the number of social users, thereby has an impact on the final SNAP. According to the proportion above, we care more about n_u

rather than b. Making $\partial P_2/\partial n_u = \varepsilon b - 4\varepsilon b n_u - a_u = 0$, we generate that when $n_u > \frac{1}{4}(1 - \frac{a_u}{\varepsilon b})$, $\partial P_2/\partial n_u < 0$, that is when the number of social users exceeds a certain 'threshold' value, the social network ad price declines as the number of users increases; and when $n_u < \frac{1}{4}(1 - \frac{a_u}{\varepsilon b})$, $\partial P_2/\partial n_u > 0$, the social network ad price increases with the number of users increases, which draws different conclusions with the analysis of traditional bilateral market.

3.3   Model Description

Without loss of generality, the paper gives an account of the R-I model with data. After analyzing the experiment conclusions with the Cox proportion hazards model, Centola [10] draws that triple stimulations of network signal can generate the most effective result of social users' spread behavior. ($Z = 1$, $P < 15\%$; $Z = 2$, $P > 30\%$; $Z = 3$, $P = 40\%$). thus we limit the studying scope within the cluster network and the strength of strong ties to be 3 ($Z = 3$). Thus, the topology of this kind of social network can be depicted as follows:

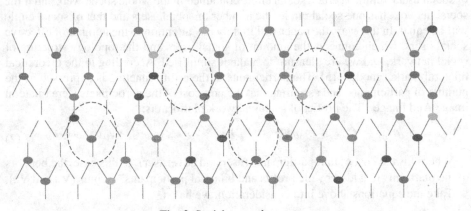

Fig. 2. Social network structure

Remarks: the number of social users ($n_u = 104$), within the structure, any user will be affected by his or her three friends effectively, which turn out to be the best impact degree of spread behavior by the study of Centola [10].

The clustering coefficient is: $C = C_G = \frac{3}{(3 \cdot 2\sqrt{2})} = 1$. Then, the strength of the social relationship is mainly measured by social users' online time: $b = f(1, T) = f(T)$, and this is set as the linear relations: $b = T/40$. While the average length of the user's online time is two hours (T = 2).

Other information is in the table below:

Table 1. XX's community information

Name	Value	Unit	Remarks
Marginal cost	f_u=1200;f_a=800	Yuan	
The number of advertisers	n_a=5	/	$n_a \ll n_u$
The proportion of pricing transformation	ε=0.2	/	
Users' cross-network intensity	a_u=-4	Yuan per advertiser	Ad-averse users
Advertisers' cross-network intensity	a_a=2	Yuan per user	User-depended

Putting these values into the R-I model we obtain:

If $b = 0.05$, $P_a = 1022 - 0.02n_u^2 + 4.01n_u$, and the threshold value is:

$$n_u^* = \frac{1}{4}(1 - \frac{a_u}{\varepsilon * b}) = 100.25 \cong 100 .$$

With $n_u = 104$, $P_a^1 = 1222.72$, That is, the final ad asking price is 1222.72.

Otherwise, the number of social users is 104($n_u > n_u^*$). Thus, the final ad asking price drops, if the amount of social users increases further; and when the number of users is less than 100, , the final asking ad price increases along with the growth of the number of users.

4   R-I Empirical Evidence

To verify the robustness of the R-I model further, the thesis takes the example of China's typical social network – Renren to draw a brief demonstration. While the key evidence to verify the conclusion is whether the impact of social user on social network advertising pricing exists a threshold value. The paper adopt the monthly amount of Unique Visitor to reflect the number of social users and take cross-quarter online advertising revenue as the income of social network platform achieved from advertisers (fixed ad proportion). Thus, the paper extracted RenRen's relative data (the number of social user (2010 Q4-2014 Q1) and social ad revenues (2009 Q1-2014Q1)) from the 199IT Internet data centers, iResearch, DCCI and so on, which is shown in Figure .2. As we can see, the data especially Ad revenues represents a seasonal fluctuation. Thus it is necessary to adjust the data to remove the influence of season, and adjusting the statistics with MA (5) is reasonable.

As is seen from the chart, the change of advertising revenue shows an oscillatory growth trend, while the number of unique visitors draws a more substantial increase trend. Furthermore, with the scatterplot composed of advertising revenue (P) and the number of monthly unique users (U) (Figure .3), it is easy to judge that advertising revenue (advertising price) presents a first-increased-then-decreased trend by the impact of the number of social users, and when N = 4000 (March 2012), the threshold value appears, which verifies the conclusions. When we explore the statistical relationship between the ad price and the number of social users by SAS 9.3 further, we are able to draw the conclusion better (Figure .4)

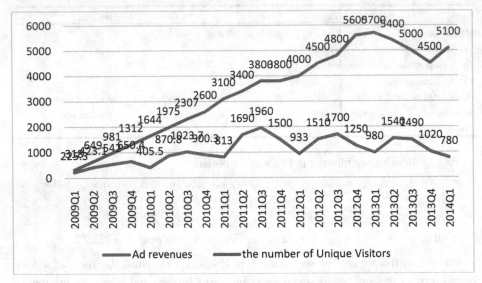

Fig. 3. The amount of the Unique Visitors and ad revenues

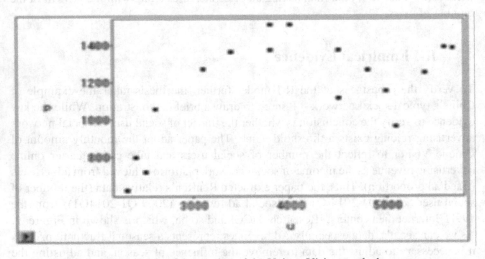

Fig. 4. Adjusted data of the number of the Unique Visitors and ad revenues

The regression fitting of parameter without intercept								
		Model			Error			
Curve	Times(polynomial)	Free degree	Mean square	Free degree	Mean square	R square	F statistics	Pr> F
	2		2	12313894.3	15	41469.5208	0.9754	296.94	<.0001

Fig. 5. The regression fitting result of their relation between Ad revenue and the number of social users

5   Conclusions and Forecasts

5.1   Conclusions

The paper draws the following conclusions by using the analysis of R-I model, and we summarize two key points as follows:

(A) The intensity of social relationships indicates user's dependence level on social networking platforms, the more clustering social relations will leads to more frequent interactions among social users and higher dependence level of the platform, the more comprehensive social relationship network, and the higher user's utility level, which attracts more users to join the network. Meanwhile, the marginal effects of one's indirect relation users (weak ties) on the social users is degressive. In other words, a weaker degree of mutual trust and intimacy appears when the social network tends to be looser.

(B) Social relationships intensity affects the final price of social network platforms by two (direct and indirect) ways. The direct way is derived from the attention of social network platforms to social users, and the indirect way lies to social relation's effects on the number and utility of social users, and affect the final pricing further. What's more, when the indirect effects surpass the direct one and the amount of social users exceeds the threshold value, the final price will decline, which turns out to be perfect for both the platform enterprise and social network participants.

5.2   Forecasts

Further analysis will focus on two aspects: the empirical test by using the big data and modify the model; thinning the SNAP, making targeted analysis of pricing model of different advertising and extracting more rigorous theoretical model. Thus, there are much more tasks for us to launch.

Acknowledgements. This research was supported by the National Natural Science Foundation of China under Grant 71162005, 71102010 & 71362016.

References

1. The 33th Statistical Report on Internet Development in China,
   http://www.cnnic.net.cn/hlwf-zyj/hlwxzbg/hlwtjbg/201401/
   t20140116_43820.htm/ (read April 16, 2014)
2. iRresearch. China's Online Ad Key Data (2013),
   http://a.iresearch.cn/others/20-140109/224661.shtml (read January 9, 2014)
3. Zeng, S.: Commodity Recommended System Based on the Social Network Trust Model. University of Technology of South, Guangzhou (2012)
4. Social Network Analysis Report in 2010-2011,
   http://report.iresearch.cn/1658.html (read February 20, 2012)

5. The Blue Paper of China's ad network,
   `http://www.dcci.com.cn/hotreport/view/id/51116-6.html`
   (read December 15, 2011)
6. Social Network Service,
   `http://zh.wikipedia.org/zhcn/%E7%A4%BE%E4%BA%A4%E7-%B6%B2%E8`
   `%B7%AF%E-6%9C%8D%E5%8B%99,2013` (read October 8, 2012)
7. Wen, C.: The Impact of Advertisement appeal, sources and type on Social Network Performance. Department of Industrial Engineering in Tsinghua University, Beijing (2011)
8. Lu, M.: The Analysis of WEB Data Mining And its Application in Social Network. University of Electronic Science and Technology, Chengdu (2012)
9. Watts, D.J., Strogatz, S.H.: Collective dynamics of 'small-world' networks. Nature 393, 440–442 (1998)
10. Centola, D.: The Spread of Behavior in an Online Social Network Experiment. Science 329, 1194–1197 (2010)
11. Borgatti, S.P., Mehra, A., Brass, D.J., et al.: Network Analysis in the Social Sciences. Science 323, 892–895 (2009)
12. Vaughn, R.: How Advertising Works: A Planning Model. Journal of Advertising Research 20, 773–779 (1980)
13. Lee, D., Kim, H.S., Kim, J.K.: The Impact of Online Brand Community Type on Consumer's Community Engagement Behaviors Consumer-Created vs. Marketer-Created Online Brand Community in Online Social-Networking Web Sites. Cyber-Psychology, Behavior and Social Networking 14(1-2), 59–63 (2011)
14. Meng Lingsheng, D.: The Analysis of online adverting strategy and pricing model. Jilin University, Changchun (2008)
15. Armstrong, M.: Competition in two-sided markets. RAND Journal of Economics (10), 668–691 (2006)
16. Rochet, J.-C., Tirole, J.: Platform competition in Two-sided Markets. Journal of the European Economic Association 1(4), 990–1029 (2003)
17. Wrigh, J.: One-sided Logic in Two-sided Markets. Review of Network Economics 3(1), 1–21 (2004)
18. Roson, R.: Two-Sided Markets: A Tentative Survey. Review of Network Economics 4(2), 142–160 (2005)
19. Kaiser, U., Wright, J.: Price structure in two-sided markets: Evidence from the magazine industry. ZEW Discussion Papers 24(1), 1–28 (2004)
20. Cheng, G., Li, Y.: The Bilateral Pricing Strategy of Media Platform with Negative Network Externalities. Journal of Shanxi University of Finance and Economics 31(4), 7–13 (2009)
21. Cheng Guisun, D.: The Analysis of Operating Mechanisms and Competition regulations in Media Industry Based of Bilateral Marker Theory. Shanghai Jiao Tong University, Shanghai (2007)
22. Wu, L.: Social Network Effects on Productivity and Job Security: Evidence from the Adoption of a Social Networking Tool. Information Systems Research 24(1), 30–51 (2013)
23. Nitzan, I., Libai, B.: Social Effects on Customer Retention. Journal of Marketing 75, 24–38 (2011)
24. Luo Jiade, M.: Social Network Analysis, 2nd edn. Social Science Document Press (2010)
25. Kleinberg, J.: The Small-World Phenomenon: An Algorithmic Prospect, pp. 12–24. ACM Press, New York (2000)
26. Boulos, M.N.K., Wheeler, S., et al.: The emerging Web 2.0 social software: an enabling suite of sociable technologies in health and health care education. Health Information and Libraries Journal (2007)

27. Richin, M.L.: Social Comparison and the Idealized Images of Advertising. Journal of Consumer Research (1991)
28. Fritsch, M., Kauffeld-Monz, M.: The impact of network structure on knowledge transfer: An application of social network analysis in the context of regional innovation networks. Jena Economic Research Papers (2008)
29. Katz, M.L., Shapiro, C.: Network externalities, competition and compatibility. The American Economic Review, 424–440 (1985)
30. Huijboom, N., van den Broek, T., Frissen, V., et al.: Key areas in the public-sector impact of social computing. European Communities (2009)

Enterprise's Online Trust Crisis Management: A Life Cycle View

Yitang Zeng[1] and Chunhui Tan[2,*]

[1] School of Information Management, Wuhan University, Wuhan, China
ztqingwen@163.com
[2] School of Information Management, Central China Normal University, Wuhan, China
tanadan@mail.ccnu.edu.cn

Abstract. Online trust is the vital mechanism for the development of e-commerce, and the significance of online trust has become a consensus. Undeniably, an obscure message may be magnified indefinitely and evolve into the enterprise's online trust crisis, which will affect the image of the enterprise, threaten the survival and development of the enterprise, or make the enterprise into a doomed situation that can never be recovered. We study the development phases of enterprise's online trust crisis life cycle, and then put forward framework of enterprise's online trust crisis management strategies based on the characteristics of online trust crisis life cycle, which may be provide some theoretical supports for enterprise's online trust crisis management.

Keywords: Online trust, online trust crisis, online trust crisis management.

1 Introduction

1.1 Background

After more than ten years of fast development, the Internet has brought reforms on people's life styles, modes of information transmission, enterprise's managing ways, and ideas of government administration. According to the "Internet World Stats" [1], by the end of June 2012, about 26% of the World's population has Internet access, which represents 566.4% more than in year 2000. Regions like North America and Europe are well above those figures, with 78.6% and 63.2% of the population having online access respectively. The "33th statistical report on the Internet Development in China" [2] showed that the number of Chinese Internet users have reached 618 million, and Internet penetration rate has raised to 39.9% by the end of 2013. The scale of e-commerce has grown substantially over the past years. Global business-to-consumer e-commerce sales will pass the 1 trillion euro mark by 2013, according to a new report by the Interactive Media in Retail Group (IMRG) [3], a U.K. online retail trade organization. The e-commerce of China has a rapid growing; the total value of E-commerce in China was 5.6 trillion Yuan (about 675 million euro) in the first three quarters of 2012, which grew 25% from the same time of 2011 [4].

---

[*] Corresponding author.

H. Li et al. (Eds.): I3E 2014, IFIP AICT 445, pp. 288–302, 2014.

Trust is the currency of all commerce. In traditional commerce, the trust is based such things as societal laws and customs, and on the intuition people tend to develop about each other during interpersonal interactions. As for Internet-based commerce, owing to lacking of the personal relationship, trust becomes even more important. Whether for the traditional entity enterprises, or for the emerging virtual enterprises, online trust is the vital mechanism for the development of e-commerce, and the significance of online trust has become a consensus [5-8]. Online trust can affect consumers' intentions to revisit the site and to recommend the site to others. Online trust plays a key role in creating satisfied and expected outcomes in online transaction. Online Trust has been found to be a significant antecedent to customer's willingness to transact with an e-vendor. Online trust serves to mitigate the perceptions of risk, uncertainty and vulnerability that are associated with the disclosure of personal and identifiable information.

The Internet has proven to be a powerful and very popular vehicle for distributing information to millions of individuals. If there were any negative information about one company in the Internet, this information might diffuse broadly in several hours or even in several seconds. The spread of Internet public opinions often plays the role of the catalyst of corporate crisis and the accelerator of dissemination. An obscure message may be magnified indefinitely and evolve into the enterprise's online trust crisis. Like viruses, online trust crises can be mutated, acquiring new and dangerous forms in, for example, social forums like Facebook, or video distribution sites such as YouTube. Once the enterprise's online trust crisis has be happened, the credibility and reputation of the enterprise will be heavily injured, the word-of-mouth and image of the enterprise will be heavily damaged, and the survival and development of the enterprise will be threaten, or make the enterprise into a doomed situation that can never be recovered. Either in China or in other countries, a lot of famous enterprises have met the online trust crises. Some of them fall in trouble and recovered from the crisis through a very long period. However, some of them were not so lucky and directly ended the corporate life cycle.

The objective of this paper is to address the following research questions:

(1) What is the life cycle of online trust crisis diffusion?

(2) What countermeasures can be used in the enterprise' online trust crisis management in the life cycle of enterprise' online trust crisis?

Hopefully, this study will help enterprises to gain some useful suggestions to manage online trust crises.

1.2   Define Online Trust and Online Trust Crisis

Trust as a social phenomenon has been studied in various disciplines and the notion of trust has been examined under various contexts over the years. Many researchers have interpreted it and made a lot of contributions from the perspective of philosophy, psychology, management, and marketing [9-12]. Among definitions, the generalized definition of trust by Rousseau et al. [13] is broadly accepted. In their opinion, "*trust is a psychological state comprising the intention to accept vulnerability based upon positive expectations of the intentions or behavior of another*"(p.395). Based on the study of trust, researchers widened the definition of trust and applied it to online trust upon Rousseau's definition, Bart et al. [14] emphasize that "online trust includes consumer perceptions of how the site would deliver on expectations, how believable the site's information is, and how much confidence the site commands" (p.140). Corritore et al. [15] define online trust as "an attitude of confident expectation in an online situation of risk that one's vulnerabilities will not be exploited" (p.740). In our opinion, online trust is a psychological affirmation of common expectations in an online environment of uncertainty that is caused by online principal parts, based on their wishes, participant's characteristics and the environmental factors, e.g. system, technology and third-party certification.

What is a crisis? The viewpoint of Seeger et al. [16] is representative for an organizational setting. They define that crises are "the specific, unexpected, and non-routine events or series of events that create high levels of uncertainty and threat or perceived threat to an organization's high priority goals" (p.235).

Based on the above notions, the enterprise's online trust crisis means that the related information about the irresponsible behaviors or negative events of the enterprise were published, disseminated and diffused via Internet, which will decline dramatically the degree of trust of consumers, suppliers, distributors, social public, government departments, thus create high levels of uncertainty and threat or perceived threat to one enterprise.

2   Review of Literature

The topic for this study cuts across more than one substantive area in the review of the literature. However, its framework is premised on a strong thematic organization based on the online trust and crisis management.

2.1   Online Trust

The literature on online trust is extensive and multi-faceted and online trust has been extensively studied.

Luo [17] examined several key mechanisms that can help increase customers' trust of e-commerce and decrease privacy concerns. These mechanisms include characteristic-based (e.g., community), transaction process-based (e.g., repeated purchases), and institution-based trust production (e.g., digital certificate). Wang and Emurian [18] pointed out that online trust is a difficult task to accomplish because it requires the establishment of trusting relationships in the online world. The study of Gefen et al. [19] showed that trust in online environments is predicated on beliefs in the trustworthiness of a trustee; trustworthiness is composed of three distinct dimensions-- integrity, ability, and benevolence. Integrity is the belief that the online merchant adheres to stated rules or keeps promises. Ability is the belief that the online merchant has the skills and competence to provide good quality products and services. Benevolence is the belief that the online merchant wants to do good to the customer without regard to making a sale. Beldad et al. [20] provided an overview of the available research into the antecedents of trust in both commercial and non-commercial online transactions and services. There are three clusters of antecedents:

customer/client-based, website-based, and company/organization-based antecedents. Benedicktus [21] offered that online trust beliefs vary positively with consensus ratings and trust is higher when ratings trends increase rather than decrease. Bock et al. [22] evaluated the effects of antecedents of online trust in the context of multi-channel retailers at different phases, taking into consideration the moderating effects of product types. With increases in product uncertainty, the effects of word-of-mouth, offline trust, and efficacy of sanctions on online trust are greater for experience products than for search products. Kim and Ahmad [23] offered a framework for modeling the trustworthiness between a content consumer and a content provider in online social media-sharing communities where users have interacted with each other as either a content consumer or a content provider.

2.2   Crisis Management

Many researchers have discussed crisis management.

Heath [24] introduced an integrated management approach in the Crisis Management Shell Structure, which includes decision component, advisory component, operation component and information component, in terms of what the components units do, why they do these activities, and who is involved. Mak et al. [25] presented a novel application of workflow technology to coordinate and disseminate tasks and related information for Crisis Management Support Systems (CMSS). Their research results indicated that the ability of workflow technology to coordinate, monitor, organize and distribute specific tasks and the associated required information in a timely and efficient manner appears to make it an ideal tool for strategic crisis management. Murray and Foster [26] provided the principles of organizing a multidisciplinary group for crisis resource management. Ryzenko and Smolarkiewicz [27] presented analysis of usefulness of space applications in crisis management activities carried out on the national level. According to them, the key to successful crisis management lays in understanding operational needs; integration into common information environment; and standardization of information exchange. The research of Pearson and Sommer [28] showed that through crisis management planning and preparation, organizational leaders do what they can to make timely decisions based on the best facts that they can gather. Speakman and Sharpley [29] proposed an alternative, chaos theory-based approach to crisis management. The elements of chaos theory include edge of chaos, the butterfly effect, bifurcation and cosmology, self-organization, strange attractors and the lock-in effect. Sardouk et al. [30] proposed a crisis management approach based on wireless sensor networks, which overcomes the problems encountered by the base stations and insures relevant, rich and real-time information about events.

From the previous survey, none of these previous articles focused on the enterprise online trust crisis management. Online trust crisis managing strategies have been substantially altered comparing with the traditional crisis management.

3   The Life Cycle Model of Enterprise's Online Trust Crisis

There is a life cycle for the development of everything in the world. The life cycle concept is originally used to describe a period of one generation of organism in a biological system [31]. Since 1960s the theory of life cycle was raised, the concept of life cycle is applied widely, especially in the fields of politics, economy, environment, technology, and society. In essence, the term life cycle can be popularly understood as a period "from Cradle to Grave", or from its birth to its end. This also applies to the enterprise's online trust crisis. Similarly, the theory of life cycle can be used to describe the stage of development of enterprise's online trust crisis.

Based on the change of individuals' attention, we investigate the stage of development of enterprise's online trust crisis from the perspective of time series, then the whole online trust crisis life cycle can be divided into four phases (see Figure 1): incubation phase, outbreak phase, diffusion phase and decline phase.

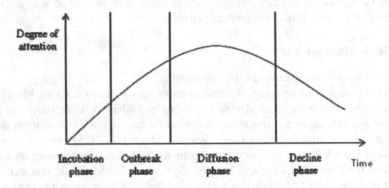

Fig. 1. The life cycle of online trust crisis

There are different characteristics and symptoms in the different stages of enterprise's online trust crisis life cycle. And these characteristics play an important role in enterprise's online trust crisis management. The characteristics of each stage of enterprise's online trust crisis life cycle are shown in Table 1.

3.1   Incubation Phase

The incubation phase includes the period from the appearing of signs of enterprise's online trust crisis to the perceived loss caused by online trust crisis. In this phase, the online trust crisis has not yet broken out; the degree of attention is low. The number of individuals involved is little, network issues are dispersed, the information of online trust crisis is relatively hidden, and the sphere of influence of online trust crisis is small. However, the signs of online trust crisis begin to appear, for example, the sporadic appearance of report on the enterprise adverse events, the sporadic negative comments against the enterprise. If these signs could be taken seriously and handled properly, then the online trust crisis will be nipped in the bud or the damages will minimize to a low degree.

3.2   Outbreak Phase

The outbreak of enterprise's online trust crisis is triggered with potential predisposing factors. There are two sources for enterprise's online trust crisis: one is internal incentive; the other is external incentive [32]. As internal incentive, the corporate itself produces negative events, such as error of strategies, unfavorable events of functions or deviation of implementation. These will cause the enterprise's online trust crisis, when these negative events are communicated and diffused with Internet news, BBS, blogs and podcasts, RSSs, SNSs, IM, or micro-blogs, just to give some examples of potential channels for the bad news. As external incentive, the negative events usually relate with production and operation. The trust crisis event of the enterprise is reported by Internet news media. The unfavorable remarks on the enterprise are then further released by individuals with social media. The evaluations and reports on the enterprise's crises may also published by other third parties. The adverse information on the enterprise might be published maliciously by competitors. These all will lead to the enterprise's online trust crisis, which is communicated and diffused with network channels.

Table 1. The characteristics of each stage of enterprise's online trust crisis life cycle

Phase	Degree of attention	Number of individuals involved	Degree of convergence on network issues	Network communication channels	Sphere of influence
Incubation phase	Very low	Few	Multipoint scatter	Single	Small
Outbreak phase	Increase rapidly	Increase gradually	Multi-aggregation	Increase gradually	Enlarge gradually
Diffusion phase	Increase stably	Continue to increase	Oligarchic gathering	Many	Enlarge rapidly
Decline phase	Decrease gradually	Decrease gradually	Scatter gradually	Reduce gradually	Diminish gradually

3.3   Diffusion Phase

During this period, the network becomes a "catalyst" of online trust crisis. The publicity of the crisis events is enlarged quickly in a short time, which is catalyzed and fermented through network environment, and attracts more individuals' concerns and participations. The number of individuals involved increase gradually, and the degree of attention increases sustainably, network medias change from niche to mass, the strong medias intervene actively, and negative information disseminating channels are formed rapidly, thus the enterprise sinks in the whirlpool of trust crises. Network public opinions are complicated, after the massive outbreak of online trust crisis, there will be "joint" effects in many fields, and increasingly more related things are involved widely. The enterprise's online trust crisis is most difficult to control in this phase.

3.4   Decline Phase

Finally the enterprise's online trust crisis enters the decline phase, mainly because of two reasons. One is that the individuals' views are shifted by new crisis events of other organizations. The other is that the online trust crisis is treated effectively. The cyber hotspot sustains for a period of time, if new events emerge and new stimulus generated, a lot of individuals will automatically shift to new network issues, and they will pay decreasingly attentions to the old network hotspot, then it will be cooled, faded and fallen. When the enterprise's online trust crisis is treated effectively, the crisis will disappear gradually and the enterprise will enter into a new development stage. However, if the treatment was invalid, the residual factors of the online trust crisis will ferment and the online trust crisis will re-enter into a new life cycle.

4   The Framework of Enterprise's Online Trust Crisis Management Strategies

There are different characteristics in different stage of the enterprise's online trust crisis life cycle, and these characteristics determine the countermeasures used in the online trust crisis management. It is vital to find the countermeasures in accordance with the online trust crisis life cycle.

The fishbone diagrams are causal diagrams created by Ishikawa [33] in the 1960s. These show the causes of a specific event and each cause or reason for imperfection is a source of variation. A fish bone diagram is a common tool used for a cause and effect analysis, and it immediately sorts ideas into useful categories [34]. Nowadays, the fishbone diagram is widely used in the field of management.

According to the characteristics of each phase of online trust crisis and the basic principles of crisis management, drawing lessons from the idea of fishbone diagram, we construct the framework of enterprise's online trust crisis management strategies, as shown in Figure 2. The framework can provide references on the enterprise's online trust crisis management decision-making.

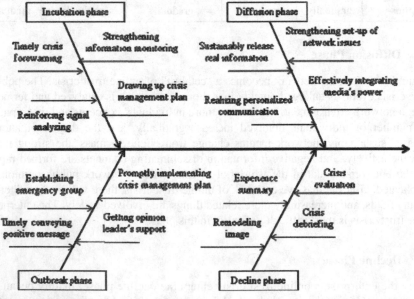

Fig. 2. The fishbone diagram of enterprise's online trust crisis management strategies

5   The Strategies of Enterprise's Online Trust Crisis Management

5.1   The Management in Incubation Phase

There are some signals to indicate the possibility of the online trust crisis though the great trust crisis events cannot yet happen during the incubation phase. At this time, the enterprise should pay more attention to the prevention of online trust crisis. If all kind of incentives which may induce the enterprise's online trust crisis could be monitored timely and the related signals could be analyzed carefully and forewarned effectively, then it could be possible to avoid the ordinary events suddenly upgrading to the burst of online trust crisis. The countermeasures in the incubation phase are discussed in more detail below.

5.1.1   Drawing up Crisis Management Plan

The enterprise needs to take the online trust crisis management into the overall work of enterprise's crisis management, and draw up the online trust crisis management plan based on the principles of integrity, predictability, initiative, operability and timeliness. It is crucial to make a plan for online trust crisis management, because it can effectively prevent the occurrence of some online trust crisis events, and avoid expanding the trust crisis as far as possible after the trust crisis is really happening. This is just a precaution for crisis management. Apart from definite management budget and segmentation of management target, there are two key points in online trust crisis management plan: one is the construction of online trust crisis management information system; the other is the construction of online trust crisis management team. The online trust crisis management information system is a subsystem of enterprise information system, which has the functions of trust crisis information collection, collation, analysis, identification, transmission, feedback, communication, publishing, recovery and assessment. The online trust crisis management team should establish the managing mechanism with centralized command and division of labor and cooperation. The online trust crisis management team should form strong professional personnel. There are some key functions of online trust crisis management team. Firstly, it should coordinate the relationships among internal departments, the enterprise and news media, and the enterprise and the social public. Secondly, it should collect, analyze, predict and evaluate various crisis-related information. Lastly, it should report the information to the enterprise decision-making setup.

5.1.2   Strengthening Information Monitoring

Although the occurrence of enterprise's online trust crisis may be sudden for unknown reasons with a certain degree of randomness, yet some information, views and attitudes will be displayed in network channels from the appearing of initial signals to the occurrence of online trust crisis. So the enterprise's online trust crisis can be monitored [35]. In order to get the signals of the enterprise's online trust crisis, the enterprise should monitor the external information related with the survival of the enterprise. The external information includes politics information, economy information, policy information, science and technology information, finance information, market information, competitor information, supply and demand information, and consumer information. Additionally the enterprise should monitor the internal sensitive causes, which result from corporate mismanagement, business decision-making errors, the problems of product and service quality, financial crisis, the poor quality of staff, and the error of PR strategies.

5.1.3   Reinforcing Signal Analyzing

The analysis of cyber information monitored is a process to find problems and solve problems. After the collection of data, the analyzing process of information arrangement, identification, filtering and evaluation should be carried out. There are differences in the methods and focus of signals processing for different information communication platforms. As for network news data, the main core elements of the original should be maintained as far as possible, such as news title, source, release time, content, click number, reviewer, comment content and the comment number. As for BBS, blog, podcast, RSS, SNS, IM data and other social media, the original element and diffusion effects should be preserved as far as possible and formed the formatting information, such as the title of the post, the spokesman, the release time and content, content and number of replies, and so on. The abnormal information should be classified and analyzed timely, the hot spots, sensitive spots and dangerous spots, which may possibly cause online trust crisis, should be researched. The successful identification of potential and possible threats will lay the foundation for the enterprise's online trust crisis management.

5.1.4   Timely Crisis Forewarning

All signals analyzed and processed should be timely submitted to the crisis management team. The crisis management team should evaluate professionally the signals to determine whether the crisis forewarning is need. After an action decision, the online trust crisis forewarning should be realized automatically with the information technologies as one module of the enterprise information system. The forewarning indicators are built according to parameters such as the news source, news authority, comment number, release time and intensive degree, the viewpoint and orientation of the article or post, the degree of concern, and so on. This module owns the functions of semantic analysis, statistical analysis, comprehensive analysis and feature extraction by making full use of natural language processing, viewpoint mining, artificial intelligence, and visualization technology. In this way, the signals can be checked if they were in a normal range, once beyond a particular scope, the alarm is submitted to the online trust crisis management.

5.1.5   The Management in Outbreak Phase

This stage is especially critical for the enterprise's online trust crisis management. If the enterprise's online trust crisis could not be disposed improperly, the consequences of the trust crisis events will be in a deteriorating direction, the "heat" of online trust crisis will increase, and more individuals will participate in the discussion and communication. Thus, cyber will become the "blower" for the enterprise's online trust crisis, and the online trust crisis will be worsening. Following actions are needed.

5.1.6   Promptly Implementing the Crisis Management Plan

Once the online trust crisis is outbreaking, the enterprise should confirm rapidly and timely the crisis, find the source of crisis, determine the potentially affected the public, and implement promptly the online trust crisis management plan, assembling all kinds of resource to control the influence power of online trust crisis within a smaller range as far as possible. According to different crisis forms, the enterprise online trust management team should format the corresponding crisis control groups to provide effective organizational support. In addition, the enterprise online trust management team should strengthen the cooperation with government agencies, NGOs and media, and rely on the opinion leaders and the public to reduce the damage to a minimum.

5.1.7   Getting the Support of Opinion Leader

Opinion leader is an active media user who interprets the meaning of media messages or content for lower-end media users [36]. There is a lot of promoting by opinion leaders almost behind every network hot issue. In some events, the viewpoints of opinion leaders can play and unexpected role on the settlement, because as the person concerned in the event, the point of view of the enterprise is difficult to get fully recognized by individuals. Therefore, the enterprise should actively win the support of opinion leaders and make full use of the third-party persuading role of opinion leaders, encouraging opinion leaders to guide online speech through blogs, podcasts, SNS, RSS, and other social media. The opinion leaders should be the strong support for the enterprise and the external opinion. At the same time, the enterprise should cultivate actively its own "opinion leaders" to form a positive guiding force in the "new opinion stratum".

5.1.8   Timely Conveying Positive Message

Winning the time is equal to winning the image. There may not be enough reliable information at the onset of the crisis. In order to avoid information vacuum and rumors flying long, the enterprise must strive for controlling the situation in a short time and endeavor to communicate with public in the first time, conveying positive messages, indicating attitudes, informing of the activities being adopted. Then the public will feel the enterprise's responsibility consciousness and public philosophy, and they will be in favor of the enterprise psychologically. The key point to keep the initiative in information release is that the enterprise itself should be the first hand information release source and provide the necessary background information to media and public as much as possible, not giving too much playing space to media. The enterprise should inform news media about the crisis basic facts and the management countermeasures in the shortest period of time, and clarify the stance and attitude to get the trust and support of the media and lead actively public opinions. In such a way, the enterprise will quickly control the situation by not giving up any rights of speech.

5.1.9   The Management in Diffusion Phase

Due to the openness, timeliness and interactivity of Web 2.0, there are cascading effects during the diffusion phase of online trust crisis. It is difficult to manage online trust crisis in the diffusion phase. In addition to adhering continuously the countermeasures mentioned above, the enterprise should also do the following works.

5.1.10 Effectively Integrating Media's Power

The enterprise online trust crisis management team should actively coordinate the relationship with the mainstream traditional media and online media. The enterprise can communicate and dialogue with individuals frankly in proper communicating atmosphere with unblocked information channels. During the communication, the positive interaction between the enterprise and the society will be formed, and the chaos of social public caused with blocking messages and uniformed information will be avoided. It is very important to ensure the rights of speech of individuals, which can overcome the desire for fear or revenge of public to reduce the instability factors. In this way, the public's resentment on the crisis event can be led to the rational thinking of the nature and solutions of the event. And then the attitudes and behaviors of individuals will be advantageous to resolve online trust crisis and reduce the negative influence and loss.

5.1.11 Sustainably Release Real Information

The enterprise should release information sincerely, meet and maintain the public's right to know, winning opportunities for subsequent disposal of the trust crisis. The online trust crisis can be led and mediated in the direction of control by disclosing relevant information, and the social anxieties will be reduced. The enterprise can sustainably release real information and enhance the transparency of information through the enterprise's website, enterprise's news spokesman, hotlines, being a guest on network media and social media.

5.1.12 Strengthening Set-up of Network Issues

The primacy effect [37], in psychology and sociology, is a cognitive bias that results in a subject recalling primary information presented better than information presented later on. In the network era, the energy of primacy effect increases in hundreds or thousands times in the process of network communication. It is an effective way to guide the cyber opinions by setting up the official network issue. When the online trust crisis is breaks out, the enterprise should try its best to set the related subjects or topics in the first time consciously. On one hand, the attentions of individuals will be directed to the special themes. On the other hand, the diffusion of online trust crisis will be decreased as much as possible. The enterprise can actively get the rights of speech and guide the development of network opinions by setting communicating topics initiatively, making the network discussion become more efficient, hierarchical, systematic and controllable.

5.1.13 Realizing Personalized Communication with Individuals

In order to achieve the communicating goals in the situation of crisis, the target audience should be divided into several types and each type should adopt different communicating method and channel. The classification standards include the value, behavior habit, life style, education degree, racial, economic status, age, gender, and information selection habit of individual. Aiming at the personalized communication needs of each type of target audience, the enterprise should choose the proper transmission time, transmission channel, information content and expression forms based on satisfying the targeted communication. Of course, these all depend on information technologies.

5.2   Management in Decline Phase

Whatever the cause, when the enterprise's online trust enters into decline phase, the enterprise should carry out crisis evaluation, crisis debriefing and image remodeling.

5.2.1   Crisis Evaluation

The crisis evaluation includes the evaluation of basic support, the evaluation of disposal process and the evaluation of settlement results. The evaluation of basic support looks for loopholes of online trust crisis monitoring, forewarning and management plan to make up the shortfalls. The evaluation of disposal process can help the enterprise to improve the online trust crisis management process and enhance the enterprise's ability to cope with the online trust crisis in the future. After the online crisis trust, the enterprise should track and mend continuously the settlement results, and think over the recovery measures. The evaluation of settlement results should be fed back to the online trust crisis management plan, perfecting the plan, and providing effective basis for the future online trust crisis.

5.2.2   Crisis Debriefing

The crisis debriefing is an inevitably phase in online trust crisis management. The work of crisis debriefing is an important mechanism the enterprise tries its best to recover the cyber order to its normal state and prevent the germination of new online trust crisis. On one hand, if the enterprise really produced the trust events and had negative effects on social public, the enterprise should take responsibility and apologize actively and initiatively to get their understanding. On the other hand, if some organizations and persons violate laws, regulations or the network obligation and cause damages to the enterprise, the enterprise should actively investigate and pursue such organizations' or persons' legal liabilities.

5.2.3   Remodeling Image

The destructiveness given by online trust crisis is mainly manifested in the deterioration of the enterprise's image. Thus, when the online trust crisis enters into the decline phase, the enterprise must try its best to reshape the corporate identity. The process of remodeling image is actually the process that the enterprise pursues self-improvement and constantly gets the public's recognition, understanding and support. On one hand, the enterprise should complete the internal works to truly provide high-quality products and service to society. On the other hand, the enterprise must maintain good relations with all kinds of social public, planning and implementing a serial of PR activities manifesting the brand reputation and corporate identity, product promotion, and public benefit activities reflected the enterprise's social responsibility.

6   Conclusions

With the rapid development of Internet and telecommunication network, the occurrences of online trust crises become increasingly more frequent. The negative influences and destructive effects caused by online trust crises are growing. We study the phases of enterprise's online trust crisis life cycle and put forward the frame of enterprise's online trust crisis management strategies based on the characteristics of online trust crisis in incubation phase, outbreak phase, diffusion phase and decline phase, which may provide some theoretical supports for enterprise's online trust crisis management. Crisis management is also called crisis communication management [4]. The emphasis of enterprise's online trust crisis management is to achieve and enhance communication with the public. Communication becomes the core of management in each stage of online trust crisis cycle. Of course, the enterprise's online trust crisis management requires a lot of manpower, material and financial resources, and thus it could be difficult to realize for micro and small enterprises (MSEs). Based on recent cases, Internet users pay more attention to the production and operation of well-known enterprises, and well-known enterprises have become the main body of the outbreak of online trust crisis. All well-known enterprises should attach importance to online trust crisis management.

References

1. Internet Users in the World Distribution By World Regions-2012 Q2 (December 26, 2012), http://www.internetworldstats.com/stats.htm
2. 33th statistical report on the Internet Development in China, http://www.cnnic.net.cn
3. Global e-commerce sales will top $1.25 trillion by 2013, http://www.internetretailer.com/2012/06/14/global-e-commerce-sales-will-top-125-trillion-2013
4. The transactions of e-commerce market in China amounted to 5.6 trillion Yuan in the first three seasons, http://finance.sina.com.cn/chanjing/cyxw/20121127/1539138136 24.shtml
5. Liu, C., Marchewka, J.T., Lu, J., Yu, C.: Beyond concern: a privacy–trust–behavioral intention model of electronic commerce. Information Management 42, 127–142 (2004)
6. Pavlou, P.A., Liang, H., Xue, Y.: Understanding and mitigating uncertainty in online exchange relationships: a principal-agent perspective. MIS Quarterly 31, 105–136 (2007)
7. Luo, X., Lib, H., Zhang, J., Shimd, J.P.: Examining multi-dimensional trust and multi-faceted risk in initial acceptance of emerging technologies: an empirical study of mobile banking services. Decision Support Systems 49, 222–234 (2010)
8. Mesch, G.S.: Is online trust and trust in social teams associated with online disclosure of identifiable information online? Computers in Human Behavior 28, 1471–1477 (2012)
9. Mayer, R.C., Davis, J.H., Schoorman, F.D.: An integrative model of organizational trust. Academy of Management Review 20, 709–734 (1995)
10. Bachmann, R.: Trust, Power and Control in Transorganizational Relations. Organization Studies 22, 337–365 (2001)

11. Devosk, T., Spini, D., Schwartz, S.: Conflicts among human values and trust in institutions. British Journal of Social Psychology 41, 481–494 (2002)
12. Alesina, A., Ferrara, E.: Who trusts others? Journal of Public Economics 85, 207–234 (2002)
13. Rousseau, D.M., Stkin, S.B., Camerer, C.: Not so different after all: a cross-discipline view of trust. Academy of Management Review 3, 393–404 (1998)
14. Beldad, A., Jong, M., Steehouder, M.: How shall I trust the faceless and the intangible? A literature review on the antecedents of online trust. Computers in Human Behavior 26, 857–869 (2010)
15. Midha, V.: Impact of consumer empowerment on online trust: An examination across genders. Decision Support Systems 54, 198–205 (2012)
16. Seeger, M.W., Sellnow, T.L., Ulmer, R.R.: Communication, organization and crisis. Praeger, Santa Barbara (2003)
17. Luo, X.M.: Trust production and privacy concerns on the Internet A framework based on relationship marketing and social. Industrial Marketing Management 31, 111–118 (2002)
18. Wang, Y.D., Emurian, H.: An overview of online trust: concepts, elements, and implications. Computers in Human Behavior 21, 105–125 (2005)
19. Gefen, D., Benbasat, I., Pavlou, P.: A research agenda for trust in online environments. Journal of Management Information Systems 24, 275–286 (2008)
20. Beldad, A., Jong, M., Steehouder, M.: How shall I trust the faceless and the intangible? A literature review on the antecedents of online trust. Computers in Human Behavior 26, 857–869 (2010)
21. Benedicktus, R.L.: The effects of 3rd party consensus information on service expectations and online trust. Journal of Business Research 64, 846–853 (2011)
22. Bock, G.W., et al.: The progression of online trust in the multi-channel retailer context and the role of product uncertainty. Decision Support Systems 53, 97–107 (2012)
23. Kim, Y.A., Ahmad, M.A.: Trust, distrust and lack of confidence of users in online social media-sharing communities. Knowledge-Based Systems 37, 438–450 (2013)
24. Heath, R.: Dealing with the complete crisis-the crisis management shell structure. Safety Science 30, 139–150 (1998)
25. Mak, H.Y., et al.: Building online crisis management support using workflow systems. Decision Support Systems 25, 209–224 (1999)
26. Murray, W.B., Foster, P.A.: Crisis Resource Management Among Strangers: Principles of Organizing a Multidisciplinary Group for Crisis Resource Management. Journal of Clinical Anesthesia 12, 633–638 (2000)
27. Ryzenko, J., Smolarkiewicz, M.: Space-enabled information environment for crisis management. Scenario-based analysis and evaluation in an operational environment. Acta Astronautica 66, 33–39 (2010)
28. Christine, M., Pearson, S.: Amy Sommer. Infusing creativity into crisis management: An essential approach today. Organizational Dynamics 40, 27–33 (2011)
29. Speakman, M., Sharpley, R.: A chaos theory perspective on destination crisis management: Evidence from Mexico. Journal of Destination Marketing & Management 1, 67–77 (2012)
30. Sardouk, A., et al.: Crisis management using MAS-based wireless sensor networks. Computer Network (2012),
    http://dx.doi.org/10.1016/j.comnet.2012.08.010
31. Chou, D.C., Chou, A.Y.: Information systems outsourcing life cycle and risks analysis. Computer Standards & Interfaces 31, 1036–1043 (2009)

32. Tan, C.H., Zeng, Y.T.: Research on the Discrimination of Enterprise Online Trust Crisis Based on "Twice Evaluation" Mechanism. Information Science 29, 1815–1819 (2011)
33. Kaoru, I.: Guide to Quality Control, Asian Productivity Organization. UNIPUB. (1976)
34. Nancy, R.: Tague's The Quality Toolbox, Second Edition. ASQ Quality Press, Milwaukee (2004)
35. Tan, C.H., Wang, X.: Research on the Model of Enterprise's Online Trust Crisis Prevention Based on Ontology and Web Mining. Information Science 29, 1559–1564 (2011)
36. Gnambs, T., Batinic, B.: Convergent and discriminant validity of opinion leader: Multitrait-multimethod analysis across measurement occasion and informant type. Journal of Individual Differences 39, 94–102 (2011)
37. Jones, E.E., et al.: Pattern of performance and ability attribution: An unexpected primacy effect. Journal of Personality and Social Psychology 10, 317–340 (1968)
38. Herrero, A.G., Pratt, C.B.: An Integrated Symmetrical Model for Crisis-Communications Management. Journal of Public Relations Research 8, 79–105 (1996)

Young People Purchasing Virtual Goods in Virtual Worlds: The Role of User Experience and Social Context

Matti Mäntymäki[1,*], Jani Merikivi[2], and A.K.M. Najmul Islam[1]

[1] Turku School of Economics, University of Turku, Finland
{matti.mantymaki,najmul.islam}@utu.fi
[2] Aalto University, School of Business, Finland
jani.merikivi@aalto.fi

Abstract. Millions of young people spend real money on virtual goods such as avatars or in-world currency. Yet, limited empirical research has examined their shopping behaviour in virtual worlds. This research delves into young consumers' virtual goods purchasing behaviour and the relevance of social context and usage experience. We assert that virtual goods purchasing behaviour is inseparable of the online platform in which it is taking place. We employ the concept of cognitive absorption to capture the user experience and examine the social context with three variables, the size of one's in-world network, trust in the other users of the online platform and social presence. We test our research model with data collected from 1,225 virtual world users and use PLS in the analysis. The results show that virtual goods purchasing behaviour is predicted by cognitive absorption, perceived size of one's in-world network as well as trust in the other users.

Keywords: virtual worlds, virtual goods, cognitive absorption, trust.

1 Introduction

Purchasing virtual goods has become increasingly pervasive among the young generations. Virtual goods (e.g., avatars) are non-physical in nature and exist in the online platforms they are created in [1]). That is, they cannot be carried off to and used in another online platform. This characteristic separates virtual goods from digital goods (e.g., audio files which work in many platforms). While virtual goods have existed as long as virtual worlds (VWs), they did not receive attention before VW operators started to sell them to users with real money.

Interestingly, many of the current VWs are targeted for users aged between 5 and 15 years, who make the majority of over 1.4 billion registered VW users [2]. The large user base has made the overall spending on virtual goods to reach $15 billion already in 2012.[1]

Despite the economic potential, the research on virtual goods purchasing behaviour

---

[*] Corresponding author.
[1] (http://www.superdataresearch.com/blog/monetization-is-a-four-letter-word/)

H. Li et al. (Eds.): I3E 2014, IFIP AICT 445, pp. 303–314, 2014.
© IFIP International Federation for Information Processing 2014

in VWs is still in its infancy– compared to the 'traditional' online shopping or that which occurs offline. To contribute to virtual goods research we seek to fill three gaps in the current literature. First, prior literature on virtual goods has focused rather heavily on adult consumers, albeit young people admittedly make a notable group of existing consumers. For example, young people have been under-investigated in information systems research [3].

With regard to the second gap, we advance virtual goods research by building on user experience. We believe this is of considerable importance since purchasing virtual goods requires engagement in online platforms where the goods are available. To this end, we employ cognitive absorption, which is an established driver of technology use, also in VWs [4; 5]. Notwithstanding, its influence on purchasing behaviour has remained poorly understood.

Third, we center on social context. While social context is demonstrated as critical for online platform success [6], studies on virtual goods fall short in examining its effect on purchasing behavior [7]. In this paper, we conceptualize social context to operate through perceived network size, social presence, and trust–all of which we consider relevant for virtual goods exchange.

By filling these gaps, we add on to three different research areas, virtual goods purchasing behavior [7; 8; 9], young users use of information technology [3], and the relationship between virtual goods and platforms where they are exchanged [10; 11].

The paper is organized as follows. It starts with a literature revive and provide a foundation for the research model. The paper will then explicate the research model and hypotheses. This is followed by the methods and results. Lastly, it concludes with a discussion, including implications, limitations and suggestions for future research.

2   Research Background and Hypotheses

2.1   Prior Literature on Virtual Purchasing Behavior

Prior VW research has largely examined user adoption   [12; 13; 14], including initial acceptance and post-adoption use   [4; 15; 16; 17; 18; 19]. Purchasing behavior, in turn, has received less empirical research attention   [20]. The prior research on the topic has found purchasing in VWs have being affected by the virtual environment [7], user motivation   [1; 8] and social influences   [8].

Here we focus on two aspects that have drawn less attention in the VW, namely user experience and social context. They supplement each other as user experience stresses the experience obtained by an individual and social context the environment which is co-created by individual users. Social context is also expected to influence the individual's behavior [21]. Given virtual goods purchasing behavior is fairly inseparable of VW use we believe social context and user experience fit in perfectly to our research goal.

2.2   The Research Model

The user experience of VWs can be characterized with three key aspects. First, VWs employ avatars as a core of the navigation mechanisms and to represent the users. Second, VWs accommodate a multi-user, 3D graphical environment that includes sounds and music. Third, the user interface is highly dynamic because of a constant influx of new features and activities to sustain users' interest. Thus, the richness of stimuli that make the user absorbed in the in-world activities lie in the core of the VW user experience. Hence, we employ the concept of cognitive absorption. Cognitive absorption consists of focused immersion, intrinsic motivation, perception of control, temporal dissociation and curiosity. We measure it as a multi-dimensional construct as it was originally developed [22].

We also scrutinize how the social context can influence virtual purchasing behaviour. The social context is essentially dependent on the number of users involved in the VW. The social interaction, and the value users derive from it, is influenced by network externalities [19]. This is articulated in Metcalfe's law that postulates that the value of a telecommunications network is proportional to the square of the number of connected users [23]. For an individual user, however, the value of interactive digital technologies is more dependent upon the presence of relevant people, i.e. the user's personal network, than the network size in general [19; 24]. From a sociological perspective, this can be explained by the concept of homophily, i.e. the tendency to bond and associate with individuals with whom one perceives similarity [25]. Prior evidence from computer-mediated communication shows that interaction that involves the use of IT is likely to occur with key interpersonal relationships [26]. Thus, network externalities stem particularly from the presence of one's key social network in the VW.

In addition to the presence of other users and an in-world social network, the social atmosphere and the relationships between users represent important aspects of the social context. For example, people tend to communicate more when they perceive human warmth and psychological presence [27]. As a result, we examine the degree of human warmth and contact associated with the VW using the concept of social presence [28]. Trust is fundamental component of interpersonal relationships and an important predictor of online purchasing [29]. Hence, we investigate the trust in other VW users as a predictor of virtual purchasing. The constructs with their definitions and references are presented in Table1.

Table 1. The Research constructs and their definitions

Construct	Definition	Source
Perceived enjoyment	The degree of enjoyment associated with using the VW.	[22]
Focused immersion	The experience of total engagement where other attentional demands are, in essence, ignored.	[22]
Perception of control	The user's perception of being in charge of the interaction.	[22]
Temporal dissociation	The inability to register the passage of time while engaged in interaction.	[22; 30]
Curiosity	The extent the experience arouses an individual's sensory and cognitive curiosity.	[22]
Perceived network size	The perception of the degree to which important others are present in the VW.	[16; 31]
Social presence	The degree of human warmth associated with the VW.	[28; 32]
User-to-user trust	The belief in the other VW users' honesty.	[29]

The research model accommodating the user experience and social context is presented in Figure 1 below.

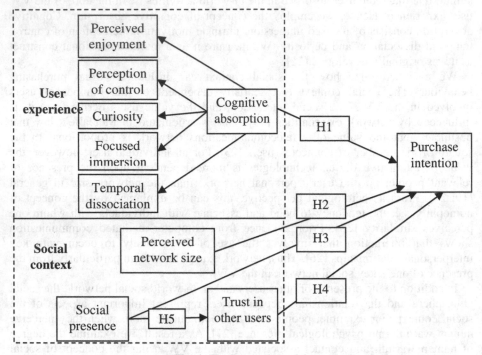

Fig. 1. The Research model

2.3   Hypotheses

Agarwal and Karahanna [22] positioned cognitive absorption as a predictor of perceived usefulness and ease of use but did not examine its direct effect on behavioural intention. Cognitive absorption is an intrinsically motivating state [33], enjoyment being one of its dimensions [22]. Intrinsic motivation, often captured with perceived enjoyment, in turn has been found to predict the intention to adopt and use various forms of IT, particularly those of hedonic nature [34; 35]. Prior VW research offers empirical support for the link between cognitive absorption and behavioural intention [4; 5]. As a result, we assume that the purchase intention is influenced by cognitive absorption and put forward the following hypothesis:

H1: Cognitive absorption has a positive effect on purchase intention.

Due to network externalities (Katz & Shapiro, 1986), the size of one's personal network inside the VW influences the amount of opportunities the user has for social interaction and communication. Furthermore, a large social circle in an VW provides more opportunities to demonstrate status through virtual purchasing or when trading virtual items with other users. Prior research on online social networking [36], instant messaging [24] and VWs [16] offers empirical evidence that the perceived size of user's network predicts the usage intention.

H2: Perceived aggregate network exposure has a positive effect on purchase intention.

Social presence has been found to have a positive effect on loyalty in the online shopping context [37]. Furthermore, previous VW research has shown a positive relationship between social presence and favourable attitudes [38] and user satisfaction [15]. However, the research has reported no relationship between social presence and behavioural intention [7; 15].

H3: Social presence has a positive effect on purchase intention.

Abundant research on e-commerce has verified a positive relationship between trust in the online merchant and user's purchasing behaviour [39]. However, considerably fewer studies have examined to what extent the trust between users affects purchasing, especially in an environment where the users are represented as avatars. Lu et al. [40] reported a positive relationship between intentions to purchase from the website and member-to-member trust.

H4: Trust in other users of the VW has a positive effect on purchase intention.

Social presence has been found to increase the number of messages exchanged in electronic communication [27]. As VWs are information-rich environments that are well capable of transmitting various non-verbal cues [15], we propose a positive relationship between social presence and trust in the other VW users. This assertion is also in accordance with the e-commerce literature that has reported social presence to have a positive effect on trust [32; 41].

H5:   Social presence has a positive effect on trust in the other users of the VW.

3   Empirical Research

3.1   Data Collection and Measurement

The data was collected through an online survey among the users of the Finnish Habbo Hotel portal in co-operation with Sulake Corporation, the Finnish company that owns and operates Habbo Hotel.

The survey was opened 8,928 times. 3,265 respondents proceeded to the final page and submitted the survey. This yielded a response rate of 36.6 percent. To further ensure the reliability of the results only fully completed questionnaires were included in the analysis. As a result, the final sample consisted of 1,225 responses. 60.8 per cent of the final sample was female.

To ensure the reliability of the measurement, the survey items were adopted from prior literature with wording adjusted to match the VW context and the target audience. The literature references of the measurement items were presented in Table 2. The items were measured with a seven-point Likert scale, anchored from strongly disagree to strongly agree – except perceived network size, which was measured with semantic scale. The constructs were modeled using reflective indicators.

3.2   Data Analysis

The data was analysed using partial least squares with smartPLS software   [42]. We began the analysis by testing the convergent and discriminant validity of the measurement model. Convergent validity was evaluated based on three criteria   [43]: firstly, all indicator factor loadings should be significant and exceed 0.70. Secondly, composite reliabilities should exceed 0.80. Thirdly, average variance extracted (AVE) by each construct should be greater than 0.5. Appendix A illustrates that the data met the criteria for convergent validity. With respect to discriminant validity, the AVE for each construct should exceed the squared correlation between that and any other construct   [43].   Table   3 shows that discriminant validity was confirmed.

Table 2. Squared correlation between constructs (AVEs in bold in the main diagonal)

	CON	CTRL	CUR	ENJ	PNS	PURC	SP	TDIS	TRU
CON	**0.749**								
CTRL	0.064	**0.812**							
CUR	0.305	0.101	**0.864**						
ENJ	0.328	0.158	0.524	**0.889**					
PNS	0.115	0.050	0.265	0.222	**0.826**				
PURC	0.182	0.077	0.282	0.303	0.257	**0.864**			
SP	0.219	0.142	0.359	0.422	0.211	0.241	**0.799**		
TDIS	0.386	0.070	0.468	0.416	0.177	0.218	0.272	**0.716**	
TRU	0.139	0.035	0.276	0.234	0.270	0.247	0.269	0.172	**0.907**

After having verified the validity and reliability of the measurement model, we proceeded to testing the structural model. According to Agarwal & Karahanna (2000), cognitive absorption was modeled as a second order construct.   Bootstrapping with 1,000 subsamples was used to estimate the significance of the path coefficients. The latent variable scores of its five constituting factors were used as an input to build the second order variable.

The R^2 of purchase intention was 42.7 per cent, which indicates that the model as a whole exerts good predictive validity. As the sample size was large, instead of looking strictly the significance of the path coefficients, we considered the value of 0.1 as a threshold to interpret that a variable exerts a substantial effect on its endogenous construct   [44]. Based on this criterion, all hypotheses were supported except H3. Age, gender and length of usage experience with the VW were included in the structural model as control variables. None of the control variables exerted a significant influence on purchase intention. Figure 2 below summarizes the results from testing the structural model.

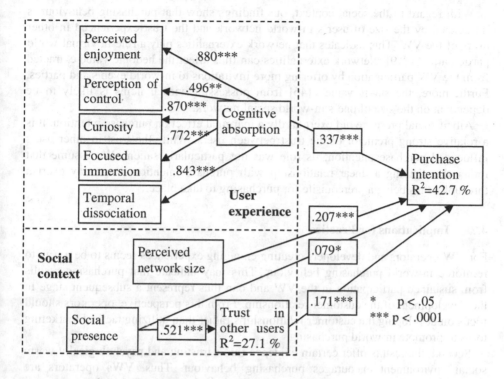

Fig. 2. Results of the PLS analysis

4   Discussion and Conclusion

The key finding of this study is that user experience captured through cognitive absorption and its first order constructs is the main driver of purchase intention. While prior research provided empirical evidence of user experience in driving usage [45] our results show that it has effect on purchase intention that takes place beyond usage. Overall, this finding implies that engaging user experience can drive VW operators' sales and thus it is critical for VWs success.

4.1   Theoretical Implications

Our results verify the importance of cognitive absorption as a component of the VW user experience and its value in predicting purchasing behaviour. On a more theoretical level, our conceptualization of cognitive absorption as a five-dimensional second order construct offers other researchers guidance how to capture the contextual characteristics of VWs. Based on our findings, virtual purchasing is substantially affected by the experiential aspects of VW usage, indicating that the user experience is a stable predictor of virtual purchasing across context [7]

With regard to the social context, our findings show that purchasing behaviour is influenced by the size of user's in-world network and the trust experienced in other users of the VW. This indicates that network externalities play a role in virtual world participation [19]. Network externalities can thus affect the hedonic value extracted from the VW participation by offering more invitations to in-world events and parties. Furthermore, the status value [46] from possessing virtual items is likely to be dependent on the size of one's in-world social network.

Albeit social presence did exert hardly a marginal effect on purchase intention, it is a relative strong predictor of the trust between users. While the trust in other users influenced purchase intention, its role was not particularly salient. We assume that rather than having a linear relationship with purchase intention, trust may exert a threshold, thus being a prerequisite for purchasing to take place.

4.2   Implications for Practice

For VW operators and developers creating engaging experiences seems to be a way to reinforce in-world purchasing behaviour. This may indicate that purchasing results from sustained participation in the VW and can thus represent a subsequent stage in the development of the customer relationship. From this perspective, operators should focus on developing the customer relationships rather than utilizing tactical marketing tools to promote in-world purchasing.

Second, the results offer certain evidence that a trusting and psychologically warm social environment encourages purchasing behaviour. Thus, VWs operators are advised to have mechanisms not only to protect users' virtual property, but to prevent aggressive behaviour and communication towards other users.

Third, we suggest operators should take a close look at how the presence and actions of other users within and beyond the VW affect users' participation and purchasing decisions. Young people have been reported to follow fads and fashion and thus be more prone to bandwagon effect than older generations [47]. This can partly explain the dynamics of the social setting and sometimes very short lifespans of trends in the VW for the young.

4.3   Limitations and Future Research

First, due to our research context, generalizability of the results is limited. Second, we examined behavioural intention instead of actual behaviour. Third, we used three constructs, perceived network size, social presence and trust in other users to examine the social context. Due to its conceptual breath, social context is very difficult to condense into a set of variables. We recommend further research to offer a richer understanding of the social context and structures behind the behavioural outcomes such as virtual purchasing. For example, future research could examine the interplay between the social context and purchasing behaviour [21].

Fourth, in our conceptualization of trust we focused only on the trustworthiness, i.e. the reliability of other users. However, prior research has highlighted the complex, multi-faceted nature of trust [29]. Moreover, in the VW context, the user may or may not trust in several entities such as the user community (or a specific subgroup within the community), the company operating the service and the service as a whole. Thus, research focusing particularly on the nature and dimensions of trust in the VW context would offer a better understanding of the social context of VWs and, at the same time, uncover the role of the avatar-centric environment in the formation of trust.

Finally, we used only the cognitive absorption to empirically examine the key aspects of the VW user experience. However, from VWs people do not necessary seek for immersion and intensive experiences but a relaxing place to spend time and socialise with other users in a casual manner. Hence, further research could examine to what extent the VW participation is perceived relaxing or stress-relieving.

References

1. Guo, Y., Barnes, S.J.: Explaining purchasing behavior within world of warcraft. Journal of Computer Information Systems 52(3), 18–30 (2012)
2. Wasko, M., Teigland, R., Leidner, D., Jarvenpaa, S.: Stepping into the internet: New ventures in virtual worlds. MIS Quarterly 35(3), 645–652 (2011)
3. Vodanovich, S., Sundaram, D., Myers, M.: Digital natives and ubiquitous information systems. Information Systems Research 21(4), 711–723 (2010)
4. Goel, L., Johnson, N.A., Junglas, I., Ives, B.: From space to place: Predicting users' intentions to return to virtual worlds. MIS Quarterly 35(3), 749–771 (2011)
5. Nevo, S., Nevo, D., Kim, H.: From recreational applications to workplace technologies: An empirical study of cross-context IS continuance in the case of virtual worlds. Journal of Information Technology 27(1), 74–86 (2012)
6. Merikivi, J., Mäntymäki, M.: Explaining the continuous use of social virtual worlds: An applied theory of planned behavior approach. In: Proceedings of the 42nd Hawaii Conference on System Sciences, Waikoloa, Big Island, Hawaii (2009)
7. Animesh, A., Pinsonneault, A., Yang, S.-B., Oh, W.: An odyssey into virtual worlds: Exploring the impacts of technological and spatial environments on intention to purchase virtual products. MIS Quarterly 35(3), 789–810 (2011)
8. Mäntymäki, M., Salo, J.: Purchasing behavior in social virtual worlds: An examination of habbo hotel. International Journal of Information Management 33(2), 282–290 (2013)
9. Guo, Y., Barnes, S.: Purchase behavior in virtual worlds: An empirical investigation in second life. Information & Management 48(7), 303–312 (2011)
10. Payne, A.F., Storbacka, K., Frow, P.: Managing the co-creation of value. Journal of the Academy of Marketing Science 36(1), 83–96 (2008)
11. Füller, J.: Refining virtual co-creation from a consumer perspective. California Management Review 52(2), 98–122 (2010)
12. Hua, G., Haughton, D.: Virtual worlds adoption: A research framework and empirical study. Online Information Review 33(5), 889–900 (2009)
13. Fetscherin, M., Lattemann, C.: User acceptance of virtual worlds. Journal of Electronic Commerce Research 9(3), 231–242 (2008)
14. Shen, J., Eder, L.: Exploring intentions to use virtual worlds for business. Journal of Electronic Commerce Research 10(2), 94 (2009)

15. Jung, Y.: Understanding the role of sense of presence and perceived autonomy in users' continued use of social virtual worlds. Journal of Computer-Mediated Communication 16(4), 492–510 (2011)
16. Mäntymäki, M., Salo, J.: Teenagers in social virtual worlds: Continuous use and purchasing behavior in habbo hotel. Computers in Human Behavior 27(6), 2088–2097 (2011)
17. Mäntymäki, M., Riemer, K.: Digital natives in social virtual worlds: A multi-method study of gratifications and social influences in habbo hotel. International Journal of Information Management 34(2), 210–220 (2014)
18. Mäntymäki, M., Merikivi, J., Verhagen, T., Feldberg, F., Rajala, R.: Does a contextualized theory of planned behavior explain why teenagers stay in virtual worlds? International Journal of Information Management 34(5), 567–576 (2014)
19. Mäntymäki, M., Islam, A.K.M.N.: Social virtual world continuance among teens: Uncovering the moderating role of perceived aggregate network exposure. Behaviour & Information Technology 33(5), 536–547 (2014)
20. Mäntymäki, M.: Continuous use and purchasing behaviour in social virtual worlds. Turku School of Economics, Series A, Turku (2011)
21. Bandura, A.: Social foundations of thought and action: A social cognitive theory. Prentice Hall, Englewood Cliffs (1986)
22. Agarwal, R., Karahanna, E.: Time flies when you're having fun: Cognitive absorption and beliefs about information technology usage. MIS Quarterly 24(4), 665–694 (2000)
23. Shapiro, C., Varian, H. R.: In: Varian, H.R. (ed.), Information rules: A strategic guide to the network economy. Harvard Business School Press, Boston (1999)
24. Lin, C., Bhattacherjee, A.: Elucidating individual intention to use interactive information technologies: The role of network externalities. International Journal of Electronic Commerce 13(1), 85–108 (2008)
25. McPherson, M., Smith-Lovin, L., Cook, J.M.: BIRDS OF A FEATHER: Homophily in social networks. Annual Review of Sociology 27(1), 415 (2001)
26. Yuan, Y.C., Gay, G.: Homophily of network ties and bonding and bridging social capital in computer-mediated distributed teams. Journal of Computer-Mediated Communication, 11(4), Article 9 (2006)
27. Gefen, D.: Gender differences in the perception and use of E-mail: An extension to the technology acceptance model. MIS Quarterly 21(4), 389–400 (1997)
28. Short, J., Williams, E., Christie, B.: The social psychology of telecommunications. Wiley, London (1976)
29. McKnight, D.H., Choudhury, V., Kacmar, C.: Developing and validating trust measures for e-commerce: An integrative typology. Information Systems Research 13(3), 334–359 (2002)
30. Ahn, T., Ryu, S., Han, I.: The impact of web quality and playfulness on user acceptance of online retailing. Information & Management 44(3), 263–275 (2007)
31. Guo, Y., Barnes, S.: Virtual item purchase behavior in virtual worlds: An exploratory investigation. Electronic Commerce Research 9(1-2), 77–96 (2009)
32. Gefen, D., Straub, D.W.: Consumer trust in B2C e-commerce and the importance of social presence: Experiments in e-products and e-services. Omega 32(6), 407–424 (2004)
33. Davis, F.D., Bagozzi, R.P., Warshaw, P.R.: Extrinsic and intrinsic motivation to use computers in workplace. Journal of Applied Social Psychology 22(14), 1111–1132 (1992)
34. van der Heijden, H.: User acceptance of hedonic information systems. MIS Quarterly 28(4), 695–704 (2004)

35. Venkatesh, V., Thong, J.Y.L., Xu, X.: Consumer acceptance and use of information technology: Extending the unified theory of acceptance and use of technology. MIS Quarterly 36(1), 157–178 (2012)
36. Aggarwal, C.C., Yu, P.S.: On the network effect in web 2.0 applications. Electronic Commerce Research and Applications (forthcoming, available online)
37. Cyr, D., Hassanein, K., Head, M., Ivanov, A.: The role of social presence in establishing loyalty in e-service environments. Interacting with Computers 19(1), 43–56 (2007)
38. Schwarz, A., Schwarz, C., Jung, Y., Perez, B., Wiley-Patton, S.: Towards an understanding of assimilation in virtual worlds: The 3C approach. Eur. J. Inf. Syst (2011)
39. Ali, N.Y., Lo, T.Y.S., Auvache, V.L., White, P.D.: Bad press for doctors: 21 year survey of three national newspaper. Br. Med. J. 323(7316), 782–783 (2001)
40. Gefen, D., Karahanna, E., Straub, D.W.: Trust and TAM in online shopping: An integrated model. MIS Quarterly; Minneapolis 27(1), 51–90 (2003)
41. Lu, Y., Zhao, L., Wang, B.: From virtual community members to C2C e-commerce buyers: Trust in virtual communities and its effect on consumers' purchase intention. Electronic Commerce Research and Applications 26(4), 346–360 (2010)
42. Hassanein, K., Head, M.: Manipulating perceived social presence through the web interface and its impact on attitude towards online shopping. International Journal of Human-Computer Studies 65(8), 689–708 (2007)
43. Ringle, C.M., Wende, S., Will, A.: Smart PLS 2.0 M3. University of Hamburg (2005), http://www.Smartpls.De
44. Fornell, C., Larcker, D.F.: Evaluating structural equation models with unobservable variables and measurement error. Journal of Marketing Research 18(1), 39–50 (1981)
45. Chin, W.W.: Issues and opinion on structural equation modeling. MIS Quarterly & The Society for Information Management (1998)
46. Islam, A.K.M.N., Mäntymäki, M.: Continuance of professional social networking sites: A decomposed expectation-confirmation approach. Paper 5 (2012)
47. Sheth, J.N., Newman, B.I., Gross, B.I.: Why we buy what we buy. A theory of consumption values. Journal of Business Research 22(2), 159–170 (1991)
48. Kastanakis, M.N., Balabanis, G.: Between the mass and the class: Antecedents of the "bandwagon" luxury consumption behavior. Journal of Business Research 65(10), 1399–1407 (2012)

Appendix A. Convergent Validity

Item	Operationalization	Load	C.R.	Alpha
CON1	I was absorbed intensively in the activity in Habbo	0,815	0.923	0.890
CON2	My attention was focused on the activity in Habbo	0.896		
CON3	I concentrated fully on the activity in Habbo	0.910		
CON4	I was deeply engrossed in the activity in Habbo	0.837		
ENJ1	It is enjoyable to use Habbo	0.931	0.960	0.937
ENJ2	It is fun to use Habbo	0.950		
ENJ3	It is entertaining to use Habbo	0.947		
TDIS1	When using Habbo, time goes by fast	0.822	0.883	0.802
TDIS2	When using Habbo, I am not aware of any noise	0.874		
TDIS3	When using Habbo, I often forget the work I must do	0.841		
CUR1	Using Habbo increases my interest in exploring things	0.932	0.927	0.842
CUR2	Using Habbo arouses my imagination	0.927		
SP1	There is a sense of human contact in Habbo	0.859	0.923	0.875
SP2	There is a sense of sociability in Habbo	0.910		
SP3	There is a sense of human warmth in Habbo	0.913		
TRU1	I believe that the other Habbo users always keep their promises	0.934	0.967	0.948
TRU3	I believe the other Habbo users are honest	0.961		
TRU3	I believe in the other users' integrity	0.960		
CTRL1	I have the resources, knowledge, and ability to use Habbo	0.891	0.896	0.769
CTRL2	Using Habbo is entirely within my control	0.911		
PNS1	How many people about your age use Habbo? none…all	0.898	0.934	0.894
PNS2	How many of your friends use Habbo? none…all	0.895		
PNS3	How many of your peers use Habbo? none…all	0.933		
PURC1	I intend to purchase Habbo items and/or Habbo Club memberships* shortly	0.939	0.950	0.921
PURC2	I predict I will purchase Habbo items and/or Habbo Club memberships in the short term	0.929		
PURC3	I will frequently purchase Habbo items and/or Habbo Club memberships in the future	0.921		

* Habbo Club membership refers to a package, which includes exclusive virtual goods such as clothing and furniture.

Insight into the Construction of Occupational Classification in E-Commerce in China

Xuewen Gui[1,*], Xiaolan Wu[2], and Song Liu[2]

[1] Electronic Commerce Research Center of Hubei Province, Information Management School,
Central China Normal University, Wuhan, China
guixuewe@mail.ccnu.edu.cn
[2] Information Management School, Central China Normal University, Wuhan, China
wuxiaolan666888@163.com, 147320325@qq.com

Abstract. The great advances on e-commerce industry have tremendously promoted the development of Chinese economy and society. However, thus far, there has been a lack of a comprehensive occupational classification system of e-commerce in China, which has inhibited the further improvement of Chinese e-commerce industry. In this article, on the basis of current situations, we highlight the importance, summarize the relevant researches and learn excellent experience of constructing e-commerce occupational classification system, from China and overseas, and thus put forward the reasonable methods and contents of the construction of e-commerce classification system in China.

Keywords: e-commerce, occupational classification system, e-commerce industry.

1 Introduction

The rapid development of Internet technology has facilitated the emergence of Chinese e-commerce era during the past two decades. As a new kind of economic style, e-commerce has attracted considerable attentions from all sectors of society in China because of its excellent abilities to enhance economic benefit and effectiveness of traditional business. Especially in recent years, e-commerce has been recognized as sunrise industry or green industry, and has become one of the most fashionable and promising industries. The statistics from China E-business Research Center has demonstrated that the total transaction amount in e-commerce industry in China has reached 10.2 trillion Yuan RMB ($1.63 trillion) in 2013 [1]; "The 12th Five-year Plan of E-Commerce Development" predicted that the total transaction amount of e-commerce in China would exceed 18 trillion Yuan RMB in 2015 [2]. Meanwhile, according to the data provided by China E-business Research Center, there would be over 2 million employees engaging in e-commerce industry by the end of the fourth quarter of 2013 and e-commerce has brought over 1.5 million job chances for employees [1]. Obviously, e-commerce industry has huge influences on society and

---

* Corresponding author.

H. Li et al. (Eds.): I3E 2014, IFIP AICT 445, pp. 315–326, 2014.

economy at present. However, thus far, there has been a lack of a comprehensive occupational classification system in China, which may lead to the poor understanding of the standards and requirements of e-commerce personnel training in universities, pose negative effects to training e-commerce professionals, and trigger the mismatch between demand and supply of e-commerce professionals. Thus it inhibits the further development of Chinese e-commerce industry. Therefore, it is highly desirable to construct a comprehensive occupational classification system, because it can provide reliable guidance for the training of e-commerce professionals, contribute to the construction of meliorating occupational classification system, and lay solid foundation for the construction and development of economy in China. This article is based on the current status of Chinese e-commerce occupational classification, sheds light on the excellent experience of the occupational classification from America and Canada as well as Chinese IT industry, subsequently demonstrates the principles and goals as well as the construction contents of e-commerce occupational classification system in China.

2   Research Background

2.1   Occupation and Its Classification

It is well known that occupation is defined as a social work with the goal of meeting material and spiritual needs by means of expertise and knowledge, in which human beings engage to gain income, achieve personal values, and realize social values as well as create wealth for society [3]. Correspondingly, an accurate description of occupational classification then refers to that following certain scientific methods and standards, the various occupations are arranged and classified systematically to form a logical and ordered system [4]. Many advances on occupational classification have been achieved in various countries, which promotes the development of economy and society. For instance, Holland thought that occupations could be classified into six major categories, such as realistic, Investigative, artistic, social, enterprising, and conventional, and subcategories on the basis of the concept of personality types with the aid of Bayesian statistics and the Dictionary of Occupational Titles (DOT) [5]. Moreover, in 2010, the Office for National Statistics (ONS) published the second version of the occupational classification, which is revised every ten years in UK by collaborating with experts and consulting with users and producers of occupational statistics [6]. Obviously, these advances on the construction of occupational classification played positive effects on the continuous development of society.

2.2   E-Commerce

Many previous reports have demonstrated that the concept of e-commerce could be defined from the following three aspects: in narrow sense, e-commerce belonged to a kind of business activity which focused on the exchange of goods and services on the basis of IT; in middle-level sense, e-commerce could be viewed as the electronization of business workflow in various industries, which was also called as e-business including a wide range of businesses and services, such as e-government, networking, the electronization of intra and inter-enterprise operating system [7]; in broad sense, e-commerce was a new model of economic activity which conducted various businesses online [8]. In general, e-commerce is a cross-discipline subject covering the fields of economy, management, information technology, and law, and is associated closely with other industries, such as manufacturing, service and public utilities ones, which requires high-quality talents with both solid professional knowledge and skilfully practical operation abilities [9]. Furthermore, the emergence of e-commerce was highly inspired by the development of IT, and both of them shared some similarities in the occupation characteristic and requirement, such as web transactions that includes advertising, buying, selling positions and the application of IT covers through almost the whole e-commerce industry. The great similarity between e-commerce and IT provides a strong indication that constructing occupational classification system in e-commerce can base on methods and principles of occupational classification of IT industry [6]. Obviously, in view of the above considerations and the highly developed feature of network technology at present, the relatively more suitable description of e-commerce should be conducted from the perspective of narrow sense. Therefore, this article attempts to give a detailed discussion on various commodity exchange dominant commercial activities by electronic means and their corresponding occupational categories from the perspective of narrow sense of e-commerce.

2.3   Research on E-Commerce Occupations

Forty years ago, China Labor & Social Security promulgated "Dictionary of Occupation in China", the first programmatic document to give an objective and comprehensive classification of Chinese occupations, which basically reflected the structures of social occupations and provided significant basis for statistic information of workers, employment guidance, and occupation introduction, etc. [10]. However, because of the development of economy, the advance of science and technology, and the improvement of industrial structure during the past four decades, the dictionary, at present, cannot meet the requirement for the construction of Chinese occupational classification in various industries, especially in newly emerging industries, such as e-commerce. Furthermore, the rapid development of e-commerce triggers the tremendous changes in its occupational structure and operation mode. Thus, these reasons result in that the principles of the occupational classification in the dictionary not only are unable to reveal the trend of the current e-commerce industry comprehensively and objectively, but also fail to provide efficient guidance for the development of the industry.

So far, there have been few research works focusing on the construction of occupational classification, even fewer in e-commerce industry. Research works on the classification of e-commerce occupations are still in their infant stage, which reserves numerous spaces for the further development. Recently some documents have generated on the construction of occupational classification of e-commerce. However, they mainly focused on elucidating the significance of occupational classification and their instructive effects on vocational education, which could not meet the demand of the development of Chinese e-commerce industry.

With the continuous development of Chinese e-commerce, researchers have gradually realized the importance of e-commerce occupational classification, which stimulated their great research interests on this theme. Liang, B. (2007) first thought that the e-commerce positions could be divided into two categories in terms of the evolution process, namely the traditional positions modified by the effect of new information technology and the situation of e-commerce booming, and the newly-emerging positions induced by computer science, Internet and communication technology, which were the pioneering works on the original classification of e-commerce positions [11]. Subsequently, Meng et al. (2009) demonstrated the four types of e-commerce positions including technology, business, management, and e-commerce engineering ones, which further promoted the construction of Chinese e-commerce occupational classification [12]. Gui (2013) from the viewpoint of the development status of commercial services and applications, classified e-commerce occupations into two categories, namely e-commerce service and e-commerce application, on the basis of job demands [13]. The iResearch Consulting Group conducted extensively interviews and investigations on e-commerce enterprises and pointed out that the occupational classification positions in e-commerce industry could be sorted into six categories, such as positions of e-business operation, marketing, network engineering, logistics, procurement and customer service; besides, with the same method, a detailed subordinate classification was conducted on the basis of the above six categories [14].

On the basis of the aforementioned research results, it can be found that as great efforts have been made in e-commerce industry, the investigations on the e-commerce occupational classification has become more and more comprehensive and deepening. However, despite of some advances, the state-of-the-art of e-commerce occupational classification cannot satisfy numerous requirements of the development status and market demands in the industry. Therefore, the construction of a maturity e-commerce occupational classification in China is highly pursuing, but still a great challenge.

3   The Experience of Occupational Classification in China and Abroad

According to the researches of the authors, there also have been few completely comprehensive e-commerce occupational classification systems in abroad, which is adverse to the improvement of e-commerce worldwide. Thus, we have to build a new system to serve for the industry. In order to construct a consummate e-commerce occupational classification in China, we should take some effective and practical experience as references from China and overseas. The following are several typical examples of occupational classification from Canada, America and Chinese IT industry.

3.1   Features of Occupational Classification in Abroad

Standard System

An occupational classification book published in Canada, is one of the most abundant, substantial and comprehensive reference tools for the classification of e-commerce industry. This reference book demonstrates that the occupations in Canada can be systematically and thoroughly classified into 10 broad occupational categories, 40 major groups, 140 minor groups and 500 unit groups. There are approximately 40,000 occupational titles classified in the 500 unit groups of the NOC (National Occupational Classification) 2011, which constitute the classification system. In the system, each major group has a unique digit code orderly, as well as minor group, unit group and occupational title [15]. This reference book featuring with extremely comprehensive and professional characteristics of occupational classification method, covers a wide range of occupational classification for almost all positions in Canada, and provides detailed description of work skills and access requirements. Inspired by its broad universality, the construction methodology of occupational classification in this book is applicable not only for Canada, but also for other countries with different social systems, as well as for the construction of Chinese e-commerce occupational classification.

Flexible Methods

There is a widely applied information system on occupational classification in America. The system classifies jobs from generality to individual into 23 major occupational groups of the revised SOC (Standard Occupational Classification) system. These major groups include 97 minor groups, 461 broad occupations, and 840 detailed occupations according to the following six criterions: requirements of knowledge, requirements of practical experience, qualification of job holders, personalities, special work requirements and professional features [16]. Meanwhile the design of the system also focuses on the quality of the users experience, allows to be classified by different methods during collecting data, and displays high flexibility and customization, which agrees well with the practical demands of users. Thus, the construction methodology of occupational classification in this system is of great value for directing the construction of Chinese e-commerce occupational classification system.

Dynamic Update of Contents

The contents in occupational classification system in America can be updated periodically, which can remove the outdated positions and simultaneously monitor newly-springing up professions on the basis of changes of social industries, such as revolution of IT, popularity of office automatic (OA) system and explosion of tertiary industry continuously. So users can make good use of the system to compare the newly developing positions with their own, and thus grasp the latest occupational information to serve for themselves. Additionally, the system gives emphasis to collecting the information of service-oriented and professional positions and eliminating the records of positions of labor-intensive industries to update the system, which is in well accordance with the laws of social development.

Adaptable for Users

The occupational classification system in America is divided into four hierarchies which are convenient to information acquisition for both organizations and individuals according to their own requirements. The users can compare the data with their own work properties to orientate themselves to make decisions or to fulfill diverse needs.

3.2   Features of IT Industry in China

A research group of IT field constructs a specific occupational classification system which is built according to two main reasons. The first reason is based on the considerations of the practical situation where the development prospects of economy, technological innovation and industries transformation present in China. The second reason is the rapid development of IT occupations with a large number of employees and positions emerging. In the scheme, the occupations in IT industries are firstly classified into 3 main categories, such as the major group, the application group and the related group. Secondly, the major group is subdivided into 5 minor groups, such as software, hardware, web, information system and manufacturing groups, which represent 5 crucial pillars in IT industry. Thirdly, the application group is subdivided into 7 minor groups, such as control, design, business, entertainment, education, and communication groups. Fourth, there are 41 unit groups formed on the basis of dividing the 12 minor groups mentioned above. Additionally, there is no subdivision in the related group [17]. The classification system data in the scheme shows that it covers almost all the positions and reflects the basic structure of occupations comprehensively in IT industry, which has been verified by its practical applications in personnel training and course arrangement.

In addition, because of the high variability of the IT occupations, the research group is currently considering constructing a long-term observable pre-warning system to monitor the variation of relevant positions, which will provide theoretical and experimental support for perfecting and updating the IT occupational classification system.

4   The Construction of Occupational Classification System in E-Commerce Industry in China

On the basis of combining the experience of the above excellent occupational classification methods with the situation of e-commerce development in China, the approach to construct occupational classification system in e-commerce industry can be divided into three parts, namely principles, aims and methods.

4.1   Principles and goals

Scientific and Standardized

Taking the development of global economy, improvement of industrial structure, and the progress of science and technology into consideration, the occupational classification system in e-commerce industry should be classified in light of properties, objects, range, and surroundings of work. Furthermore, the system should be constructed rigorously in line with rules of "Dictionary of Occupation in China". On the basis of the above considerations, the occupation system of Chinese e-commerce industry should be divided into four categories including major groups, minor groups, unit groups and detailed occupations. Of these classifications, the major groups are categorized according to working properties and capability requirements; the minor groups are classified further on the basis of working tasks and division of labor within the respectively major groups; the unit groups are divided and aggregated further on the basis of objects, circumstances and demands of work in the same minor groups; the detailed occupations are classified further according to the tools, equipment, techniques, etc. within the same unit groups; besides, those immature and indefinite positions should also be placed and explained properly. Moreover, if the method is in line with the traditional classification standards and rules, it may lead to the classification system failing to cover all the occupations in e-commerce industry accurately, because of the rapid, variable, and time-depending development of the e-commerce occupations, and as a result, the classification may be repeated or neglected. Therefore, it is highly desirable to classify e-commerce occupations flexibly to cover all their corresponding aspects to meet the criterion of national occupational classification.

Rational and Applicable
As the ultimate aim of classifying the occupational system of e-commerce is to boost the development of e-commerce industry and to provide efficient assistance for the economy development, so the system should agree well with the rational and applicable rules. During the classification process, the core concept for the construction species of each hierarchy should keep integrity, the standards of classification level should be identical, and the structure of the occupational system should be logical, which can ensure the rationality of occupational classification in e-commerce industry. Besides, the expressions in the system should be as concise and clarified as possible, which can sufficiently reveal the job properties, activity modes and work requirements, and also be applicable for business and national administration.

Comprehensive and Long-Acting
The construction of occupational classification system of e-commerce should be comprehensive and long-acting throughout its every aspect including data collection, position setting and position description. Moreover, it should combine the advantages of all the similar researches and display distinct characteristics. Except for the rules mentioned above, the system should be advanced in its framework and long-acting in function, and in the top list among various occupational classification systems. Only by using this method can the functions of the system be comprehensive and long-lasting.

Dynamic and Open
As e-commerce occupations develop rapidly and variably, the system of occupational classification in e-commerce industry should display the feature of dynamic update.

Once new positions emerge, they should be first subjected to classification positioning timely and effectively and subsequently, match automatically on the basis of their working properties, ability requirements. Then, their corresponding affiliation classifications should be confirmed according to the above positioning results and this operation process needs to develop into a continuing system, which can facilitate the dynamic update and timely addition or deletion of e-commerce occupations. By adopting this sort of method, this system not only can reflect the state and trend of occupations scientifically, objectively, completely and accurately, and additionally, but also obeys the rules of development and meets practical demands in e-commerce industry.

Moreover, the construction of e-commerce occupational classification system should meet the relevant criterions and absorb advanced experience in China and overseas. And then, we should select the concrete e-commerce positions as the main constituents, and distinguish the difference among various e-commerce positions by analyzing their contents and styles to form a scientific, rational, ordered, open and applicable system of occupational classification.

4.2 The Construction Approaches of E-Commerce Occupational Classification System

Meeting Requirements and Standards

Through investigating and interviewing those relevant e-commerce enterprises in China to collect information from e-commerce market, we should grasp the situation of positions setting and then analyze the category of e-commerce corresponding to "Dictionary of Occupation in China" to place the e-commerce occupational classification system and finally construct the system in line with related standards and requirements strictly. Also, the system should meet requirements of Chinese current market system, and simultaneously be in accordance with national circumstances. Furthermore, the system should exhibit unique characteristic and also provide references to the revision and improvement of the "Dictionary of Occupation in China".

Learning Experience

The authors hold that the systematically occupational classification results carried out previously in China and abroad should be studied thoroughly, and the statistics of those relevant positions in e-commerce industry should be paid more attentions. Moreover, after comparative researches, we should draw a comprehensive summary of all relevant experience, adopt excellent methods and eliminate outdated approaches.

Constructing a Dynamical System

The completed e-commerce occupational classification system should meet the standards of Chinese occupational classification, and in the meantime, it can develop into an independent system. In the vertical direction, the system should establish a logical framework including major groups, minor groups, unit groups, detailed occupations, and further form the main part and related part occupations in the classification, while in the horizontal direction, the definitions and descriptions of positions should be accurate and complete, which consist of basic skills, basic knowledge, work environment, relevant training, professional qualifications etc.

In addition to accurately revealing the state of e-commerce occupations, the system should predict the development trend of e-commerce industry. Therefore, it is suggested to build monitoring systems in universities or e-commerce institutes to track the trend of occupations, to detect emerging and fading ones, so that the contents and structures of the system can be adjusted and updated timely to reflect the e-commerce development authentically.

4.3   The Construction of E-Commerce Occupational Classification System

The system of e-commerce occupational classification should mirror the contents in the form of the catalog table, and can be divided into vertical and horizontal parts which interconnect and correspond with each other. In the catalog table, by taking management responsibilities and technical operations abilities as references, the vertical part is classified into four grades including high level, middle level, primary level and bottom level, all of which have an ordered arrangement. This construction method of e-commerce occupational classification, enables its pyramid-like characteristic, is well arranged and able to comprehensively cover the whole e-commerce industries. While the horizontal part of the system is classified according to the sequence modes of occupations, which can reflect the positions types comprehensively and avoid the repeating classification of emerging positions induced by monotonous arrangement. Additionally, because of the rapid development of e-commerce industry, the vertical-horizontal classification method can timely update the state of positions variation and maintain the system open. Furthermore, the, single existence of every position ensures the independence and integrity of the system and the methods for maintaining and adding new classifications are very simple and convenient, which provides reliable references for the construction of other industries.

Below is the construction method of detailed e-commerce occupational classification system (as shown in Table 1)

Table 1. E-commerce occupational classification contents

Rank	Type	Occupations contents	
Senior manager		General manager/ Vice general manager	
Middle-level manager		Executive officer, Chief financial officer, Director of operation, Chief procurement officer, Director of safety management	Director of marketing, HR director, Director of equipment, Director of IT,
Comprehensive primary-level staff	Business	Online marketer, Business operator, Logistics staff, Administrative staff, Market developer,	Salesclerk, Secretary, Financial officer, HR clerk, Information manager

Table 1. (*Continued*)

	Technology	Website designer/ engineer, Website art editor, Maintainer, Mobile phone terminal developer, New media developer,	Picture /Video maker, Software designer, Database constructor, App developer, Online stores director
	Service	Information service personnel, Training service personnel, After-sales service personnel,	Consumer-service staff, Casher, CRM personnel
General staff		Securityguard, Mechanic, Sanitation worker, Operator	

It can be seen from table 1 that in the vertical direction, the e-commerce occupational categories consist of 4 major groups: senior manager, middle-level manager, comprehensive primary-level staff and general staff representing four ranks which share similarities with the form and the personnel structure of traditional enterprises. At the same time, the comprehensive primary level staff contains three types of occupations which are business type, technology type and service type, however other ranks are not obviously with such types. In the horizontal direction, the construction of the system is on the basis of the four occupational categories, and highlights the vocational characteristics of e-commerce industry. Moreover, the method for the contents arrangement can basically meet the criterion of "Dictionary of Occupation in China", which is beneficial for the human resource management and labor statistics in China.

Additionally, the system covers not only the majority of traditional occupations, but also some ones associating with newly emerging technology and information. Among the system, the whole senior management positions, parts of middle level and comprehensive primary-level staff positions and almost all the general staff ones originated from the evolution of traditional industries, and are still the dominating part of occupational categories in e-commerce industries and maintaining their characteristics and skills of original vocations, which also endows e-commerce industries with unique characteristic and novel definition, such as General manager in the rank of Senior manager, Executive officer in the rank of Middle-level manager, Financial officer and Casher in the rank of Comprehensive primary-level staff and Securityguard in the rank of General staff, etc.. Parts of middle-level and comprehensive primary-level staff positions associating with lately-emerging technology and information, have generated following the development of information technology, especially the e-commerce industry such as Director of IT, Information manager, App developer, CRM personnel, etc. Besides, all the positions of the system listed above are categorized basing on the narrow sense of the e-commerce concept, and basically include all of them in the system of the industry, which induces the formation of a comprehensive system.

Obviously, the classification system established herein not only gathers all occupations in e-commerce industry, but also endows them with the specific industry characteristics. However, the occupations in American and Canadian classification systems are not classified according to industry content, which poses great obstacles to the development of those industries. Compared to Canadian and American occupational classification systems whose aims are to facilitate the management, the occupational classification systems of Chinese e-commerce established in this study helps the enhancement of e-commerce knowledge and professional qualities and has a goal of optimizing allocation of human resources and promoting the development of e-commerce, though both kinds of occupational classification methods share the same theories. More importantly, the construction methods originate from the combination between the consideration of current situations in Chinese e-commerce and the references of advanced occupational classification experience from China and abroad, which resulted in the formation of e-commerce occupational classification with Chinese characteristics.

5   Conclusion

In summary, through basing on the current state, obeying the rules of the development of relevant industry, following the corresponding classification modes and standards and learning advanced experience of classification system in China and abroad, we propose the construction methods and principles, and finally realize the reasonable construction of Chinese e-commerce occupational classification system. We believe that the successful construction of occupational classification system in e-commerce will have a far-reaching impact on the development of society and economy in the following aspects. First, the proposed methods and principles of classification can enrich fundamental theories of occupational classification. Second, constructing Chinese e-commerce occupational classification system can perfect the classification framework of the whole industries in China. Third, this event can undoubtedly boost the development of e-commerce industry. Fourth, the established theory of occupational classification in this study can offer efficient guidance, not only for the classification construction of other Chinese industries, but also for the construction of occupational classification system in other countries. However, some problems still remain unsolved, such as the classification of position properties, salary, job requirements, which encourage us to conduct a further exploration on these issues from a broader and deeper viewpoint in future.

References

1. Report on Monitoring Data of Chinese E-Commerce Market in 2013 (2013), http://b2b.toocle.com/zt/2013ndbg/
2. The 12th Five-year Plan of E-Commerce Development, http://www.miit.gov.cn/n11293472/n11293832/n11293907/n11368223/14527814.html
3. Li, H.Y.: Methodology and Functions of Occupational Stratification. Academic Exchange 12, 113–118 (2004)
4. Gui, X.W.: Theory and Method on Effect Measure of E-commerce Promoting Economic Development: Relevant Theory and Method on Effect Measure of E-commerce Promoting Economic Development. Dissertation, University of Central China Normal University studies in information management. Central China Normal University (2011)

5. Viernstein, M.C.: The Extension of Holland's Occupational Classification to all Occupations in the Dictionary of Occupational Titles. Journal of Vocational Behavior 2, 107–121 (1972)
6. Elias, P., Birch, M.: SOC2010: Revision of the Standard Occupational Classification. Economic & Labour Market Review 4, 48, 52 (2010)
7. Feng, G.L.: Analysis on Basic Principles of Occupational Classification Standard in China. China Training 3, 25–26 (1996)
8. Eleventh Five-year Plan on E-Commerce development, http://www.gov.cn/ztzl/2007-06/25/content_661213.htm
9. Chen, X.H.: Research on Cross-border E-commerce Talents Cultivating Pattern—Analysis based on Yiwu. Prices Monthly 3, 66–67 (2014)
10. Dictionary of National Occupational Classification and Vocational qualification Authentication Committee, Dictionary of Occupation in China: Introduction. China Labor Press, Beijing (1999)
11. Liang, B.: Analysis on Professional Orientation and Personnel Training of E-Commerce. Journal of Anyang Institute of Technology 5, 118–121 (2007)
12. Meng, Z.X., Wu, H.Z.: Analysis on E-Commerce Personnel Training and Employment. E-Business Journal 10, 83 (2009)
13. Gui, X.W.: New Impetus of Economic Booming—Mechanism and Effect of E-Commerce Measurement: E-Commerce Occupations. Science Press, Beijing (2013)
14. Report on Occupational Development and Salaries of E-Commerce in China, http://b2b.toocle.com/detail-6096637.html
15. National Occupational Classification (NOC) 2011 (2011), http://www5.statcan.gc.ca/access_acces/alternative_alternatif.action?l=eng&loc=http://www.statcan.gc.ca/pub/12-583-x/12-583-x2011001eng.pdf&teng=National%20Occupational%20Classification%20(NOC)%202011&tfra=Classification%20nationale%20des%20professions%20(CNP)%202011
16. Cosca, T., Emmel, A.: Revising the Standard Occupational Classification system for 2010. Monthly Labor Review 133, 37 (2010)
17. Report on Occupational Classification Research in IT Industry, http://www.chinajob.gov.cn/NewsCenter/content/2006-07/13/content_530717.htm

Investigating Key Antecedents of Customer Satisfaction in B2B Information Service Firms

Vikas Kumar[1,*], Archana Kumari[1], Ximing Ruan[1], Jose Arturo Garza-Reyes[2], and Supalak Akkaranggoon[3]

[1] Bristol Business School, University of the West of England, Bristol, BS16 1QY, UK
{Vikas.Kumar,Archana.Kumari,Ximing.Ruan}@uwe.ac.uk
[2] Centre for Supply Chain Improvement, The University of Derby, Derby, DE22 1GB, UK
J.Reyes@derby.ac.uk
[3] Khon Kaen University, Nong Khai Campus, 112 M. 7 A., Muang, Nong Khai,
43000, Thailand
supakk@kku.ac.th

Abstract. Service sector has grown significantly over the years and now is one of the major contributors of the Gross Domestic Product (GDP) of most of the developed and developing economies. Within the service sector, information economy has grown significantly with the rapid developments in the Internet and Communication Technologies (ICTs). However, the research so far has focused on the manufacturing sector rather than the service sector. This paper, therefore, aims to fill this void by testing the service management and Service Quality (SERVQUAL) theories in Businesses to Business (B2B) information services context. An empirical investigation with secondary data is carried out to explore the relationship between three key antecedents of Customer Satisfaction (CS) namely; Functional Service Quality (FSQ), Technical Service Quality (TSQ) and Corporate Image (CI). This re-search also aims to investigate the interrelationship between the three key antecedents of customer satisfaction. The findings show that FSQ, TSQ, and CI are positively correlated with customer satisfaction. Results also show that CI is positively correlated with TSQ and FSQ.

Keywords: Information Services, FSQ, TSQ, Corporate Image, Customer Satisfaction, B2B, Service Management e-business, e-government.

1 Introduction

Over the years, the manufacturing sector has dominated the world economy; however, the world economy is now slowly transforming to a service based economy. The growing importance of services to the economy can be realized by the fact that service sector nowadays is a major contributor to the Gross Domestic Product (GDP) of most of the developed and developing nations [1] [2]. For example, the contribution of services to the U.S. GDP is around 80%, UK around 73%, Japan around 74%, France around 73% and Germany is around 68% [1] [2] [3]. The growth

---

* Corresponding author.

H. Li et al. (Eds.): I3E 2014, IFIP AICT 445, pp. 327–337, 2014.
© IFIP International Federation for Information Processing 2014

in service sector is primarily attributed to the rapid advancements and implementations of the Internet and Communication Technologies (ICTs). ICT growth has led to the emergence of the information sector as a strong contributor in the service dominant economy. The significance of information economy was firstly highlighted by Machlup [4] and then later by Porat and Rubin [5] who attempted to measure the US information economy. In the late 70s, Porat and Rubin [5] measured the size and structure of the US information economy. Their findings reveal that the information sector contributed to around 46% of the US Gross National Product (GNP) in 1967. A research by Apte and Nath [6] concluded that the contribution of the information sector to US GNP rose to around 63% in 1997. The significance of information economy was further verified by [7] and [8], who point out that during the last fifty years, developed or developing economies have evolved from a goods or manufacturing-oriented economy to a service oriented and now have moved on to the information-oriented economy.

Information economy has become integral part of the economy and their contribution cannot be ignored. However, the research so far has primarily focused on the manufacturing sector. In light of the growing significance of information services in the worldwide economy and the lack of enough research evidence, this paper sets out to test some of the service management and SERVQUAL theories in the context of information service settings, particularly, in Business to Business (B2B) information service firms. Literature identifies Service Quality and Corporate Image as key indicators of Customer Satisfaction [9] [10] [11] [12] [13]. Nevertheless, quality is often treated as multidimensional construct in services which can be understood from two main dimensions: Technical Service Quality (TSQ) and Functional Service Quality (FSQ) [8] [9] [14] [15] [16]. Existing literature has identified a number of antecedents of customer satisfaction, such as; quality, trust, dependability, corporate image and waiting time [17] [18] [19]. However, this research will primarily investigate the linkages between three key antecedents of customer satisfaction: Technical Service Quality, Functional Service Quality, and Corporate Image and their impact on customer satisfaction in the context of B2B information service organizations. The choice of three antecedents is restricted by the availability of the secondary data provided by the case organization. In addition, the interrelationship between the three key antecedents of customer satisfaction has not been fully explored in literature and this study aims to also bridge this gap.

The rest of the paper is organized as follows. Next section reviews a number of research papers to set out the context of the research and identify the linkages among the variables. Section 3 delineates the research objectives and proposes the research framework. Section 4 elaborates the research methodology. The findings of the research are presented in section 5. Section 6 concludes this study and sets out the direction for future research.

2   Theoretical Background

Earlier discussion indicates that with the rapid advancement in ICTs, economy is slowly moving towards a more information based economy. However, very few researchers have looked at the information service sector [1] [2] [6] [7] [8]. Literature also indicates that the growth in information technology, inter-net and web technologies has revolutionized the way in which Business to Business (B2B) and Business to Customers (B2C) organizations interact and offer services to their customers [20] [21]. B2B and B2C firms have different characteristics that set them apart from each other, for example, different types of purchases and authorizations; unique contracts, terms, and conditions for different business customers; variety of customer sizes, demands and requirements; and different level of participation in customer's supply chain [8] [22] [23] [24]. Due to these differences, the strategies adopted by B2B and B2C firms vary [25]. A study by Rauyruen and Miller [26] further suggests that B2B service providers need to understand the nature and the circumstances of their customers. Therefore, B2B service-providers should pay attention to the quality control of their service delivery systems and must put a lot of effort into creating high-level perception of the service quality [8]. Tang et al. [27] argues that perceived service quality is an important factor in the B2B. This was also noted in the study of Bhappu and Schultze [21], where they found that in the B2B environment, both the relational (soft quality) and operational performance (hard quality) are of prime significance particularly for achieving customer satisfaction. However, apart from the study by Bhappu and Schultze [21] there are limited studies that aim to investigate the significance of the service quality dimensions in the B2B environment [27] [28] [29]. This study, therefore, aims to empirically provide further evidence of the relationship between FSQ, TSQ and Customer Satisfaction in a B2B service setting.

The drivers of customer satisfaction are well highlighted in the operations management literature, which identifies service quality, speed, flexibility, cost, corporate image and dependability as critical drivers [9] [18] [19] [30]. One of the popular frameworks that identify the linkage between quality and customer satisfaction is the service profit chain proposed by Heskett et al. [18]. In brief, it proposes a positive linear relationship between staff satisfaction, service quality and customer satisfaction leading, ultimately, to profitability [16]. Gonzalez et al. [30] also identify perceived service quality as an antecedent of satisfaction. This was also further verified in the work of [2] [19] [31] [32]. These SERVQUAL studies show that a strong link exists between customer satisfaction and service quality. Parasuraman et al. [33] differentiate the service quality construct as Functional Service Quality (FSQ) (doing things nicely) and Technical Service Quality (TSQ) (doing things right). Grönroos [34] suggests that dissatisfaction occurs if expectations are greater than actual performance. As a result, evaluations are not based solely on the outcome of the service, the technical quality; they also involve the process of service delivery or functional quality. A number of published researches emphasize the relationship between the two dimensions of service quality (i.e. technical and functional) and customer satisfaction [16] [35] [36] [37]. A study by Rosenzweig and Roth [38] shows that an interrelationship exists between the two dimensions of service quality, i.e., TSQ and FSQ. They further provide empirical evidence of their impact on profit-ability. Lai and Yang [39] also demonstrate that TSQ affects user satisfaction positively. Apart from the service quality dimensions, corporate image has also emerged as one of the drivers of customer satisfaction and loyalty. Andreassen and Lindestad [40] based on data from 600 individual customers, reported that corporate image impacts customer satisfaction and loyalty. Their study also reports that corporate image impacts customer loyalty directly. This relationship was also reported in the study of Martenson [41] where they showed that corporate image is a key driver of customer satisfaction. Cameran et al. [42] in their study of the professional service firms show that corporate image and service quality impacts customer satisfaction. Wu et al. [43] show that higher perceptions of service quality have a positive influence on corporate image. The study also indicates that higher perceptions of corporate image influences customer satisfaction positively. The literature review clearly indicates that service quality dimensions (FSQ and TSQ) and corporate image influence customer satisfaction and customer loyalty positively. Moreover, there is an interrelationship between service quality dimensions (FSQ and TSQ) and corporate image.

The information sector has shown rapid growth in the last few decades and has started showing the dominance in today's economy. The literature review further indicates that FSQ, TSQ, and Corporate Image influences Customer Satisfaction. However, literature investigating these relationships in a B2B information service environment is scarce. Realizing the gap in the literature, this paper sets out to fill this void by empirically investigating the service management and SERVQUAL theories that links these variables together in B2B information services. The next section provides the research framework and propositions to be tested in this research.

3   Research Model and Propositions

The literature review highlights that functional service quality (FSQ); technical service quality (TSQ); and corporate image (CI) affect customer satisfaction (CS) and loyalty. This investigation is a confirmatory study which aims to test the findings of service management and SERVQUAL literature in the context of B2B information service settings. This study looks beyond the traditional linkages between these variables and sets out to understand their interrelationships. Apart from highlighting the linkages, this study also stresses that performing well on these dimensions can help B2B information service firms to successfully meet customer satisfaction levels which in the longer run can lead to customer loyalty and ultimately benefit firms to achieve sustainable competitive advantage. The study primarily focuses on identifying the significance of Functional Service Quality, Technical Service Quality, Corporate Image and Customer Satisfaction. The investigation involves assessing these relationships in a large B2B information intensive firm operating in the UK. Customer loyalty is not included in the framework due to the restrictions imposed by the secondary data. This research sets out to test the following relationships in the context of information service settings:

- Functional Service Quality (FSQ) affects Customer Satisfaction
- Technical Service Quality (TSQ) affects Customer Satisfaction
- Corporate Image (CI) affects Customer Satisfaction
- An interrelationship exists between FSQ, TSQ, and CI

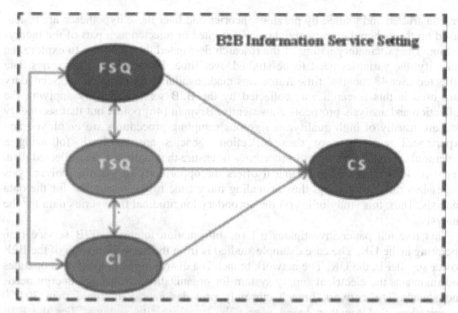

Fig. 1. Research Framework

The research framework tested in this research comprises of three key antecedents of customer satisfaction namely, FSQ, TSQ, and CI (see Fig. 1). As mentioned above, the literature review indicates that a number of papers link these individual key indicators with customer satisfaction, however, studies investigating their interrelation-ships in information service setting is limited. The study proposes a number of propositions derived from the framework based on the review of literature. This study sets out to test following propositions:

P1: Functional Service Quality affects with Customer Satisfaction positively

P2: Technical Service Quality affects Customer Satisfaction positively

P3: Corporate Image affects Customer Satisfaction positively

P4: Functional Service Quality and Technical Service Quality are positively correlated

P5: Functional Service Quality and Technical Service Quality are also positively correlated with Corporate Image

The next section elaborates the research methodology and data collection method followed in this research.

4   Research Methodology

In management research, positivism has traditionally been much more dominant than phenomenology [8], [44], [45]. In this study the selection of the quantitative approach is not motivated by its dominance in management research; rather that the choice has been based purely on the research aims. This research therefore supports the positivist approach. A hypothetico-deductive approach has been followed where the hypotheses are constructed and guided by previous theories and later these hypotheses are tested, based on the gathered empirical data, and accepted or rejected as a part of the theory-testing. The explanatory nature of the research demanded that, in order to explore the causality, the variables need to be studied over time. Longitudinal time series data collected over 48 months' time frame was made available for researchers. Secondary data used in this research was collected by the B2B service firm by applying the collection and analysis protocols consistently. Bryman [46] points out that secondary data are usually of high quality, as rigorous sampling procedures are employed and experienced researchers or data collection agencies are involved, following a structured approach and control procedures to ensure the quality of the collected data. Bryman [46] further identifies that it offers the opportunity of longitudinal analysis and is less time-consuming, thus providing more time to the researcher for the data analysis. Thus, this study relies on the secondary longitudinal time series data for the analysis.

This research paper investigates a large information intensive B2B service firm operating in the UK. The case example studied is from the network branch of the B2B power supplier in the UK. The network branch is a distribution company that operates and maintains the electrical supply system for organizations with large energy needs and spends. The network branch currently provides energy to more than 30,000 organisations that together spend over £2bn on electricity and gas. For the B2B network branch the customers are mainly the Small and Medium scale Enterprises (SMEs) and large business firms. The firm offers a range of agreements for their business customers, such as fixed term contract, flexible purchasing contracts and specialist customer contracts. The firm measures and collects data through telephone interviews and surveys to monitor its performance over time. The secondary data set for a 48 months' time frame (48 data points) was made available for this research that comprised monthly measurements of variables considered as proxies for Functional Service Quality, Technical Service Quality, Corporate Image and Customer Satisfaction.

In this research, Functional Service Quality (FSQ) refers to soft quality or relational element of the quality whereas Technical Service Quality (TSQ) is referred as a hard quality or an ability to perform the promised service dependably and accurately, including time commitments [10] [15] [16]. The B2C service firms used a multi-dimensional scale to measure the FSQ and TSQ. The firm did not employ SERVQUAL scale to measure the quality construct. However, the measures used by the firm resemble to some of the items of the SERVQUAL scale including the empathy, access, assurance, and responsiveness dimensions. Corporate Image was also measured on a multi-dimensional scale. How-ever, Customer Satisfaction was measured on a single item scale as an overall satisfaction with the level of services provided. The constructs studied in this study were measured on a 5 point Likert scale and conform to the reliability and validity tests. FSQ, TSQ and CI were finally converted to a single scale item as Cronbach's Alpha value was > 0.70. The detailed scales of the variables studied are presented below:

FSQ (Cronbach's Alpha = 0.90)
FSQ_1: Staff were polite and helpful
FSQ_2: They were able to provide sufficient information about the problem

FSQ_3: I was able to easily contact them to find out about the problem

TSQ (Cronbach's Alpha = 0.81)

TSQ1: Overall, how would you rate the non-half hourly bills and billing service?

TSQ2: Overall, how would you rate the non-half hourly meter reading service?

TSQ3: Overall, how would you rate the half hourly statements and invoicing service?

TSQ4: Overall, how would you rate the half hourly meter reading service?

CI (Cronbach's Alpha = 0.80)

CI. They are well-known

CI. They have a good reputation

CI. They are environmentally responsible

CI. They support local communities

CI. They recognize and reward your loyalty

CI. They offer competitive prices

CS: Overall, how would you rate their level of customer service?

The next section discusses the data analysis findings.

5   Research Findings

Data analysis firstly involved running a correlation analysis (see Fig. 2). The analysis shows there is a positive correlation (0.615) between functional service quality and customer satisfaction with a significance level of P<0.01. This verifies our first

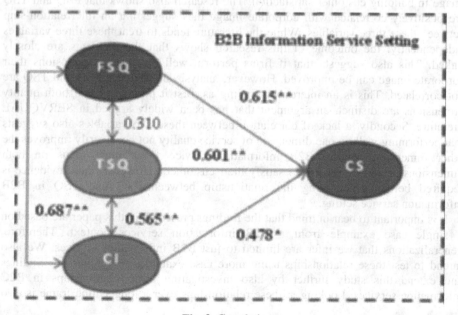

Fig. 2. Correlations

proposition (P1). A positive correlation (0.601) was also evident between technical service quality and customer satisfaction significant at P<0.01 level. This verifies our second proposition (P2). Corporate image was also positively correlated with customer satisfaction (0.478) significant at P<0.05 level. This verified our third proposition (P3). The next step of the analysis was aimed at investigating the interrelationship between the three key antecedents studied in this research. A positive correlation was found between TSQ and CI (0.565) as well as between CI and FSQ (0.687) both significant at P<0.01 level. This verifies our fifth (P5) proposition. Interestingly, no significant correlation (0.310) was found between FSQ and TSQ. Thus, our proposition P4 could not be verified.

6   Conclusions

The paper shows that Functional Service Quality (FSQ), Technical Service Quality (TSQ) and Corporate Image (CI) are key antecedents of Customer Satisfaction (CS). The findings are interesting as these relationships were tested in a new context of B2B information services. The outcome shows that for B2B information service firms performing well on both quality dimensions (FSQ and TSQ) is very important to maintain a satisfied customer base. This counter argues some of the SERVQUAL findings (such as Grönroos [9]) that prioritize FSQ over TSQ in achieving customer satisfaction by indicating that both dimensions are equally important. This verifies the earlier findings of Bhappu and Schultze [21], who emphasized the significance of both quality dimensions. Findings suggest that failure to meet quality expectations of customers can lead to dissatisfaction and ultimately customers can move on to other competitors. Another important finding of this study is the importance of corporate image in building customer satisfaction. The research also shows that FSQ and TSQ are positively correlated with corporate image thus suggesting an interrelation-ship between these three variables. While the literature tends to treat these three variables independently, the outcome of this research shows that the elements are closely linked. This also suggests that if firms perform well on quality dimensions their corporate image can be improved. However, analysis showed that FSQ and TSQ are not correlated. This is an interesting finding as, first it points out that both quality dimensions are distinct, an argument that has been widely argued in SERVQUAL literature. Secondly, a lack of correlation between these two variables also suggests that performing well on one dimension of service quality not necessarily improves the other dimension. Hence, B2B information service firms must focus on both dimensions of service quality to satisfy their customers. However, more evidence is required before generalizing the relationship between FSQ and TSQ in B2B information service setting.

It is important to bear in mind that the findings presented in this paper are based on a single case example from a B2B information services context. Therefore, generalizations that we infer are limited to just B2B information services. We also intend to test these relationships using more case examples to support our findings and extend this study further by also investigating these relationships in B2C information services. Looking at these relationships over a longer time frame is also vital and this requires alternative methods of research. Longitudinal studies are necessary to test out some cause-effect phenomena.

Future studies should involve testing these relation-ships using more robust statistical methods such as multiple regressions and structural equation modeling in B2C information sector. Additionally, more case examples should be investigated to broaden the generalizations of the findings across the information services sector. Moreover, future studies should also aim to investigate the differences in the relationships among the variables between the SMEs and the large business firms.

References

1. Lovelock, C., Wirtz, J.: Services Marketing: People, technology, strategy, 6th edn. Pearson Prentice Hall, US (2007)
2. Kumar, V., Kumari, A., Garza-Reyes, J.A., Lim, M.: Dependability a Key Element for Achieving Competitive Advantage: A Study of Information Service Firms. In: Prabhu, V., Taisch, M., Kiritsis, D. (eds.) APMS 2013, Part I. IFIP AICT, vol. 414, pp. 493–500. Springer, Heidelberg (2013)
3. Paulson, L.D.: Services Science: A New Field for Today's Economy, Industry Trends, pp. 14–17. IEEE Computer Society (August 2006)
4. Machlup, F.: The Production and Distribution of Knowledge in the United States. Princeton University Press, Princeton (1962)
5. Porat, M.U., Rubin, M.R.: The Information Economy (9 Volumes), OT Special Publication, US Department of Commerce, 77-12 (1-9), Washington, DC (1977)
6. Apte, U.M., Nath, H.K.: Size, structure and growth of the US information economy. In: Managing in the Information Economy, pp. 1–28. Springer US (2007)
7. Godin, B.: The Information Economy: the History of a Concept through it Measurement, 1949-2005, Project on the History and Sociology of S & T Statistics Working Paper No. 38, 1–85 (2008)
8. Kumar, V.: An Empirical Investigation of the Linkage between Dependability, Quality and Customer Satisfaction in Information Intensive Service Firms, Doctoral dissertation, University of Exeter, UK (2011)
9. Grönroos, C.: Service quality: the six criteria of good perceived service quality. Review of Business 9(3), 10–13 (1988)
10. Newman, K.: Interrogating SERVQUAL: a critical assessment of service quality measurement in a high street retail bank. International Journal of Bank Marketing 19(3), 126–139 (2001)
11. Srivastava, K., Sharma, N.K.: Service Quality, Corporate Brand Image, and Switching Behavior: The Mediating Role of Customer Satisfaction and Repurchase Intention. Services Marketing Quarterly 34(4), 274–291 (2013)
12. Wu, H.C.: An Empirical Study of the Effects of Service Quality, Perceived Value, Corporate Image, and Customer Satisfaction on Behavioral Intentions in the Taiwan Quick Service Restaurant Industry. Journal of Quality Assurance in Hospitality & Tourism 14(4), 36 (2013)
13. Ishaq, M.I., Bhutta, M.H., Hamayun, A.A., Danish, R.Q., Hussain, N.M.: Role of corporate image, product quality and customer value in customer loyalty: Intervening effect of customer satisfaction. Journal of Basic Applied Science Research 4(4), 89–97 (2014)

14. Levesque, T., McDougall, G.H.G.: Determinants of customer satisfaction in retail banking. International Journal of Bank Marketing 14(7), 12–20 (1996)
15. Jamal, A., Naser, K.: Factors influencing customer satisfaction in the retail banking sector in Pakistan. International Journal of Commerce and Management 13(2), 29–53 (2003)
16. Kumar, V., Smart, P.A., Maddern, H., Maull, R.S.: Alternative Perspectives on Service Quality and Customer Satisfaction: The Role of BPM. International Journal of Service Industry Management 19(2), 176–187 (2008)
17. Chi, C.G.Q., Qu, H.: Examining the structural relationships of destination image, tourist satisfaction and destination loyalty: An integrated approach. Tourism Management 29(4), 624–636 (2008)
18. Heskett, J.L., Jones, T.O., Loveman, G.W., Sasser Jr., W.E., Schlesinger, L.A.: Putting the service-profit chain to work. Harvard Business Review, 164–174 (March/April 1994)
19. Kumar, V., Batista, L., Maull, R.: The impact of operations performance on customer loyalty. Service Science 3(2), 158–171 (2011)
20. Zhao, J., Wroe, C., Goble, C.A., Stevens, R., Quan, D., Greenwood, M.: Using Semantic Web Technologies for Representing E-science Provenance. In: McIlraith, S.A., Plexousakis, D., van Harmelen, F. (eds.) ISWC 2004. LNCS, vol. 3298, pp. 92–106. Springer, Heidelberg (2004)
21. Bhappu, A.D., Schultze, U.: The Role of Relational and Operational Performance in Business-to-Business Customers' Adoption of Self-Service Technology. Journal of Service Research 8(4), 372–385 (2006)
22. Minett, S.: B2B Marketing, Financial Times/Prentice-Hall, London (2002)
23. Barschel, H.: B2B Versus B2C Marketing - Major Differences Along the Supply Chain of Fast Moving Consumer Goods (FMCG). Grin Verlag, Germany (2004)
24. Homburg, C., Fürst, A.: How organizational complaint handling drives customer loyalty: an analysis of the mechanistic and the organic approach. Journal of Marketing 69(3), 95–114 (2005)
25. Mithas, S., Krishnan, M.S., Fornell, C.: Why Do Customer Relationship Management Applications Affect Customer Satisfaction? Journal of Marketing 69(4), 201–209 (2005)
26. Rauyruen, P., Miller, K.E.: Relationship quality as a predictor of B2B customer loyalty. Journal of Business Research 60(1), 21–31 (2007)
27. Tang, Y.C., Liou, F.M., Peng, S.Y.: B2B brand extension to the B2C market—The case of the ICT industry in Taiwan. The Journal of Brand Management 15(6), 399–411 (2008)
28. Molinari, L.K., Abratt, R., Dion, P.: Satisfaction, quality and value and effects on repurchase and positive word-of-mouth behavioral intentions in a B2B services context. Journal of Services Marketing 22(5), 363–373 (2008)
29. Rauyruen, P., Miller, K.E., Groth, M.: B2B services: linking service loyalty and brand equity. Journal of Services Marketing 23(3), 175–186 (2009)
30. Gonzalez, A.E.M., Comesana, R.L., Brea, F.A.J.C.: Assessing tourist behavioural intensions through perceived service quality and customer satisfaction. Journal of Business Research 60, 153–160 (2007)
31. Sohail, S.M.: Service quality in hospitals than you might think. Managing Service Quality 13(3), 197–206 (2003)
32. Chiou, S.J., Dorge, C.: Service Quality, Trust, Specific Asset Investment, and Expertise: Direct and Indirect Effects in a Satisfaction-Loyalty Framework. Journal of Academy of Marketing Science 34(4), 613–627 (2006)
33. Parasuraman, A., Zeithaml, V.A., Berry, L.L.: A conceptual model of service quality and its implications for future research. Journal of Marketing 49, 41–50 (1985)

34. Grönroos, C.: A service quality model and its marketing implications. European Journal of Marketing 18(4), 36–44 (1984)

35. Lassar, M.W., Manolis, C., Winsor, R.D.: Service quality perspectives and satisfaction in private banking. Journal of Services Marketing 14(3), 244–271 (2000)

36. Luk, S.T.K., Layton, R.: Managing both Outcome and Process Quality is Critical to Quality of Hotel Service. Total Quality Management & Business Excellence 15(3), 259–278 (2004)

37. Bell, S.J., Auh, S., Smalley, K.: Customer Relationship Dynamics: Service Quality and Customer Loyalty in the Context of Varying Levels of Customer Expertise and Switching Costs. Journal of the Academy of Marketing Science 33(2), 169–183 (2005)

38. Rosenzweig, E.D., Roth, A.V.: Towards a theory of competitive progression: Evidence from High-Tech Manufacturing. Production and Operations Management 13(4), 354–368 (2004)

39. Lai, J.Y., Yang, C.C.: Effects of employees perceived dependability on success of enterprise applications in e-business. Industrial Marketing Management 38(3), 263–274 (2009)

40. Andreassen, T.W., Lindestad, B.: Customer loyalty and complex services: the impact of corporate image on quality, customer satisfaction and loyalty for customers with varying degrees of service expertise. International Journal of Service Industry Management 9(1), 7–23 (1998)

41. Martenson, R.: Corporate brand image, satisfaction and store loyalty: A study of the store as a brand, store brands and manufacturer brands. International Journal of Retail & Distribution Management 35(7), 544–555 (2007)

42. Cameran, M., Moizer, P., Pettinicchio, A.: Customer satisfaction, corporate image, and service quality in professional services. The Service Industries Journal 30(3), 421–435 (2010)

43. Wu, J.H.C., Lin, Y.C., Hsu, F.S.: An empirical analysis of synthesizing the effects of service quality, perceived value, corporate image and customer satisfaction on behavioral intentions in the transport industry: A case of Taiwan high-speed rail. Innovative Marketing 7(3), 83–100 (2011)

44. Meredith, J.R., Raturi, A., Amoako-Gyampah, K., Kaplan, B.: Alternative Research Paradigms in Operations. Journal of Operations Management 8(4), 297–326 (1989)

45. Riege, A.M.: Validity and Reliability Tests in Case Study Research: a Literature Review with "hand-on" Applications for each Research Phase. Qualitative Market Research: An International Journal 6(2), 75–86 (2003)

46. Bryman, A.: Social Research Methods. Oxford University Press, New York (2004)

A Research of Taobao Cheater Detection

Baohua Dong[1], Qihua Liu[2], Yue Fu[1], and Liyi Zhang[1,*]

[1]School of Information Management, Wuhan University, Wuhan 430072, P.R. China
{baohua,lyzhang}@whu.edu.cn, fuyue412@gmail.com
[2]School of Information Technology, Jiangxi University of Finance and Economics,
Nanchang 330013, P.R. China
qh_liu@163.com

Abstract. This paper focuses on Taobao cheater detection. At present the phenomenon of fake trading is widespread in Taobao, which makes it difficult for consumers to distinguish between true and fake product reviews. To solve this problem, we collect a total number of 50,285 historical review data from 100 cheaters and 100 real buyers to create a dataset. By using these data, we extract 8 features from three dimensions that are reviewer, commodity, and review. Then we use the SVM algorithm to construct the classification model and choose the RFB kernel function, which has a better performance to identify the cheater. The precision of the final classification model we built to identify the cheater reaches up to 89%. The experimental result shows that extracting features from the historical review data can recognize the cheaters effectively. It can be applied to the recognition of the cheaters in Taobao.

Keywords: Fake trading, Product review, SVM, Cheater detection.

1   Introduction

According to the 33rd China Internet network development state statistic report issued by China Internet Network Information Center (CNNIC), by the end of December 2013, the number of online consumers in China has reached 302 million [1]. Such a huge online shopping market makes numerous entrepreneurs see business opportunities, thousands of new stores set up on Taobao every day. However, many of them have little online traffic due to low credibility. Therefore, many newly opened stores try to improve their reputation in various ways. The most typical one is fake trading. Fake trading refers to the cheating behavior that some merchants in e-commerce platform improve sales, store ratings and credit score by improper means. Fake trading has developed into a huge industry. We can find hundreds of third party fake trading platforms and thousands of QQ or YY groups that are serviced for fake trading.

Consumers usually read the reviews before buying. However, due to the existence of fake trading, it is difficult for them to judge the reality of reviews. Thus, reading reviews while shopping online is a double-edged sword, real reviews make them understand the products better, but when a spam review comes, they will be misguided.

---

[*] Corresponding author.

H. Li et al. (Eds.): I3E 2014, IFIP AICT 445, pp. 338–345, 2014.
© IFIP International Federation for Information Processing 2014

Thus, finding spam reviews timely is of great significance. This paper adopts the method of identifying cheater to identify spam reviews. If a reviewer is cheater, the reviews he has published can be regarded as spam reviews.

2   Related Work

Online reviews research has been a hot topic in recent years, especially spam reviews that attract attention of many scholars at home and abroad. Researches on spam reviews mainly concentrate on text classification and consider the recognition process as the classification of spam and true reviews through manual label, extracting features from the text and using machine learning methods to identify spam reviews automatically. However, due to the judgment of true or fake of product reviews is relevant to psychology, philosophy and many other fields, in addition, it involves the process of natural language understanding and opinion extraction of the review text, therefore, it is difficult to detect fake reviews based on review content, the effect is not very good either. Thus, scholars began to focus on the behavioral characteristics of reviewers based on which we can determine whether a review is a spam review. This method is often used to find spammers. It holds that, if a user is a spammer, then his reviews have strong possibility to be spam reviews. Jindal et al [2] take reviewer's behavior into consideration and analyze the possibility of the reviewer to be a spammer through finding abnormal review patterns. If a user repeatedly publishes positive reviews, then there is strong possibility that he is a spammer. Wu et al [3, 4] use the proportion of positive singletons (singleton refers to the only review that a reviewer had published) in all the reviews and the time aggregation degree of these singletons to analyze reviewer's suspicious behavior. Wang et al [5] discover the reinforcement relations of reviews' trustiness, reviews' honesty, and stores' reliability. They use such relations to discover suspicious spammers. Mukherjee et al [6] use frequent pattern mining to find groups of reviewers who frequently write reviews together, and then they construct features to find the most likely groups of spammers. They also construct a graph modeling the relations between groups of spammers, spammers and products for group spammer ranking [7]. Zhang [8] focuses on detecting the credibility of customers by analyzing online shopping and review behavior, and then they re-score the reviews for products and shops. Spammers can be taken as the special case of their work, which had very low credibility. Lim et al [9] propose a behavioral approach to detect review spammers who try to manipulate review ratings on some target products or product groups.

However, because these studies basically define the behavior of spam reviewers artificially, then determine the spammers or spam reviews. The precision of this method is hard to estimate. In addition, the former studies mainly focus on reviews of one or several products; few studies focus on the historical review data of the reviewers that exist on a platform such as Taobao. For this reason, we collect cheaters' account information from the third party fake trading platforms and real buyers' account information from popular stores on Tmall. We get the historical review data of these buyers on Taobao using crawler software, and then extract features from three aspects including reviewer, commodity, and review to identify the cheaters.

3   Feature Extraction

There are many forms of fake trading; currently the most used one is fake trading on third-party platform, which links both side (sellers who need fake trading and cheaters who engage in fake trading) through an intermediate platform. Both sides publish relevant information on the middle platform, and platform provides guarantee as a third party for both sides to reach an agreement. There are several typical fake trading platforms, such as hiwinwin.com, shuaxinyong.com, shuakewang.com, etc.

Many scholars have adopted the machine learning method, establishing spam review feature library for identification of spam reviews. Among them, most of the scholars extracted corresponding features according to the content of review, such as review sentiment polarity, scoring etc. However, since the review content belongs to natural language, which is difficult to process, and there is no significant difference between true reviews and spam reviews. Therefore, it is almost impossible to identify spam reviews from the comment content. For these reasons, many scholars begin to look for features from other sources. Mukherjee et al [10, 11] believe that the factors such as content similarity, maximum number of reviews, ratio of first reviews and review length have a significant effect on identifying spam reviews. In another paper, Mukherjee et al [12] report that opinion spammers are usually not longtime members of a site. Real reviewers, however, use their accounts from time to time to post reviews. These features have played a certain effect on identifying artificially labeled spam reviews, but whether they can be used to identify the real spam reviews remains to be verified. More importantly, the previous research object is the data of Amazon, but there are many acts of fake trading and spam reviews on Taobao. In order to identify cheaters on Taobao, this paper absorbs the previous research results, analyzing the information of cheaters on third party spam trading platform and information of normal reviewer from Tmall, and we summarized the specific characteristics of cheaters on Taobao. Review centric features include gender (F1), number of registration days (F2) and identity authentication status (F3). Commodity centric features include total number of commodity categories (F4), number of categories purchased in a single day (F5). Review centric features include average length of reviews (F6), daily number of reviews (F7), and no repetition rate of reviews (F8).

Using the above features of reviewers to constitute an eigenvector for every reviewer, calculation methods of the eigenvector are as follows:

$$F1 = n, \ (n = 0, 1)(0: \textit{female} ; 1 : \textit{male})$$

$$F2 = n, \ (n = 0, 1, 2 \ldots\ldots) \ (n : \textit{number of registration days})$$

$$F3 = n, \ (n = 0, 1) \ (0 : \textit{no authentication}; 1 : \textit{authentication})$$

$$F4 = n, \ (n = 12 \ldots\ldots, 16)(n : \textit{total number of commodity categories which are divided into 16 classes according to the Taobao classification})$$

$$F5 = \frac{\text{sum of } \textit{daily number of commodity categories}}{\textit{total number of days that purchase behavior happened}}$$

$$F6 = \frac{number\ of\ words\ in\ all\ reviews}{total\ number\ of\ reviews}$$

$$F7 = \frac{sum\ of\ daily\ number\ of\ reviews}{total\ number\ of\ days\ that\ purchase\ behavior\ happened}$$

$$F8 = \frac{number\ of\ no\ repetition\ reviews}{total\ number\ of\ reviews}$$

4   Methodology

4.1   SVM Model

In our study, we focus on distinguishing cheaters from real buyers. In order to solve this problem, we treat the task as a binary classification problem. Given a training data set $D = \{xi, yi\}_1^n$, we can build a model that can minimize the error in prediction of y given x (generalization error) [13]. Here $x_i \in X$ and $y_i = \{cheater, real\ buyer\}$ represents a buyer and a label, respectively. The model predicts the corresponding y and outputs the score of the prediction when it is applied to a new instance. We use SVM (Support Vector Machines) [14] as the model of classification since it is very effective to solve the binary classification problem. If an instance x (reviewer) is given, SVM assigns a score to it according to

$$f(x) = w^T x + b \tag{1}$$

where w denotes a vector of weights and b denotes an intercept. The value of $f(x)$ presents the quality of the instance x, the higher value of $f(x)$ is, the higher quality of the instance x is. The sign of $f(x)$ is employed in our classification. If the sign is positive, then x is classified into the positive category (real buyer), otherwise into the negative category (cheater). The building of SVM needs labeled training data (in our case, the categories are "real buyer" and "cheater"). Briefly, the learning algorithm creates the "hyper plane" in (1), and the hyper plane separates the positive and negative instances in the training data with the largest margin.

4.2   Experimental Program

This study selects historical review data of cheaters and real buyers as object, using LIBSVM tool for training and testing experimental data. Specific programs are as follows: Firstly, data processing and format adjusting are done on historical review data of reviewers. We use the crawler software crawling the basic personal information and review information of reviewers on Taobao. These data are consolidated, and we build a data set of historical reviews for each reviewer. Elements and the format of all data sets are the same. Secondly, we assign features according to reviewer, commodity, and review to construct a standard data set. We extract features from historical review of each reviewer and construct a data set using feature data of all reviewers. All data are labelled completed. Thirdly, we call LIBSVM tool and use radial basis kernel function to train samples and generate a model file. We construct a classification model using the data of 70 cheaters and 70 normal reviewers for testing. Lastly, we test and forecast the generated model using the test data set. We use the classification model to test the remaining 30 cheaters and 30 normal reviewers, in order to testing the effectiveness of the classification model.

4.3   Evaluation Indexes

The most commonly used classification indexes for text classification are recall, precision and F1-measure [15]. The recall of a class X is the ratio of the number of users correctly classified to the number of users in class X. The precision of a class X is the ratio of the number of users classified correctly to the total predicted as users of class X. The F1-measure is the harmonic mean between both precision and recall, and it is usually reported to evaluate classification effectiveness. Our research can also be seen as a classification problem. To assess the effectiveness of our classification strategies, we use the standard classification index of recall, precision and F1-measure.

(1) Recall
Recall refers to the ratio of the cheaters correctly classified to the number of cheaters in our dataset. The calculation formula is as follow:

$$Recall = \frac{|TP|}{|TP| + |FN|}$$

(2) Precision
Precision refers to the probability that cheaters are correctly predicted to be cheaters. The calculation formula is as follow:

$$Precision = \frac{|TP|}{|TP| + |FP|}$$

(3) F1-measure
F1-measure refers to the harmonic mean between both precision and recall. The calculation formula is as follow:

$$F1 = \frac{2 * Recall * Precision}{(Recall + Precision)}$$

|TP| refers to the number that cheaters are correctly predicted to be cheaters, |FN| refers to the number that cheaters are incorrectly predicted to be real buyers, |FP| refers to the number that real buyers are incorrectly predicted to be cheaters.

5   Experiments and Result Analysis

5.1   Introduction to Dataset

In this study, we collect nearly 50285 data about historical review information of cheaters and real buyers.

The data collection method of cheaters is as follows: Firstly, we get merchants' fake order information and specified requirement from the professional third-party fake trading platforms, such as shuaxinyong.com, hiwinwin.com, shuakewang.com and so on. Secondly, we find out the cheaters' account information by comparing the specified requirements with the order information. Thirdly, we collect the cheaters' personal information by the Taobao inquiry website taodake.com. Lastly, we collect each cheater's data by using the web crawler (The system default reviews and anonymous reviews are not included in the data).

The data collection method of real buyers is as follows: Firstly, we choose the stores with great influence and high credit which don't need to improve credit ratings through fake trading, such as the official flagship store of Xiaomi. Secondly, we find out the real buyers' account information by judging the behavior of the anonymous reviewers. If the reviewer does not have abnormal behavior, that is, the reviews are objective, it can be concluded that the reviewer is a real buyer. Thirdly, we collect the real buyers' personal information through the Taobao inquiry website taodake.com. Lastly, we collect each real buyer's data by using the web crawler (The system default reviews and anonymous reviews are not included in the data).

5.2   Experiments

Our experiments take matlab2010b (matlab7.1) as the experimental platform, the support vector machine is professor Lin's libsvm-mat-3.1 version. We adjust the format of data first, and then import it into the matlab and conduct experiments. The training set includes 140 samples (70 cheaters and 70 real buyers), the test set includes 60 samples (30 cheaters and 30 real buyers), and all the samples have been labeled. The experiment uses a RBF SVM model to identify cheaters out of the data set. When the data is normalized, we use grid search and ten-fold cross validation to optimize the SVM parameters. Then we use the optimized parameters to construct a classification model that can be used to predict the test set. The predict result is shown in Figure.1.

We can see that the model established using the data of 70 cheaters and 70 normal reviewers for training have better ability to identify cheaters, 30 cheaters were successfully predicted. Considering our ultimate goal is to kick out the spam reviewers to reduce the harm caused to consumers, although some normal reviewer are mistakenly identified as cheaters, the error is still acceptable. Overall, the classification model we developed has good prediction ability, and it can well separate cheaters and normal reviewers.

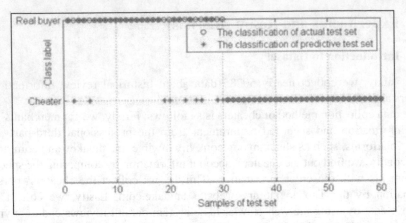

Fig. 1. The actual classification and predictive classification of the test set

5.3   Result Analysis

The experimental result of all samples is as follow:

Table 1. The final results of all samples classified by the classification model

Cheaters					
The number of cheaters correctly identified	The number of cheaters identified by the model	The actual number of cheaters	recall	precision	F1-measure
93	104	100	93%	89%	91%

As we can see, the better cheater recognition effect is due to the choice of 8 characters. The characteristic dimension is relatively high and the RBF kernel function effectively maps it into a high dimensional feature space, so the cheaters and real buyers are certainly linearly separable. What's more, the effect is better.

6   Conclusion

This paper identifies the spam reviews by identifying cheaters. We choose the history review data of 100 cheaters and 100 real buyers as our research object. From these data we extract 8 features using SVM algorithm to construct a classification model, and the precision of the final classification model we built to identity the cheater reaches up to 89%. The experimental result shows that, extracting features from the history review data can identify the cheaters effectively. It can be applied to the recognition of the cheaters in Taobao.

Acknowledgement. The work is supported by the Natural Science Foundation of China (No. 71373192, 71363022) and the MOE Project of key research Institute of Humanities & Social Science in Chinese Universities (No: 14JJD870002)

References

1. China Internet Network Information Center. The 33rd Statistical Report on Internet Development in China, pp. 38–39 (2014)
2. Jindal, N., Liu, B., Lim, E.-P.: Finding unusual review patterns using unexpected rules. In: CIKM 2010, pp. 1549–1552. ACM, New York (2010)
3. Wu, G., Greene, D., Smyth, B., Cunningham, P.: Distortion as a validation criterion in the identification of suspicious reviews. In: Proceedings of the First Workshop on Social Media Analytics, SOMA 2010, pp. 10–13. ACM, New York (2010)
4. Xie, S., Wang, G., Lin, S., Yu, P.S.: Review spam detection via temporal pattern discovery. In: KDD, pp. 823–831 (2012)
5. Wang, G., Xie, S., Liu, B., Yu, P.S.: Identify Online Store Review Spammers via Social Review Graph. Journal of ACM Transactions on Intelligent Systems and Technology (TIST) 3(4) (September 2012)
6. Mukherjee, A., Liu, B., Wang, J., Glance, N., Jindal, N.: Detecting Group Review Spam. In: Proceedings of the 20th International Conference on World Wide Web. ACM (2011)
7. Mukherjee, A., Liu, B., Glance, N.: Spotting fake reviewer groups in consumer reviews. In: Proceedings of the 21st International Conference on World Wide Web. ACM (2012)
8. Zhang, R.: Exploiting shopping and reviewing behavior to re-score online evaluations. In: Proceedings of the 21st International Conference Companion on World Wide Web (2012)
9. Lim, E.-P., Nguyen, V.-A., Jindal, N., Liu, B., Lauw, H.: Detecting Product Review Spammers using Rating Behaviors. In: Proceedings of the 19th ACM International Conference on Information and Knowledge Management (2010)
10. Mukherjee, A., Kumar, A., Liu, B., et al.: Spotting opinion spammers using behavioral footprints. In: Proceedings of the 19th ACM SIGKDD International Conference on Knowledge Discovery and Data Mining. ACM (2013)
11. Mukherjee, A., Liu, B., Glance, N.: Spotting fake reviewer groups in consumer reviews. In: Proceedings of the 21st International Conference on World Wide Web. ACM (2012)
12. Mukherjee, A., Venkataraman, V., Liu, B., et al.: What Yelp Fake Review Filter Might Be Doing? In: ICWSM (2013)
13. Liu, J., Cao, Y., Lin, C.Y., Huang, Y., Zhou, M.: Low-Quality Product Review Detection in Opinion Summarization. In: EMNLP-CoNLL, pp. 334–342 (2007)
14. Vapnik, V.N.: An overview of statistical learning theory. IEEE Transactions on Neural Networks 10(5), 988–999 (1999)
15. Yang, Y.: An evaluation of statistical approaches to text categorization. Information Retrieval Journal (1999)

Analysis of Two Methods of Systemic Risk Measurement Based on Option

Peng Wang[1],* and Mingjia Xie[2]

[1] School of Economic Information Engineering, Southwestern University of Finance
and Economics, Chengdu, China
wangpeng596@gmail.com
[2] Research Institute of Economics and Management, Southwestern University of Finance
and Economics, Chengdu, China
mingjiax@qq.com

Abstract. This paper makes a comparison and analysis about those two main systemic risk measurement methods based on option. Through the comparison and analysis in 5 aspects, more knowledge about each method can be discovered including its advantages and disadvantages. Some conclusions about extensions of credit risk measurement methods can also be included in this paper.

Keywords: Systemic risk measurement, contingent claim analysis, option-iPoD.

1 Introduction

Data gains its dominance in researches in different fields with the development of efficient accessibility of massive data. As one of the most important fields – financial industry, data from option and future markets offer basis for various theoretical approaches that cannot be applied without high frequency data or sufficient panel data.

The current crisis demonstrates the need for tools to detect systemic risk in financial industry. There are two main systemic risk measurement methods which are based on data from financial market, Contingent Claim Analysis (CCA)[1,2] and option-iPoD[3]. Despite the similar dependence on market data, the different methods show differences in design of methods, in/out parameters, empirical researches, etc. This paper makes a comparison between the two methods in four different aspects, which provides more information about each method that is based on high frequency data. The results show that neither of two methods is suitable for all kinds of analysis while advantages come with disadvantages.

2 Summarization of CCA and Option-iPoD

2.1 About Two Methods

A. Contingent Claim Analysis
Contingent claim are the assets whose value depends on other assets' value, for example, options. Contingent Claim Analysis (CCA) is a widely used analytical method

---

* Corresponding author.

H. Li et al. (Eds.): I3E 2014, IFIP AICT 445, pp. 346–357, 2014.

in risk measurement. While contingent claim analysis is applied in credit risk analysis, it is normally called Merton Model. The key to understand CCA is the Risk-Adjusted Balance Sheet, which comes from traditional balance sheet with adding "market risk factor". As for elements in balance sheet, this analysis transfers the book value into market value, which can't be observed directly.

The risk-adjusted balance sheet is:

$$Assets = Equity + RiskyDebt$$
$$A(t) = J(t) + D(t)$$

While

$$RiskyDebt = DefaultFreeDebt - debtguarantee$$
$$D(t) = Be^{-rT} - P(t)$$

Assets A(t) is the market value of assets and the equity J(t) is the market value of equity. Debt guarantee P(t) can be expressed as the expected loss of the debt because the debtor may default at due time. B is the promised payments at time T and r is the risk-free rate. There are two important conclusions as follows:

- Equity is a call option whose underlying assets is Assets and strike price is the promised payments
- Debt guarantee is a put option whose underlying assets is Assets and strike price is the promised payments

According to BSM option pricing model,

$$J(t) = A(t)N(d_1) - Be^{-rT}N(d_2) \tag{1}$$

with $d_1 = \frac{\ln(A/B) + (r + \sigma^2/2)T}{\sigma\sqrt{T}}$, $d_2 = d_1 - \sigma\sqrt{T}$, σ equals to the standard deviation of the asset return. N(X) is the cumulative standard normal distribution.

Main Risk exposure indicators

- Risk-neutral default probability: With expect assets return rate equals the risk-free rate r, $N(-d_2)$ is the risk-neutral default probability
- Credit spread: $s = y - r, y = \frac{\ln(B/D)}{T}$
- Risk-adjusted balance sheet elements: Assets, Equity, Risk Debt
- Delta, it measures the non-linear change in the value of an option per unit change in the value of the underlying asset

B. Option-iPoD

Buchen and Kelly (1996) propose a numerical method to get probability density distribution function of assets. Capuano (2008) proposes option-iPoD based on former function, which is a systemic financial risk measurement method.

Entropy measures the uncertainty of a variable and maximum entropy shows the most uncertainty of a variable. Principle of Maximum Entropy, that is PME, is ideally suited to estimating the probability distribution of an asset on which a derivative security or contingent claim is written[1].

---

[1] Peter W. Buchen, Michael Kelly, "The Maximum Entropy Distribution of an Asset Inferred from Option Prices".

The problem is following:

$$min_D \left\{ min_{f(V_T)} \int_{V_T=0}^{\infty} f(V_T) \log \left[\frac{f(V_T)}{f^0(V_T)} \right] dV_T \right\} \tag{2}$$

While $f^0(V_T)$ is the prior probability density distribution function, $f(V_T)$ is the true probability density distribution function.

The constraints are following:

a) Constraints based on Merton Model

$$C_T^K - max(E_r - K; 0) - max(max(V - D; 0) - K; 0) - max(V_r - D - K; 0) \tag{3}$$

C_T^K: The price of an option whose strike price is K and due time is T.

V: the market price of assets

D: the market price of risky debt

E: the market price of equity

r : the risk-neutral rate

b) Constraints based on option market price

$$C_0^i = e^{-rT} \int_{V_T=0}^{\infty} max(V_T - D - K_i; 0) f(V_T) dV_T = e^{-rT} \int_{V_T=0}^{\infty} (V_T - D - K_i) f(V_T) dV_T \tag{4}$$

c) Constraints based on probability density function

$$1 = \int_{V_T=0}^{\infty} f(V_T) dV_T \tag{5}$$

We treat the problem as solving an optimization problem for $f(V_T)$ with 3 constraints. $f(V_T)$ Will be solved first and given the optimal $f(V_T)$ solve for D. The Larangian is:

$$L = \int_{V_T=0}^{\infty} f(V_T) \log \left[\frac{f(V_T)}{f^0(V_T)} \right] dV_T + \lambda_0 \left[1 - \int_{V_T=0}^{\infty} f(V_T) dV_T \right]$$

$$+ \lambda_L \left[E_0 - e^{-rT} \int_{V_T=0}^{\infty} (V_T - D) f(V_T) dV_T \right]$$

$$+ \sum_{i=1}^{n} \lambda_{2,i} [C_0^i - e^{-rT} \int_{V_T=0}^{\infty} (V_T - D - K_i) f(V_T) dV_T] \tag{6}$$

We can solve the function by numerical method and get the probability distribution function of assets by constraints mentioned above.[2] When the function is solved, some risk exposures, such as default probability[3], are measured.

---

[2] Standard Newton numerical method for example. The constraints are constraints in Lagrangian which can be the functions to solve unknown parameters.

[3] Default probability equals $\int_0^D f(V_T)$.

2.2   Data Source

Not only the algorithm but also the data source are based on financial markets especially option & future market. Data concerning about "level" and "volatile" are required to construct risk detecting indicators, which means a higher frequency data are always better because the second moment of data series are needed. The data required can be divided into following categories:

A. Data from stock market

Market value of equity is required. In most cases, prices of stocks are proxy variables of equity that are unobservable directly. Daily data or even weekly data can be used to construct risk detecting indicators while other frequency are available depending on different scenarios and purposes. When necessary, frequency higher than daily, such as hourly, can be applied to measure a more precise process of changing of risk.

Risk-free interest rate is also needed for both two methods. Government bond or other composite indexes can be used as this indicator.

B. Data from option market

Information concerning about the details of contracts in the option market is used to construct the constraint equations in option-iPoD. Precisely, expiration time of options, prices of options, are required in the algorithm. Analogous to the frequency issue in stock market, higher frequency data are always better.

3   Comparison and Analysis

Two methods mentioned above have different features. This part focuses on the difference in 5 fields: thinking of designation, in/out variables, mathematical methods, empirical analysis approaches, and extensions of methods. It is expected that more info about individual method will be discovered through comparison.

3.1   Thinking of Designation

I) Contingent Claim Approach

Thinking of design of CCA can be divided into 2 parts.

- Risk-adjusted Balance Sheet in theory
- Calibration to risk-adjusted Balance Sheet for different sectors

CCA is an extension of BSM model that assumes a stochastic process of Assets. It is reasonable because of the "Weak Axiom of Efficient Market". CCA assumes that default happens when asset is lower than debt. The core function is following:

$$Assets = Equity + Defaultfree - DebtGuarantee$$

Equity is a call option whose underlying asset is Assets and strike price is the promised payments. Debt guarantee is a put option whose underlying asset is Assets and strike price is the promised payments. This function is a risk-adjusted traditional balance sheet. Scalars in traditional balance sheet are transferred into time variables with "risk factors". Several risk exposure indicators can be calculated including elements in risk-adjusted balance sheet, risk-neutral default probability, Delta[4].

Another innovation is the calibration to risk-adjusted balance sheet for different sectors. For example, some financial institutions are "too-big-to-fall", so guarantees from government are often existed, which adds the assets of financial sectors. In empirical analysis, some calibrations to equations are made to insure rationality of the methods. Merton et al. (2007) divide the economy into 4 sectors: corporate sector, financial sector, households sector, and sovereign sector, so there are 4 equations for different sectors.

II) Option-iPoD

CCA requires many assumptions, such as constant rate, distribution function of assets, which Option-iPoD is designed to release. Too many assumptions are the barrier to suitable empirical solutions. Christian Capuano (2008) proposes option-iPoD, which releases two main assumptions: probability distribution function of Assets and the default barrier[5].

The key to option-iPoD is Maximum Entropy Distribution of Assets, which is the result of Principle of Maximum Entropy (PME). The PME is ideally suited to estimating the probability distribution of an asset which a derivative security or contingent claim is written. With constraints from option market prices, probability distribution of an asset can be calculated.

The unknown coefficient in Lagrangian function and default barrier D can be calculated by means of numerical mathematical methods. When probability distribution of an asset is calculated, some risk exposure indicators can be calculated. "Less assumptions" is a feature of option-iPoD. Exception for prior probability distribution, all constraints are from real option market prices, which do no harm in the freedom of variables.

There are also "CCA factors" in option-iPoD. Market value of assets, partial probability distribution, delta, and gamma can be calculated through option-iPoD.

Table 1. Comparison in design of thinking

	CCA	Option-iPoD
Role of option market	Thinking of BSM option pricing model	Data source of constraints
Focus	Risk-adjusted balance sheet	Probability distribution of an asset
Data	Traditional balance sheet	Option market
similarity	Coming from option market, either thinking or data	

---

[4] Equation: debt guarantee/Assets, measures the non-linear change in the value of an option per unit change in the value of the underlying asset.

[5] CCA assumes that default happens when asset is lower than promised payments, but promised payments are not a certain volume. In KMV model, default barrier equals 50% long-term debt plus 100% short-term debt.

3.2   In/Out Variables

Comparison between two methods tells attitudes to "variable freedom" from two methods.

Table 2. Comparison in in/out variables

	CCA	Option-iPoD
Input parameters	Market value of equity; Standard variation of equity; Risk-free interest rate; Default barrier; Expect return rate of asset;	Expiration time of options; Stock price; Risk-free interest rate;
Main output variables (relate to risk-adjusted balance sheet)	Default probability; Market value of asset; Expected loss of debt; Equity;	Default probability; Market value of asset; Expected loss of debt; Equity; Default barrier;
Main output variables (relate to risk exposure indicators)	Standard variation of asset; Delta; Gamma;	Standard variation of asset; Probability distribution of asset;
	Vega	Delta; Gamma; Vega

CCA has more input parameters than option-iPoD does while the two methods have nearly same main output variables. Option-iPoD has two distinct outputs: probability distribution of asset and default barriers that are two prior assumptions of CCA. To some degree, CCA has more "constraints" than option-iPoD does, which leads to deviation from real market.

Table 3. Comparison in Assumption & in/out variables

	CCA	Option-iPoD
In/out variables	More input parameters	Less input parameters but two more distinct variables
Similarity	Number of input variables is larger than number of output variables. Fewer assumptions are better for methods.	

3.3   Mathematical Methods

Analytic expression and numerical solution are two main mathematical methods used in empirical analysis. CCA applies analytic expression while option-iPoD uses numerical method. Different methods lead to different results that generate different understanding about risk measurement.

I) Mathematical methods in CCA

The main mathematical method in CCA is analytic expressions used in BSM option pricing model. While equity and debt guarantee are regarded as two options whose underlying assets are assets. Through mathematical solution, elements in risk-adjusted balance sheet can be solved and relational expression of elements can also be solved.

It is clear that absolute value of elements from risk-adjusted balance sheet and its relationship can be used to do further research through this mathematical method.

II) Mathematical methods in option-iPoD

To obtaining the distribution of equation (2), it is necessary to first find the Lagrange parameters, $\lambda_D, \lambda_1, \lambda_{2,1} \dots \lambda_{2,n}$, which are determined by the constraints entailed by equation (3), (4) and (5). These equations are nonlinear and have to be solved numerically. A standard Newton method, for example, in Djafari (2000). Buchen and Kelly (1996) and Avellaneda (1998) show that the objective function is convex, and that the solution is unique.

Then the last unknown parameter default barrier D, also has to be solved numerically with following constraint:

$$\lim_{\Delta \to 0} \frac{L\{f(V_T, D+\Delta)\} - L\{f(V_T, D)\}}{D+\Delta} = 0 \tag{8}$$

The whole probability distribution function can be established.

It is of importance that the solved probability distribution function is a "partial function" with "partial domain of definition". Because constraints of option-iPoD come from real option market, which shows that constraints are meaningful if option market's price information is effective. Stock prices and option prices are insignificant when corporate defaults. For investors of stock or option, there is no trading of stocks, no trading of options in default state. Equity options are not suited to describe the market value of asset in default state.

While equity options don't contain information on shape of the probability density function in the default state, they do contain information on the cumulative distribution function, the probability of default. As it can be found in equation (7), $f^R(V_T)$ has same information as $f(V_T)$ when $V_T < D$, so it provides sufficient "cumulative information" in $V_T - D$. For example, default probability and other risk exposure indicators can be measured.

Table 4. Comparison in mathematical methods

	CCA	Option-iPoD
Emphasis of mathematical methods	Analytic expression of all elements of risk-adjusted balance sheet	Solution to probability distribution of asset and default barrier
Similarity	Mathematical methods are regarded as tools for solution in different purpose	

3.4   Empirical Analysis with Data

I) CCA

There are 3 important basic input variables: market value of asset, standard variation of assets return rate, default barrier. Unlike trading equity, market value of asset and its variation can't be observed directly. So two equations are needed to solve this problem.

$$S_t = h(V_t, \sigma_V, r, D, T)(9)$$

$$\sigma_s = g(\sigma_V)(10)$$

S_t is equity's market price, V_t is market value of asset, σ_V is standard variation of asset, r is risk-free interest rate, and D is market value of default barrier.

Equation (9) is BSM option pricing model and equation (10) can be expressed as following:

$$\eta_{S,V} = \frac{\Delta S}{S} / \frac{\Delta V}{V} \Longrightarrow \sigma_s = \eta_{S,V} \sigma_V (11)$$

$$\eta_{S,V} = \Phi(d_1) \frac{V}{S} (12)$$

As for market value of default barrier D, it's always be set manually. Taking KMV for example, default barrier equals 100% short-term debt plus 50% long-term debt.

The economy-wide CCA can be used with scenario, simulation, and stress-testing analysis. The level of analysis depends on practical issues related to data availability, data reliability, and goals of the analysis.

a)   Risk exposure indicators from CCA can be analyzed with other macroeconomic factors. Factor model is a suitable model[6]. The time pattern of asset returns of each financial institution can be used as the dependent variable in a factor model.
b)   "Sub risk-adjusted balance sheet for household sectors"[7].
c)   Stress-testing and assessing capital adequacy using CCA model of financial institutions.
d)   Making connections between traditional monetary policies based on interest rate & money supply and CCA risk exposure indicators.

II) option-iPoD

One of main empirical analysis of option-iPoD is what Capuano (2008) has done with Citi Bank. In that research, option-iPoD is well used to estimate how default barrier and leverage ratio generate through financial crisis in 2008, which proves that option-iPoD is suitable for systemic risk measurement. Based on results from option-iPoD, some advices, such as "de-leveraging" should be applied for Citi Bank, are suggested.

---

[6] Based on data from Chile central bank, Dale et al (2008) build a factor model with 4 main economic factors and asset return rate calculated by CCA. This factor model well simulates how the systemic risk generates
[7] See more details in Dale F. Gray et al, "NEW FRAMEWORK FOR MEASURING AND MANAGING MACROFINANCIAL RISK AND FINANCIAL STABILITY".

At the same time, Capuano observes how Moody's KMV EDF[8] generates, which shows what the systemic risk Citi Bank faces.

Table 5. Comparison in empirical analysis

	CCA	Option-iPoD
Application fields	Wider than option-iPoD does including macro-economy, financial sector, households sector	Fields with trading equity and options
Data	Suitable balance sheets for different sectors	Price information from option market
For risk supervisors	All four sectors[9]	Corporate and financial sector
Similarity	Time pattern series are usually used in analysis. It is expected that risk indictors can well predict breakout of crisis	

3.5   Extensions of the Methods

I) relaxing assumptions of CCA
Assumptions lead to gap between theory methods and reality. Extensions of CCA focus on its assumption's relaxation.

● Some researchers have established a relationship between implied volatility of two equity options, leverage and implied asset volatility. In fact, this is another way of implementing Merton's model to get spreads and risk-neutral default probabilities directly from the implied volatility of equity options
● Shimko, Tejima, and Van Deventer (1993) include a Vasicek interest rate term structure model which allows interest rates and term structure of interest rates to vary

II) More constraints about option-iPoD
More constraints from markets make option-iPoD more accurate. Capuano (2008) adds zero-coupon bonds constraints to option-iPoD.

Table 6. Comparison in extension of methods

	CCA	Option-iPoD
Extension	Relaxing more assumptions	Adding more constraint from market
Similarity	Fewer manual assumptions and more constraints from market	

Out of the potential participants (249 members), 118 filled the questionnaire, resulting in a response rate of 47%. Table 1 contains the non-respondent analysis results.

---

[8] Estimated Default Frequency: a measure of probability of default given distance to default.
[9] Corporate sector, financial sector, households sector, sovereign sector.

Table 7. Reasons for non-participation in the survey

Reason for non-participation	N:o of persons (total = 131)	Percentage
Too busy or forgot	27	21%
Dislike of such surveys or not interested	10	8%
Technical difficulties with the survey	19	15%
Did not feel competent enough to fill the survey despite being a health journalist	7	5%
Unknown or person could not be contacted	67	51%

4   Conclusion

4.1   Data from Different Fields of the Economy Solve Different Problems

Based on Merton's analysis, economy can be divided into 4 parts: corporate sector, financial sector, households sector and sovereign sector. Different sector prefers different risk measurement approaches because of the accessibility to different data.

For sectors with trading equities and options, some corporates, some financial institutions, even some public sectors, option-iPoD provides a more accurate and effective systemic risk measurement approach. Compared with CCA, option-iPoD gets fewer manual assumptions which makes it closer to real market. Option-iPoD is more suitable in this situation for its data used in option-iPoD comes from high-frequency option market. More constraints come from markets makes option-iPoD more accurate for various goals of analysis.

CCA shows its wide range of application because of Merton's risk-adjusted balance sheet. For sectors like sovereign sector, with no trading options, CCA is better at estimating systemic risk. The core function[10] is calibrated to correspond with various sectors thus CCA can be used in all sectors, even in public sector or households sector. CCA works at sectors with trading equity or "analogous equity" and option is not necessarily needed.

All above, CCA has wider range of application than option-iPoD does but option-iPoD has less manual assumptions, which makes it closer to reality.

4.2   CCA Needs Release of Assumptions and Option-iPoD Needs More Constraints

Assumptions simplify the problem in theory but deduce the freedom of variable, which leads to inconsistency with reality. CCA will be better if more assumptions are released. In classical Merton Model, constant interest rate and probability distribution are required. For example, Shimko et al. (1993) include a Vasicek interest rate term

---

[10] Assets = Equity + Risk Debt

structure model which allows interest rates and term structure of interest rates to vary, which is a release to interest rate assumption.

Option-iPoD needs more constraints of reliable, available market. More reliable constraints makes prior probability distribution function closer to real probability distribution function by Lagrangianoptimization process. Christian Capuano (2008) adds zero-coupon bonds constraints to option-iPoD which makes option-iPoD more accurate.

Above all, subjectivity should be reduces and objectivity should be increased both in CCA or option-iPoD.

4.3 The Improvement of the Quality of Data Tells More

CCA and option-iPoD are two main credit risk measurement methods. These two methods are better than traditional FSIs[11] at solutions to deal with high-frequency market value. CCA applies the thinking of option while option-iPoD applies the price information of options.

The two methods focus on option market. On one perspective, option market is a highly financial market whose information if effective. On another perspective, price information from option market is easy to get. With the development of option market, it can provide more information in financial market which are constraints for credit risk measurement model.

Information from option market and theories of option market can work in more sectors not only in financial and corporate sectors. Like CCA, contingent claim analysis works well in sovereign sector that represents macroeconomic systemic risk exposure. Contingent claim analysis can also be used in key banks to simulate systemic risk in financial sector after weighting process.

Above all, macro financial engineering is one of extensions of credit risk measurement method.

4.4 Future Work

This paper makes comparison and analysis between two main systemic risk measurement methods based on option in 5 perspectives. But effectiveness of two methods in empirical analysis is not estimated which requires more related researches to support.

Acknowledgement. The research work was supported by the National Social Science Foundation of China (Grant No. 13CJY121), by the Fundamental Research Funds for the Central University under Grant No. JBK130503.

---

[11] FSIs: Financial Stability Indicators system, it is a system of indictors measuring financial risk.

References

1. Gray, F., Merton, R., Bodie, Z.: A new framework for analyzing and managing macrofinancial risks of an economy. National Bureau of Economic Research working paper 12637 (2006)
2. Gray, F., Merton, R., Bodie, Z.: New framework for measuring and managing macrofinancial risk and financial stability. National Bureau of Economic Research working paper 13607 (2007)
3. Capuano, C.: The option-iPoD—the probability of default implied by option prices based on entropy. IMF working paper WP/08/194 (2008)

Author Index

Printed in the United States
By Bookmasters